Analyzing the Physics of Radio Telescopes and Radio Astronomy

Kim Ho Yeap
Universiti Tunku Abdul Rahman, Malaysia

Kazuhiro Hirasawa
University of Tsukuba, Japan

A volume in the Advances in Environmental Engineering and Green Technologies (AEEGT) Book Series

IGI Global
DISSEMINATOR OF KNOWLEDGE

Published in the United States of America by
IGI Global
Engineering Science Reference (an imprint of IGI Global)
701 E. Chocolate Avenue
Hershey PA, USA 17033
Tel: 717-533-8845
Fax: 717-533-8661
E-mail: cust@igi-global.com
Web site: http://www.igi-global.com

Library of Congress Cataloging-in-Publication Data

Names: Yeap, Kim Ho, editor. | Hirasawa, Kazuhiro, 1941- editor.
Title: Analyzing the physics of radio telescopes and radio astronomy / Kim Ho Yeap and Kazuhiro Hirasawa, editors.
Description: Hershey, PA : Engineering Science Reference, [2020] | Includes bibliographical references and index. | Summary: "This book examines the underlying operational principles of radio telescopes and illustrates how signals from the cosmic sources are detected and analyzed. It also expounds on the physics of the celestial objects detected by radio telescopes"-- Provided by publisher.
Identifiers: LCCN 2019039180 (print) | LCCN 2019039181 (ebook) | ISBN 9781799823810 (hardcover) | ISBN 9781799823827 (paperback) | ISBN 9781799823834 (ebook)
Subjects: LCSH: Radio telescopes. | Radio astronomy.
Classification: LCC QB479.2 .A53 2020 (print) | LCC QB479.2 (ebook) | DDC 522/.682--dc23
LC record available at https://lccn.loc.gov/2019039180
LC ebook record available at https://lccn.loc.gov/2019039181

This book is published in the IGI Global book series Advances in Environmental Engineering and Green Technologies (AEEGT) (ISSN: 2326-9162; eISSN: 2326-9170)

British Cataloguing in Publication Data
A Cataloguing in Publication record for this book is available from the British Library.

Advances in Environmental Engineering and Green Technologies (AEEGT) Book Series

Sang-Bing Tsai
Zhongshan Institute, University of Electronic Science and Technology of China, China & Wuyi University, China
Ming-Lang Tseng
Lunghwa University of Science and Technology, Taiwan
Yuchi Wang
University of Electronic Science and Technology of China Zhongshan Institute, China

ISSN:2326-9162
EISSN:2326-9170

Mission

Growing awareness and an increased focus on environmental issues such as climate change, energy use, and loss of non-renewable resources have brought about a greater need for research that provides potential solutions to these problems. Research in environmental science and engineering continues to play a vital role in uncovering new opportunities for a "green" future.

The **Advances in Environmental Engineering and Green Technologies (AEEGT)** book series is a mouthpiece for research in all aspects of environmental science, earth science, and green initiatives. This series supports the ongoing research in this field through publishing books that discuss topics within environmental engineering or that deal with the interdisciplinary field of green technologies.

Coverage

- Cleantech
- Contaminated Site Remediation
- Industrial Waste Management and Minimization
- Radioactive Waste Treatment
- Green Technology
- Pollution Management
- Waste Management
- Air Quality
- Water Supply and Treatment
- Electric Vehicles

IGI Global is currently accepting manuscripts for publication within this series. To submit a proposal for a volume in this series, please contact our Acquisition Editors at Acquisitions@igi-global.com or visit: http://www.igi-global.com/publish/.

The Advances in Environmental Engineering and Green Technologies (AEEGT) Book Series (ISSN 2326-9162) is published by IGI Global, 701 E. Chocolate Avenue, Hershey, PA 17033-1240, USA, www.igi-global.com. This series is composed of titles available for purchase individually; each title is edited to be contextually exclusive from any other title within the series. For pricing and ordering information please visit http://www.igi-global.com/book-series/advances-environmental-engineering-green-technologies/73679. Postmaster: Send all address changes to above address.

Titles in this Series

For a list of additional titles in this series, please visit:
https://www.igi-global.com/book-series/advances-environmental-engineering-green-technologies/73679

Climate Change and Anthropogenic Impacts on Health in Tropical and Subtropical Regions
Ahmed Karmaoui (Southern Center for Culture and Sciences, Morocco) Abdelkrim Ben Salem (Southern Center for Culture and Sciences, Morocco & Research Center of Plant and Microbial Biotechnologies, Biodiversity, and Environment (BioBio), Faculty of Sciences, Mohammed V University, Rabat, Morocco) and Ashfaq Ahmad Shah (National University of Sciences and Technology, Pakistan)
Engineering Science Reference • © 2020 • 350pp • H/C (ISBN: 9781799821977) • US $225.00

Physiological Aspects of Imprinting and Homing Migration in Salmon Emerging Research and Opportunities
Hiroshi Ueda (Hokkaido University, Japan)
Engineering Science Reference • © 2020 • 200pp • H/C (ISBN: 9781799820543) • US $175.00

Artificial Intelligence and IoT-Based Technologies for Sustainable Farming and Smart Agriculture
Pradeep Tomar (Gautam Buddha University, India) and Gurjit Kaur (Delhi Technological University, India)
Engineering Science Reference • © 2020 • 350pp • H/C (ISBN: 9781799817222) • US $215.00

Predicting, Monitoring, and Assessing Forest Fire Dangers and Risks
Nikolay Viktorovich Baranovskiy (National Research Tomsk Polytechnic University, Russia)
Engineering Science Reference • © 2020 • 417pp • H/C (ISBN: 9781799818670) • US $235.00

Effects of Emerging Chemical Contaminants on Water Resources and Environmental Health
Victor Shikuku (Kaimosi Friends University College, Kenya)
Engineering Science Reference • © 2020 • 350pp • H/C (ISBN: 9781799818717) • US $225.00

Waste Management Techniques for Improved Environmental and Public Health Emerging Research and Opportunities
Sang-Bing Tsai (Zhongshan Institute, University of Electronic Science and Technology of China, China & Wuyi University, China) Zhengxi Yuan (Zhongshan Institute, University of Electronic Science and Technology of China, China) Jian Yu (Research Center for Environment and Sustainable Development of China Civil Aviation, Civil Aviation University of China, China) and Xuexin Liu (Capital University of Economics and Business, China)
Engineering Science Reference • © 2020 • 229pp • H/C (ISBN: 9781799819660) • US $190.00

IGI Global
DISSEMINATOR OF KNOWLEDGE

701 East Chocolate Avenue, Hershey, PA 17033, USA
Tel: 717-533-8845 x100 • Fax: 717-533-8661
E-Mail: cust@igi-global.com • www.igi-global.com

Editorial Advisory Board

List of Reviewers

Table of Contents

Detailed Table of Contents

In this chapter, the design and construction of a simple home-brew radio telescope are illustrated. The radio telescope is built essentially from commercial off-the-shelf components. The main components include an offset prime-focus Ku-band reflector antenna, RF detector circuit, Arduino microcontroller, and a computer. To demonstrate the viability of the telescope, a drift-scan of the sun was performed. From the measurement, it is observed that the radiation signal from the sun has a peak power of –34.3 dBm, half power of –35.4 dBm, and a beamwidth of 3.13o. The impact of rain on the signal from the sky was also investigated. The result shows that there was a noticeable degradation of the power received. A significant amount of energy is absorbed by the dense water vapour in the atmosphere.

Electromagnetic theory covers the basic principles of electromagnetism. This chapter explores relationships between electric and magnetic fields. The chapter describes the behaviour of electromagnetic wave. The four sets of Maxwell's equations which underpin the principles of electromagnetism are briefly explained. An illustration on wave polarization and propagation is presented. The author describes the classification of waves according to their wavelengths (i.e. the electromagnetic spectrum).

Electromagnetic wave propagation is an invisible phenomenon that cannot be detected by the human senses. To understand wave propagation, one must first learn what wave propagation is and the basic principles that affect wave propagation. This chapter introduces the atmospheric windows which allow electromagnetic radiation from bands to penetrate Earth. Helmholtz equations, i.e. the equations which govern wave propagation, and the properties of waves (such as propagation constant and characteristic impedance) are then briefly explained. When waves encounter different media during its propagation, they may be reflected, refracted, or diffracted. These phenomena are also covered. The last part of this chapter concisely explains the terminologies commonly used to describe electromagnetic radiation.

Physical optics (PO) is one of the fundamental and powerful high-frequency theories for electromagnetic scattering and radiation. The total field of a source (antenna) which radiates in the presence of a perfectly conducting surface may be expressed as a superposition of the incident and the scattered fields. The current fields which exist everywhere are chosen in PO to denote the electric and magnetic fields of the source, i.e., they exist as if the scatterer was "absent"; this is unlike the geometrical optics (GO) incident field, which exists in the presence of the surface of the scatterer. The scattered fields in this case can be expressed in terms of the radiation integrals over the actual currents induced on the surface of the scatterer. These currents also radiate the scattered fields in the absence of the scatterer. This chapter shows the fundamental PO formulation and calculated results, and some topics which improve the conventional PO to the extended PO such as "physical theory of diffraction (PTD)" and "PO with transition current (PTD-TC)".

Analysis of Terahertz waves comes in three main forms, physical optics, geometrical optics, and Gaussian optics. Physical optics has the highest accuracy but it is time consuming when it is applied in the design of large radio telescopes. Also, it is only capable of computing radiation characteristics. Geometrical optics, on the other hand, reduces computational time significantly. But it does not give accurate results when designing telescopes which are to operate at Terahertz frequencies. Gaussian optics is a good trade-off between these two methods and it is a popular approach used in the design of large radio telescopes — particularly those which operate near/in the Terahertz band. Since it accounts for the effects of diffraction, this method produces reasonably accurate results. This chapter describes Gaussian optics, with emphasis given on its application in the design of radio telescopes.

One of the simple and most widely used microwave antennas is the horn as a feed element for large radio telescopes, satellite tracking, and communication reflector, which are found installed throughout the world. In addition to its utility as a feed for reflectors and lenses, it is a common element of phased arrays and serves as a universal standard for calibration and gain measurement of other high gain antennas. Its widespread applicability stems from its simplicity in construction, ease of excitation, large gain, wide-band characteristics, and preferred overall performance. An electromagnetic horn can take many different forms, such as basic pyramidal, conical, corrugated, double-ridged, and dual polarized horns, as well as horns with lens and so on. The horn is nothing more than a hollow pipe of different cross-sections, which has been tapered to a larger opening aperture. This chapter explains the fundamentals of the pyramidal horn antenna in detail using aperture field method. Numerical and measured examples, are also shown.

This chapter introduces some experiments on meteor burst communications conducted in Japan and the method of QSO by meteor scatter communications conducted by amateur radio operators.

Section 6
Cosmic Sources

The radio emissions from the Sun and planets provides a new window to explore and improve our understanding of these celestial objects and its magnetic activities. Radio emissions allow us to see things that are not seen in other wavelengths. The radio emissions observed from the Sun and planets can generally be divided into thermal and non-thermal. The thermal emission is related to the temperature of the objects, i.e. the blackbody radiation and the non-thermal emission is usually associated with the magnetic fields.

With the discovery of radio emission of extraterrestrial origin, it had opened a new and broader window in the electromagnetic spectrum to observe the sky. Two of the common sources of radio emissions are active galactic nuclei (AGN) and pulsars. AGN are very luminous at radio wavelength and are powered by the accretion disk surrounding supermassive black holes at the center. AGN are useful as an alternative standard ruler to determine cosmological parameters. Pulsars, on the other hand, is a type of neutron star that is highly magnetized and rotates at enormous speed. This cosmic lighthouse produces a very precise pulsation period that can be used as cosmic clock.

The cosmic microwave background (CMB) holds many secrets of the origin and the evolution of our universe. This ancient radiation was created shortly after the Big Bang, when the expanding universe cooled and became transparent, sending an afterglow of light in all directions. It is a pattern frozen in place that dates back to 375,000 years after the birth of the universe. Numerous experiments and space missions have made increasingly higher resolution maps of the CMB radiation, with the aims to learn more about the conditions of our early universe and the origin of stars, galaxies, and the large-scale cosmic structures that populate our universe today.

Preface

Radio astronomy studies the properties of cosmic sources and phenomena at the radio wave spectrum. In radio astronomy, radio telescopes are built and used to observe naturally occurring signal emission from celestial objects, such as planets, stars, galaxies, pulsars, quasars, etc. The electromagnetic waves emanated from cosmic sources carry spectral and spatial information which is important in the field of astrophysics. The wave signal radiated from dust clouds in the interstellar medium (ISM), for instance, allows astrophysicists to probe the origins of stars, how they are formed, as well as their underlying physical and chemical properties. The cosmic microwave background (CMB) radiation is believed to be the relic radiation of the Big Bang. Hence, by detecting and analysing CMB radiation, the information of the early universe when it was merely 375,000 years old could be unveiled.

A typical radio telescope is made up of a main and a sub-reflector, a mixer circuit, a local oscillator, a detector circuit, and a data processing unit (Yeap, Tham, Yassin, & Yeong, 2011). The latest generation of radio telescopes is designed to collect wave signals from multiple frequency bands (Yeap & Tham, 2018). Hence, each cartridge at the cryostat comprises several receivers (each fitted in a different cartridge) which are used to detect signals at different bands. The electromagnetic beam radiated from cosmic sources is first detected by the main reflector. The signal is then scattered to the sub-reflector. This is then followed by the propagation of the beam through focusing elements, which are usually in the form of mirrors or lens, to the feed horn. The electromagnetic signal at the aperture of the horn is then channelled via waveguides to a mixer circuit. The signal at the mixer is modulated with the signal from the local oscillator and is then down converted to a lower frequency. After going through numerous stages of amplification, the signal is eventually fed to a data analysis system, such as a spectrum analyser (Yeap, Tham, Yassin, & Yeong, 2011).

This book can be broadly classified into two sections – the first section illustrates the engineering of radio telescopes; while the second section explains the science of radio astronomy. In the first section, an extensive description of the underlying theories and approaches involved in the design of radio telescopes is given. The working principles of the astronomical instruments involved in the construction of a radio telescope are also highlighted. In the second section, the physics of astronomical sources, such as stars and galaxies, are explained in detail.

OBJECTIVES OF THE BOOK

This book correlates the engineering aspect (i.e. knowledge involved in the construction of a radio telescope) with the science (i.e. the astrophysics of cosmic sources) of radio astronomy. In short, the objectives of this book are to

(i) provide readers with the underlying operation principles of radio telescopes
(ii) illustrate how signals from cosmic sources are detected and analyzed
(iii) explain the nature of the universe and the cosmic sources

TARGET AUDIENCE

This book starts with elementary notions in radio astronomy and radio telescopes. It then progressively advances to more complicated concepts. Hence, it provides a step-by-step guide for novices who wish to delve into the fields of radio astronomy and astrophysics. Experts who have already possessed certain knowledge in these fields can also consider this book as a technical resource for their researches. The potential readers are, but not limited to:

(i) Engineers, in particular, electrical and electronic engineers
(ii) Astronomers as a whole and radio astronomer in particular
(iii) Physicists
(iv) Astrophysicists
(v) Cosmologists
(vi) Researchers (including Master and PhD students) in the fields of radio astronomy and astrophysics
(vii) Amateur astronomy hobbyists

ORGANIZATION OF THE BOOK

The goal of this book is to provide a complete coverage of both common and technical knowledge on radio telescopes and radio astronomy. The chapters are organized in a structural and coherent manner. Upon finishing this book, readers shall be able to appreciate radio astronomy from both the engineering and science perspectives. A brief description of each of the chapters is given below:

- **Chapter 1:** This chapter provides a historical walkthrough on the development of radio astronomy and radio telescopes. Significant discoveries and works pioneered by scientists in radio astronomy are highlighted here. At the end of this chapter, Nobel laureates who have led to ground breaking contributions in this field are listed as well.
- **Chapter 2:** This chapter provides a complete overview of the working principle of the astronomical instruments involved in the construction of a radio telescope. The underlying physics of the components in a radio telescope, ranging from the antenna to the front and back-end systems are illustrated.

- **Chapter 3:** In this chapter, the design and construction of a simple home-brew radio telescope are illustrated. The main components include an offset prime-focus Ku-band reflector antenna, RF detector circuit, Arduino microcontroller, and a computer. The viability of the telescope is demonstrated at the end of the chapter.
- **Chapter 4:** In this chapter, the relationship between electric and magnetic fields shall be explained. The topics covered in this chapter include Maxwell's Equation, wave polarization and propagation, as well as, the electromagnetic spectrum.
- **Chapter 5:** Astronomy can only be studied by analysing the signals (i.e. electromagnetic waves) emanated from cosmic sources. Hence, the basic principles which govern wave propagation is elaborated in more detail in this chapter.
- **Chapter 6:** In this chapter, the fundamental physical optics (PO) formulation and calculated results are elaborated in detail. Methods which improve the conventional PO to the extended PO such as "physical theory of diffraction (PTD)" and "PO with transition current (PTD-TC)" are also highlighted.
- **Chapter 7:** Apart from physical optics, Gaussian optics is an alternative method popularly used by researchers in the design of radio telescopes. This chapter describes Gaussian optics, with emphasis given on its application in designing radio telescopes.
- **Chapter 8:** The horn is commonly used as a feed element for large radio telescopes. In this chapter, the fundamentals of the pyramidal horn antenna are explained in detail. Numerical and measured examples, as well as codes used to compute the geometry of a horn are also shown here.
- **Chapter 9:** In the receiver of a radio telescope, the incoming wave at the horn is channelled to the mixer block via waveguide-to-probe transition (Yeap, Tham, Nisar, & Loh, 2013; Yeap et al., 2018). The probe used to couple the signal can be a microstrip antenna (Yassin & Withington, 1996). This chapter gives a detail elaboration on microstrip patch antennas. The topics covered in this chapter include the models employed in designing and analysing patch antennas, the important parameters used in characterizing the antenna and the classification of the antenna radiation regions.
- **Chapter 10:** This chapter is a sequel of its preceding (i.e. chapter 9). In order to improve the performance of an antenna (such as a microstrip antenna), an active device, such as an amplifier, could be integrated with it. This chapter discusses the active integrated antenna.
- **Chapter 11:** When space dust rushes into the atmosphere, meteor bursts are formed. Being a good reflector for radio waves, meteor bursts are used in the field of communication. In this chapter, the basics of meteor burst communication and its applications are outlined.
- **Chapter 12:** While the previous chapter (i.e. chapter 11) outlines the basic concepts of meteor burst communication, this chapter emphasizes on the applications of it. It introduces some experiments on meteor burst communications conducted in Japan and the method of a contact (QSO) by meteor scatter communications conducted by amateur radio operators.
- **Chapter 13:** The radio emissions from the Sun and planets provide a new window to explore and improve our understanding of these celestial objects and its magnetic activities. This chapter explains radio emissions observed from the Sun and planets which can generally be divided into thermal and non-thermal forms.
- **Chapter 14:** Two common sources of radio waves in the universe are pulsar and radio galaxies. Pulsar is a type of neutron star that is highly magnetized and rotates at enormous speed. Radio galaxies, on the other hand, are active galactic nuclei that are very luminous at radio wavelength.

Like its preceding chapter (i.e. chapter 13), this chapter illustrates these two cosmic sources observed at radio frequencies.

- **Chapter 15:** Being the remnant of the early universe, the cosmic microwave background (CMB) comprises the secrets of the origin and the evolution of the universe. These secrets are unravelled in this chapter.
- **Chapter 16:** The constant depictions of contact with extra-terrestrial life and their constant basic presence in science fiction show the deep human desire for connection and transcendence with other life forms. This chapter describes the endeavours that have been made to locate outer space intelligence.

CONCLUSION

The study of astronomy has had centuries of history since mankind started to gaze at the stars to appease their curiosity about the universe. However, our understanding about the universe only burgeons when the epiphany that radio waves conceal a plethora of the information struck us less than a century ago. Radio telescopes have been erected and exploited to collect radio signals from astronomical sources since then.

This book gives a detail elucidation on the operation of radio telescopes and the revelation of astronomy through the vista of radio waves. The chapters that compose this book were written by experts and/or researchers in their diverse fields. Besides providing a detail explanation on the topics related to radio astronomy, the authors attempt to share with the readers some of the recent advancements in those domains as well.

Although it may not be possible to cover all the bases in this field, we hope that the selected topics in this book could help to inspire researchers or scientists working in this field in some way or another and to motivate amateur astronomers to explore further.

Kim Ho Yeap
Universiti Tunku Abdul Rahman, Malaysia

Kazuhiro Hirasawa
University of Tsukuba, Japan

REFERENCES

Yassin, G., & Withington, S. (1996). Analytical expression for the input impedance of a microstrip probe in waveguide. *International Journal of Infrared and Millimeter Waves*, *17*(10), 1685–1705. doi:10.1007/BF02069582

Yeap, K. H., Goh, Z. Y., Lai, K. C., Nisar, H., Teh, P. C., & Rizman, Z. I. (2018). Analysis of hammerhead probes in a rectangular waveguide. *Journal of Telecommunication, Electronic, and Computer Engineering*, *10*(1-4), 115–118.

Yeap, K. H., & Tham, C. Y. (2018). Optimization of an offset receiver optics for radio telescopes. *Journal of Infrared, Millimeter and Terahertz Waves*, *39*(1), 64–76. doi:10.100710762-017-0449-z

Yeap, K. H., Tham, C. Y., Nisar, H., & Loh, S. H. (2013). Analysis of probes in a rectangular waveguide. *Frequenz Journal of RF-Engineering and Telecommunications*, *67*(5-6), 145–154.

Yeap, K. H., Tham, C. Y., Yassin, G., & Yeong, K. C. (2011). Propagation in Lossy Rectangular Waveguides. In A. A. Kishk (Ed.), *Electromagnetic Waves Propagation in Complex Matter* (pp. 255–272). UK: IntechOpen.

Section 1

An Overview of Radio Astronomy and Radio Telescopes

Chapter 1
Introduction to Radio Astronomy and Radio Telescopes

Kim Ho Yeap
https://orcid.org/0000-0001-7043-649X
Universiti Tunku Abdul Rahman, Malaysia

Kazuhiro Hirasawa
https://orcid.org/0000-0002-1510-1141
Universiti of Tsukuba, Japan

Humaira Nisar
https://orcid.org/0000-0003-2026-5666
Universiti Tunku Abdul Rahman, Malaysia

ABSTRACT

Radio astronomy studies the physics and chemistry of cosmic sources and cosmic phenomena at the radio wave spectrum. Radio telescopes are built to detect radiation within this spectrum. This chapter provides a historical walkthrough on the development of radio astronomy and radio telescopes. Significant discoveries and works pioneered by scientists in radio astronomy are highlighted. Nobel laureates who have led to groundbreaking contributions in this field are listed.

INTRODUCTION TO RADIO ASTRONOMY

Astronomy is the branch of science which studies the physical and chemical properties of cosmic sources and phenomena. It applies the laws of physics and chemistry in an attempt to explain the behavior of these sources and phenomena. Being one of the oldest fields of science, the history of astronomical study can probably be dated back to as early as 3000 years ago, when the ancient Egyptians, Chinese, Mayans and Babylonians studied the sky by observing the stars at night with their naked eyes. The observa-

DOI: 10.4018/978-1-7998-2381-0.ch001

tions were believed to serve the purposes of religious ceremonial and agricultural activities. The oldest evidence of astronomical observation is the Nebra sky disk which was found in northern Europe. The 30 cm circular bronze disk is believed to be constructed in the 1600s BC. Images of the sun, crescent moon, and stars were engraved on the disk. Hence, most conjectured that the disk is an instrument once used for astronomical purposes.

Although Galileo Galilei was not the inventor of optical telescopes (the invention of which is credited to German-Dutch spectacle-maker Hans Lippershey, Dutch spectacle-maker Zacharias Janssen, and Dutch lens-maker Jacob Metius), the Italian astronomer and physicist was known to be the person who has enhanced the design of the optical telescope so that it was sophisticated enough for astronomical observation. Being the pioneer of astronomical observation in the modern era in early 1600s, Galileo earned himself the name “the father of observational astronomy”. Since the advent of optical telescopes, a wealth of information related to cosmic sources and events – particularly those in our solar system have been discovered.

The optical telescope, however, renders limitations. The signals detected by the optical telescope are only restricted to those within the visible region of the electromagnetic spectrum. It is to be noted that cosmic sources, such as stars, galaxies, planets, quasars, pulsars, etc. do not only radiate naturally occurring signal within this region; important spectral and spatial information is also found in the radio wave region. The radio image of the Centaurus A galaxy at 1.4 GHz is depicted in Figure 1 (Feain, Cornwell, Ekers, Morganti, & Junkes, n.d.). Since radio waves can penetrate molecular dust clouds and that cosmic sources emit more radio waves than visible light, the detection of signals in the radio wave region very often unravels information failed to be accounted for by the visible light. The signal emission detected from the dust clouds in the interstellar medium (ISM), for example, allows astrophysicists to study the physical and chemical conditions during the formation of a star. Also, the cosmic microwave background (CMB) radiation, which is believed to be the remnant of the Big Bang, consists of information of the early universe when it was merely 375,000 years old (Planelles & Biffi, 2017). Studying cosmic sources at radio waves therefore opens the realm of radio astronomy.

Radio telescopes, which consist primarily of a parabolic reflector antenna, a receiver, a detector circuit, amplifiers, and a data processing unit, are built to detect radio waves. Figure 2 depicts the structure of a 70 m radio telescope antenna at the Canberra Deep Space Communication Complex (CDSCC) (Kerton, 2009). Since the wavelengths of radio waves are larger than the visible light, i.e. larger than 1 mm, the size of a radio telescope is much larger than its optical counterpart. As a matter of fact, the larger is the size of the parabolic reflector antenna, the higher is the sensitivity and resolution of the telescope. This is to say that, if the antenna is to collect more energy radiated from a faint object or to resolve very close objects, then the reflector of which has to be sufficiently large.

HISTORICAL HIGHLIGHTS OF RADIO ASTRONOMY

The history of radio astronomy has gone a long way since the framework of electromagnetism was formed more than 100 years ago. In this section, the work and discovery made by the pioneers of radio astronomy throughout all these years are progressively highlighted.

Figure 1. Radio waves radiated from the Centaurus A galaxy (Feain, Cornwell, Ekers, Morganti, & Junkes, n.d.)

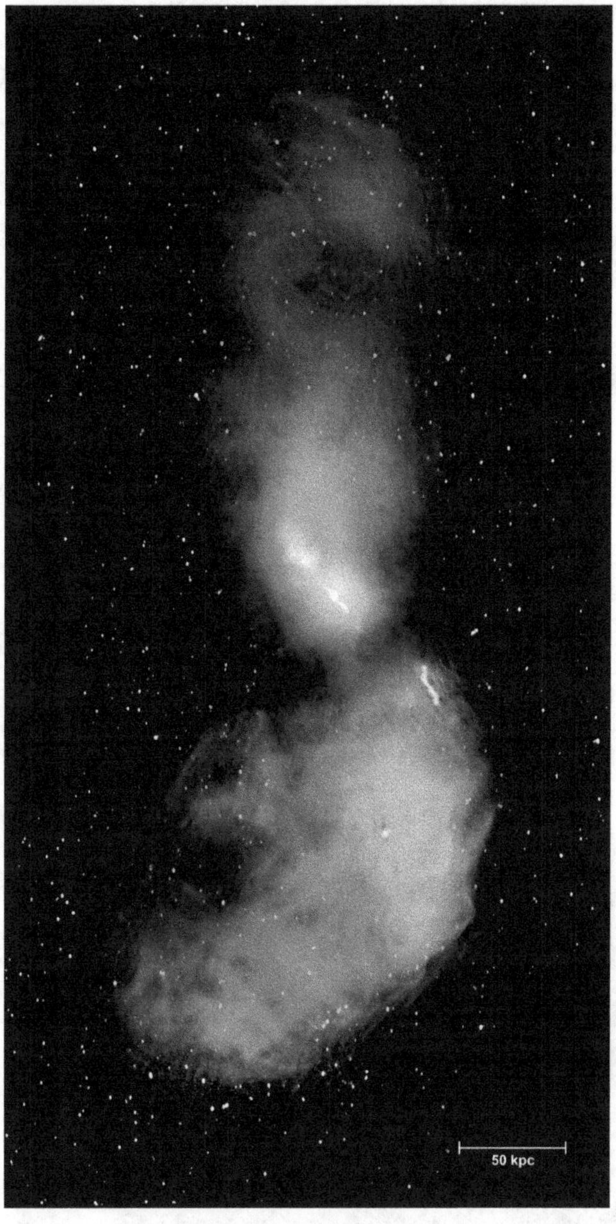

Formulation of Maxwell's Equations in 1881

Inspired by the experimental observations reported independently by three scientists (i.e. French physicist Andre-Marie Ampere, British physicist Michael Faraday, and German physicist Johann Carl Friedrich Gauss), Scottish physicist James Clerk Maxwell established the unified theory of electricity and magnetism. In 1873, Maxwell published his formulations in his textbook "A Treatise on Electricity and Magnetism". The complexity of the formulations in the textbook was later reduced by English

Figure 2. The 70-m antenna at CDSCC (Kerton, 2009)

physicist Oliver Heaviside in 1881 to four sets of differential equations (Yeap & Hirasawa, 2019). These four sets of notable mathematical equations outline the fundamental principles of electromagnetism and have contributed significantly towards our understanding on electromagnetic fields and waves, as well as, the rapid advancement of telecommunication technology today.

Experimental Validation of Waves in the Late 1800s

In the late 1800s, German physicist Heinrich Rudolf Hertz conducted a series of experiments which validated the presence of electromagnetic waves theorized by Maxwell. Among his experimental works, there was one in which he constructed a Ruhmkorff coil-driven spark gap and one-meter wire pair as a transmitter at one end, and a half-wave dipole antenna as a receiver at the other. A micrometer spark gap was left opened between the transmitter and receiver. Hertz was able to show in his experiment the propagation of radio waves. In 1886 to 1889, he was also able to show that waves propagating in free space are in the form of transverse electromagnetic TEM mode. He had also devised methods to measure the magnitude of electric field, as well as, the polarization and reflection of waves.

Discovery of Cosmic Radio Waves in 1931

In the early 19th century, scientists were already aware that the visible light is part of the electromagnetic spectrum. They also knew that radio waves and x-rays are present in the spectrum and that these two types of electromagnetic waves have, respectively, longer and shorter wavelengths in comparison to visible light. But no one realized that cosmic sources radiate radio waves, not until the serendipitous discovery made by American radio engineer Karl Guthe Jansky in August 1931. Hence, Jansky is regarded as one of the founders of radio astronomy. When working for Bell Telephone Laboratories in Holmdel, New Jersey in 1931, Jansky was delegated the task to investigate the effect of wave interference caused by thunderstorm towards trans-atlantic telecommunications systems. He built a rotatable Bruce array antenna to detect the signal (see Figure 3). The antenna was connected to a receiver which was able to display the magnitude of the static signals on a strip-chart. After performing his measurements in all directions for a few months, Jansky was able to classify the recorded signals into three types – two of which were concluded to be caused by the distant and nearby thunderstorms. However, there was one where he was not able to identify the source of it and was rather perplexed with it. For over a year, Jansky attempted to investigate the unknown signal. He learned that the signal was not constant. It repetitively rose to its peak intensity and then gradually fell at every interval of 23 hours and 56 minutes. The cycle of the unknown signal was in agreement with the period of the earth's rotation with respect to the stars (which is more technically referred to as the sidereal day by astronomers). By mapping the direction of the signal with the astronomical maps, Jansky concluded that the signal had its origin from the Milky Way and it was the strongest in the constellation of Sagittarius.

First Survey of the Radio Sky from 1937 to 1943

The advancement in the field of radio astronomy came close to a hiatus during World War II. American amateur astronomer Grote Reber was most likely the only person who persisted to work in this field during that period. Having learned of Jansky's work, Reber constructed his own radio telescope in his backyard in Wheaton, Illinois in September 1937. Reber was the first who used a parabolic reflector

Figure 3. Jansky's merry-go-round antenna

antenna when constructing his telescopes. Due to its high directivity, radio astronomers have since resorted to the reflector antenna for radio astronomy. During his initial attempts to detect radio waves from the sky, Reber encountered several setbacks. Since radio waves at higher frequencies carry higher energy, Reber tried to detect the extraterrestrial signals at 3.3 GHz and then 900 MHz at the beginning. However, both attempts ended in vain. In his third attempt, Reber modified his receiver to detect signals at 160 MHz and was finally successful in detecting the radiation along the Milky Way, confirming the discovery made by Jansky. During World War II, Reber spent his time drawing out the radio sky map. He completed the contour radio map in 1941 and the extended version in 1943. In the map, several areas were revealed to have high emissions of radio sources – the one with the highest intensity was found to be at the centre of the Milky Way. Reber's map also revealed discrete sources which were later associated with the supernova remnant Cassiopeia A and galaxy Cygnus A (Kellermann & Moran, 2001). Reber was also the first scientist who detected the emissions from the Andromeda galaxy.

Discovery of the Sun as a Radio Source in 1942

During World War II, English physicist James Stanley Hey was hired by the Army Operational Research Group (AORG) to work on radar anti-jamming methods. In February 1942, Hey detected significant interference at radar stations along the south coast of England. The direction of the interference appeared to be following the sun as it traversed across the sky. Hey consulted London's Royal Observatory and was informed that there was an active sunspot traversing the solar disc which was emitting streams of energetic ions and electrons of approximately 100 G of magnetic fields. This led to Hey's conclusion that the Sun was a radio source itself.

Improvements in Radio Astronomical Instruments in 1946

Although there was not much progress in the development of radio astronomy during World War II, knowledge in this field grew in leaps and bounds after the war. This, to a certain extent, can be attributed to the radio equipment and military devices developed during the war – in particular, the radar system. From the outbreak of the war in 1939 until the end of it in 1945, English physicist Sir Alfred Charles Bernard Lovell worked for the Telecommunications Research Establishment (TRE) to develop radar systems for the detection and navigation of aircrafts. The development of radar led to the improvements in antennas and electronics. After the war, Lovell acquired an ex-military radar detector set. To avoid interference from the surroundings, he moved to Jodrell bank, an open area 30 km away from Manchester. Using his radar set there, he was able to obtain the radar echoes reflected from the ionized trails of meteors. In 1946, the first radar reflections from the moon were detected at 111 MHz.

Discovery of the Hydrogen Line from 1944 to 1951

Being enshrouded by dust clouds along the galactic plane, Dutch astronomer Jan Hendrik Oort was frustrated that observation could only be performed up to a few thousand light years towards the galaxy centre before the energy from visible light was completely absorbed by the dust particles. Knowing that radio waves are capable of penetrating dust clouds, he postulated that detecting radiation from the radio wave spectrum would allow him to learn more about the galactic centre and possibly the opposite side of the galaxy. When visiting Leiden observatory from January to April 1944, Oort discussed with his student Hendrik Christoffel van de Hulst the detection of radio waves radiation from the Milky Way. Since hydrogen is the most abundant element in the universe, both Oort and Hulst devoted concerted effort on the search for the spectral lines of hydrogen. Meanwhile, having heard of Hulst's proposal in 1949, American physicist Edward Mills Purcell and his student, Harold Irving Ewen in Harvard, decided to join the search for the hydrogen spectral line as well. On 10th March, 1950, Oort and Hulst's laboratory in Kootwijk was badly struck by a fire. The disaster had delayed their progress on the search significantly. On 25th March, 1951, Purcell and Ewen discovered the 21 cm (or at 1.42 GHz frequency) radio emission from the ground state of neutral hydrogen atoms; while Oort and Hulst, with the assistance from the newly employed engineer Lex Muller, detected the line almost 6 weeks later after them (Strom, 2013). Ewen completed his doctorate study with the detection of the hydrogen line and he and Purcell made no further attempts on the study. For Oort and Hulst, however, the successful discovery of the hydrogen line motivated them to pursue further. From July 1952 to June 1953, the Dutch team in Kootwijk worked on the mapping of the neutral hydrogen in the Milky Way galaxy. Completing the map and having it published, Hulst, Muller and Oort revealed the spiral arms structure of the galaxy in 1954. By combining with the southern part of the galaxy mapped by Australian astronomer Frank John Kerr and his team, a more complete map of the Milky Way galaxy was produced and published in 1958 (Oort, Kerr, & Westerhout, 1958).

Discovery of Quasars in 1960

After World War II, English astronomer Sir Martin Ryle who, also worked for the TRE during the war, formed the Cambridge Radio Astronomy group. In late 1940s, Ryle developed the radio interferometer (Sullivan, 1991). By 1950, Ryle and his team had spawned a list of 50 radio stars using the interferometer

(Sullivan, 1991). They published the first catalogue of radio sources in the Monthly Notices of Royal Astronomical Society in 1950, which was referred to as the 1C catalogue (Ryle, Smith, & Elsmore, 1950). Due to its poor resolution, however, the exact positions of the radio sources failed to be accurately identified in the catalogue. In 1955 and 1959, respectively, the improved second (2C) and third (3C) catalogues were subsequently published in the Memoirs of the Royal Astronomical Society (Shakeshaft, Ryle, Baldwin, Elsmore, & Thomson, 1955; Edge, Shakeshaft, McAdam, Baldwin, & Archer, 1959). In 1962, A. S. Bennett revised the 3C catalogue and published it in the Monthly Notices of the Royal Astronomical Society, i.e the 3CR catalogue (Bennett, 1962). In early 1960s, Ryle and his colleagues introduced the aperture synthesis technique – by moving portable antennas, a large aperture could be synthesized (Kellermann & Moran, 2001). By incorporating aperture synthesis into the interferometer, they produced the 4C catalogue. Subsequently in mid 1960s, the 5C catalogue was produced using the One-Mile telescope (Kellermann & Moran, 2001).

In 1960, two American astronomers, Thomas A. Matthews and Allan Rex Sandage discovered a faint blue star-like object in the 3C catalogue and by 1963, they had found three of these objects. The spectrum of these objects was found to be very different from that of the usual blue stars. In 1963, Dutch astronomer Maarten Schmidt found that these objects had extraordinarily large red-shifts. These quasi-stellar objects or quasars were later learned to be the most distant cosmic sources. During this period, there existed two contradicting theories which explained the origin of the universe. The first one was the steady state theory proposed by Anglo-Austrian cosmologist Sir Hermann Bondi, Austrian astrophysicist Thomas Gold and British astronomer Fred Hoyle. In the steady state theory, space and time are postulated to be homogeneous and eternally remains unchanged. The other one was the big bang theory proposed by Belgian Roman astronomer Georges Henri Joseph Edouard Lemaitre. The big bang theory hypothesizes that the universe was expanded from a high density and high temperature state and it is constantly expanding. This theory enjoyed more popularity then, since it was observationally confirmed by American astronomer Edwin Hubble who, using the Hooker optical telescope, demonstrated that galaxies were moving away from each other. The discovery of the quasars further shows that the universe is evolving in time, instead of being held constantly in a steady state. It, therefore, provides further support to the validity of the big bang theory.

Discovery of the CMB in 1965

The discovery of the cosmic microwave background (CMB) marked a quantum leap in the development of physical cosmology. Like Jansky's, however, the discovery was also made through sheer luck. While working on the detection of neutral hydrogen in July 1965, American astronomers, Arno Allan Penzias and Robert Woodrow Wilson were puzzled by the persistent noise interference picked up from all directions by their 6 m horn-shaped antenna in Holmdel, New Jersey. They ruled out all possible sources of interference that they could think of. They suppressed the noise caused by radar and radio broadcasting and cooled down their receiver to -269°C to avoid thermal noise. They also cleaned the pigeons' nesting and droppings in the antenna. However, the noise remained. Although the big bang theory was gradually gaining traction during this period, the competition between it and the steady state theory was still quite stiff. American astrophysicist Robert Henry Dicke, Canadian-American cosmologist Phillip James Edwin Peebles, and American cosmologist David Todd Wilkinson from the Princeton University believed that if the big bang theory was valid, then radiation from the explosion, albeit being red-shifted to the microwave region, should be permeating the space. Having heard of this prediction and after carefully confirming that the characteristics of the noise fitted exactly those described by the scientists from Princeton University, Penzias and Wilson finally realized that the noise was actually the remnant survived from the early stage of the universe. Although the afterglow was initially coined the primeval fireball, it is more commonly referred to as the cosmic microwave background (CMB) or the cosmic microwave background radiation (CMBR) nowadays. This cosmic source has since become the most convincing evidence for the big bang theory.

Discovery of Pulsars in 1967

Yet another scientific breakthrough in the field of radio astronomy was made in 1967. British astronomer Antony Hewish and his PhD student Jocelyn Bell Burnell built a dipole array antenna which was initially meant for the study of quasars. On 28th November, 1967, Bell observed an unusual signal which emitted pulses at a regular interval of 1.3373011 s at the same location of the sky, kept to the sidereal time. Puzzled with the strange behavior of the signal and carefully not ruling out the possibility of extraterrestrial life, she and her supervisor nicknamed the signal LGM-1, for "little green men". Not long after, however, they found a second pulsing signal in a different part of the sky. This time the signal was emitting at an interval of 1.2 s. This led them to drop the ludicrous idea of the source originating from intelligent life forms in space. They then changed the name of the signals from LGM to CP, which denotes Cambridge pulsars. Astrophysicists believe that pulsars are rapidly rotating neutron stars which emit intense radio waves from the north and south poles. Figure 4 shows the artist's impression of the XTEJ1810-197 pulsar which lies 10,000 light years away in the constellation of Sagittarius (Rowe, n.d.). The discovery of the pulsars was important as it was the first clue which hinted the existence of black holes.

Discovery of Binary Pulsars in 1974

The first binary pulsar was discovered by American astrophysicist Joseph Hooton Taylor, Jr. and his PhD student Russel Alan Hulse in 1974. Taylor sent Hulse to study pulsars using the largest radio telescope in the world, the Arecibo telescope in Puerto Rico. On 2nd July, 1974, Hulse detected a pulsar which behaved in a peculiar manner. Instead of pulsing at a constant rate, the pulse appeared to have varying frequencies. However, he also noticed that the pulsar exhibited a periodic pattern – it would repetitively return to its original frequency at every 7.75 hrs before increasing and then decreasing at a fixed rate of frequencies again. Both Taylor and Hulse came to the conclusion that the only explanation to this bizarre phenomenon was that the pulsar itself was orbiting another companion star, which was most likely a neutron star. The changes of frequencies were caused by the Doppler Effect – the frequencies red-shifted and then blue-shifted when it moved, respectively, away from and towards earth. The discovery made quite an impact in the astrophysics community as it provided a firm validation to Einstein's general theory of relativity, as well as the first experimental proof for the existence of gravitational waves. According to the general theory of relativity, when a pair of binary stars accelerates in a strong gravitational field, they will emit energy in the form of gravitational waves. As the objects lose energy, they will inspiral towards each other, gradually reducing the orbital distance, and eventually combining together. The measurement that Taylor and Hulse worked out showed that the orbits are shrinking at 1 cm each day which was in agreement with that predicted by Einstein's theory. It is to be noted that the Arecibo telescope played a very significant role in this discovery. Had it not been for the high sensitivity of the radio telescope, the binary pulsar (which was rather faint) may not have been discovered by the two scientists.

Figure 4. An artist impression of Magnetar XTE J1810-197 (Rowe, n.d.)

Launch of COBE in 1989

When the CMB was discovered in the 1960s, the radiation appeared to be uniformly distributed in all directions. Scientists were rather bothered by the uniformity of the radiation. Theoretically speaking, if the universe was highly uniform, then matters would have failed to coalesce into stars and galaxies through gravitational effect. This would then refute the foundation of the big bang theory. To find out the distribution of the CMB across the sky in more detail, scientists concluded that a more sensitive detector was necessary. On 18th November, 1989, the National Aeronautics and Space Administration (NASA) of the United States launched the Cosmic Background Explorer (COBE) satellite in California. The COBE mission was dedicated primarily for the development of cosmology. COBE carried three instruments, namely the Far Infrared Absolute Spectrophotometer (FIRAS), the Differential Microwave Radiometers (DMR), and the Diffuse Infrared Background Experiment (DIRBE). COBE collected data in space with a resolution of about 7° (about 420 arcmin) from 1989 to 1993. Measurements from the FIRAS showed that the CMB radiation follows closely a blackbody spectrum; whereas those from DMR indicated that small random fluctuations of temperature (i.e. anisotropy) on the order of 1 part in 100,000 exist throughout the sky (Mather & Hinshaw, 2008). Also, from the blackbody curve, the temperature of the CMB was determined to be at an average of 2.725 ± 0.002 K. The DIRBE was successful in mapping the faint infrared and submillimeter background light radiated from the early stars and galaxies.

Launch of WMAP in 2001

Building on the success of the COBE mission, NASA launched a second microwave space telescope on 30th June, 2001. Initially known as the Microwave Anisotropy Project (MAP), the telescope had its name changed later to Wilkinson Microwave Anisotropy Project (WMAP) as a tribute to American physicist David Todd Wilkinson who died in 2002, for his immense contribution to both COBE and MAP. As compared to COBE, WMAP had much improved temperature sensitivity and resolution (about 13.5 arcmin). The satellite telescope operated for nine years from 2001 to 2010. The data obtained from WMAP revealed that a temperature variation of 0.0002 K exists in the CMB, the leftover glow was emitted about 375,000 years after the inflation, the age of the universe is about 13.77 billion years old, and stars formed about 400 million years after the big bang. WMAP also measured the E-mode polarization of the CMB and the composition of the universe. It also showed that the universe was ever expanding and that the observable edge (which is technically referred to as the particle horizon) of it was 14,000 Mpc (about 45.66 billion light years) away (Halpern & Tomasello, 2016).

Launch of Planck in 2009

The successor of WMAP is the European Space Agency's Planck satellite which was launched on 14th May 2009 and ceased operation on 23rd October 2013. The satellite was formerly known as the Cosmic Background Radiation Anisotropy Satellite and Satellite for Measurement of Background Anisotropies (COBRAS/SAMBAS). However, it was renamed in 1996 in tribute of German physicist Max Karl Ernst Ludwig Planck. The purpose of the Planck satellite was to map the sky with an even higher sensitivity and resolution than that of WMAP. To suppress thermal noise, the detectors were cooled close to absolute zero. Planck observed from the microwaves to the very far infrared range. It consisted of two instruments, i.e. the Low Frequency Instrument (LFI) which operated from 30 GHz to 70 GHz and the High Frequency Instrument (HFI) which operated from 100 GHz to 857 GHz. With much better sensitivity and resolution (about 5 arcmin), Planck was able to produce more accurate readings than its two predecessors. From the data obtained from Planck, scientists found that the universe is actually expanding more slowly than that predicted previously, the composition of the universe contains of more matter, the universe is flat, and that the universe is 13.8 billion years old, which is much older than that predicted by WMAP; while the particle horizon is 0.7% smaller, i.e. 13,900 Mpc (about 45.34 billion light years) away (Halpern & Tomasello, 2016).

HISTORICAL DEVELOPMENTS OF GROUND-BASED RADIO TELESCOPES

Since the first discovery of cosmic radio waves made by Jansky, there has been significant advancement in the technology of radio telescopes. The development of ground-based radio telescopes according to chronological time is summarized in this section.

Figure 5. Reber's radio telescope

First Radio Telescope in 1930

The first ever built radio telescope could be dated back to the fall of 1930. It was the antenna that Jansky used to detect signals at 20.5 MHz. As illustrated in Figure 3, the antenna was simply a 30 m long and 6 m tall Bruce array, made of brass pipes. In order to receive signals from all directions, the scaffolding was mounted onto a rotatable platform, with four Ford model-T tyres affixed on the platform. The antenna was dubbed Jansky's merry-go-round due to the fact that it could be rotated in a circular fashion.

Second Radio Telescope in 1937

The second radio telescope was built by Reber in Illinois in 1937. It was the first telescope which used a parabolic reflector to receive signals from the sky. Modern radio telescopes today resemble the one that Reber built. Although reflector antennas consume relatively larger space, they are able to obtain larger collecting areas and higher angular resolution over a wider range of frequencies. They are, therefore, the preferred option when it comes to radio astronomy. As can be seen in Figure 5, Reber's adopted the prime focus configuration for his antenna design, i.e. the receiver is mounted on top of the reflector with the aperture of the feed facing the reflector. The reflector dish was made of sheet metal, the diameter of which was approximately 9.5 m in size. It was used to focus radio waves to the feed, mounted around 6 m above it. Radio signals from cosmic sources are usually very faint. Hence, an amplifier circuit was installed in the receiver to increase the magnitude of the signal. One drawback of Reber's telescope was that it could only be rotated upward and downward in the elevation direction but could not be turned in the azimuth direction.

Ryle's Radio Interferometer in 1946

The development of the aperture synthesis and the idea of incorporating it into interferometers are credited to Ryle. In late 1940s, Ryle came up with the idea of an interferometer (Sullivan, 1991). In Ryle's design, the signals collected from two separate antennas could be combined via aperture synthesis to provide excellent angular resolution for the telescope. By increasing the spacing to the point where the fringe amplitudes became weaker, the size of the radio source could then be estimated. Although the first optical interferometer built for astronomical purposes was invented by American physicist Albert Abraham Michelson in 1890, Ryle denied that the idea was inspired by Michelson's design. Nevertheless, the basic principal of Ryle's radio interferometer was somewhat similar to Michelson's optical interferometer (Kellermann & Moran, 2001). In 1946, Ryle and his team members conducted the first astronomical observation using a radio interferometer. The interferometer was an array of eight halfwave dipole antennas, mounted onto a support covered with a chicken-wire reflecting screen (Sullivan, 1991). The antennas were connected by coaxial cable to a receiver. Ryle named the receiver the "cosmic radio pyrometer". However, it is more affectionately known as the Ryle-Vonberg receiver today (Sullivan, 1991; Ryle & Vonberg, 1948). At the receiver, the signal collected from the antenna was compared with a reference noise source. The magnitude of the noise was continually varied in correspond to the changes in the antenna signal. Hence, as long as the source could track gain changes accurately and quickly enough, the impact caused by variations in the rest of the receiver parameters would become minimal (Sullivan, 1991). In 1948, Ryle became interested with Hey's discrete source of emission in the constellation Cygnus. When studying the source at 80 MHz using his interferometer, not only had he and his then research student Sir Francis Graham Smith detected the emission from the Cygnus source, they had also inadvertently discovered Cassiopeia A, the strongest radio source in the sky (Sullivan, 1991).

The 66-m Transit Telescope in 1947

When working in Jodrell Bank, Lovell built the 66 m Transit telescope in 1947. The parabolic reflector was supported by 7.3 m long of scaffolding poles and the focal point of which was 38 m above the ground. The structure of the telescope was quite rigid and was non-steerable. The telescope was swiftly replaced 10 years later by the Lovell telescope.

First Earth Rotational Synthesis in the 1950s

The first earth rotational synthesis was applied in the 1950s for the purpose of solar radio astronomy (Wendt, Orchison, & Slee, 2008). The solar grating array was built by Australian astronomer Wilbur Norman Chirstiansen in Potts Hill, Australia. It comprised two rows of arrays, along the east-west (E-W) and north-south (N-S) baselines. Prime focus reflector antennas with a diameter of 1.7 m were adopted. The E-W array had a total length of 213 m and it comprised 32 antennas evenly separated apart at an interval of 7 m; whereas, the N-S array was slightly shorter, with a total length of 160 m and it comprised only 16 antennas. This is to say that, the angular resolution of the N-S baseline was relatively lower than that of the E-S baseline. A dipole probe was mounted at the feed of the prime focus.

The antenna outputs were combined using a branching system of transmission lines (Wendt, Orchison, & Slee, 2008). A superheterodyne receiver was connected to the array transmission lines via a radio-frequency (rf) switch which operated at a rate of 25 Hz. The modulated signal was directed to a mixer which was coupled to an oscillator and a 30 MHz amplifier with 4 MHz bandwidth (Wendt, Orchison, & Slee, 2008). Subsequently after the 30 MHz amplifier, the signal was fed to a further detector, a 25 Hz amplifier, a phase-sensitive detector, and finally a recording milli-ammeter (Wendt, Orchison, & Slee, 2008). In order to map the two-dimensional image, Fourier analysis was to be performed on each of the one-dimensional distribution, for the different scanning angles. As computer technology was relatively obsolete during that period, it took nearly six months of manual calculation for Christiansen and another Radiophysics staff Joe Warburton to complete an image then (Frater, Goss, & Wendt, 2017). In 1955, Christiansen decided to abandon Potts Hill grating array, as he planned to construct a new array proposed by Australian engineer Bernard Yarnton Mills at Fleurs. With the operation of the more sophisticated 21 cm 64 elements Fleurs array (which was more popularly known as the "Chris-Cross", since it constituted a cross shape and was developed by Christiansen) in 1957, the Potts Hill array finally completed its journey of operation.

The 25-m Dwingeloo Telescope in 1956

After the successful mapping of the Milky Way galaxy, Oort felt that there was a need to build a more sensitive radio telescope which could produce higher angular resolution. With the financial support from Netherlands Foundation for Radio Astronomy (NFRA) and Post, Telephone and Telegraph service (PTT), Ben G. Hooghoudt was appointed to design and supervise the new telescope. As soon as Hooghoudt came out with the design in 1954, construction process immediately took place at Dwingeloo, Netherlands. On 3rd and 30th November 1955, however, construction work was temporarily interrupted. The partially completed telescope was used to observe the occultation of the Crab nebula by the moon at 75 cm wavelength (Woerden & Strom, 2007). In April 1956, the telescope was inaugurated by Queen Juliana. The telescope was of prime focus configuration, the reflector of which had a 25 m diameter and 12 m focal length. It was covered with a wire mesh of 15 mm spacing, had a surface accuracy of 1 cm, and the beamwidth at 21 cm was 36 arcmin (Woerden & Strom, 2007). Before the inception of the Lovell telescope, the Dwingeloo telescope used to be the largest in the world. Throughout more than 40 years of its operation, the telescope had brought vast amount of scientific discoveries. Using the telescope, scientists were able to map the hydrogen lines with greater detail, discover the high velocity clouds, and observe galaxies behind our Milky Way, just to name a few of its contributions. It is to be noted that, galaxies Dwingeloo 1 and Dwingeloo 2 were named after the telescope (Woerden & Strom, 2007). Dwingeloo was decommissioned in 1999. Being designated as a protected national monument today, the telescope, as shown in Figure 6, is still being properly preserved (and operated) by the C. A. Muller Radio Astronomy Station (CAMRAS), a volunteer organization of amateur astronomers and ham radio enthusiasts (Baars & Karcher, 2018).

The 75-m Lovell Telescope in 1957

Due to the limitation of the Transit radio telescope (i.e. it was non steerable), Lovell decided to build a better replacement for it. Although the rigid Transit telescope was designed and built by astronomers, Lovell realized that a fully steerable one would simply be too challenging for them. He gathered that

Figure 6. The 25-m Dwingeloo radio telescope

an engineer with the appropriate professional experience and knowledge should be assigned to handle it. In September 1949, British civil engineer Henry Charles Husband took up the task to build the new telescope. In January 1950, Husband produced the initial drawings of the design and in September 1952, the construction of the telescope commenced. The construction process encountered delay in the work progress which was partially due to the changes proposed by Lovell in an attempt to improve the design. Nevertheless, the first fully steerable radio telescope managed to be completed in 1957. Upon its completion, the size of the telescope reflector was 75 m, surpassing that of the Dwingeloo's and rendering it the largest radio telescope of its time. Indeed, the telescope is still one of the largest of its kind today. On 12th June 1957, the antenna was rotated in the azimuth direction for the first time, and on 20th June, it was tilted in the elevation angle for the first time. Initially called the Mark I telescope, it was renamed to Lovell telescope on its 30th anniversary in 1987. The telescope became an iconic symbol of British pride and both Lovell and Husband were knighted for that reason. Over more than six decades of operation, Lovell telescope is still in good shape today (see Figure 7) and it is still being used to make forefront scientific discoveries.

Owens Valley 27-m Telescopes in 1960

Owing to their preoccupation with the Cold War, the development of radio astronomy in the United States was generally recognized to be lagging behind Australia and Europe (including the United Kingdom) (Cohen, 1994; Bowen, 1981). However, things started to make a difference when American physicists Lee Alvin DuBridge and Robert Fox Bacher, as well as, American astronomer Jesse Leonard Greenstein joined the California Institute of Technology (Caltech). Being ardently interested with radio astronomy,

Figure 7. The 75-m Lovell radio telescope

they were determined to set up a radio astronomy program in Caltech. They were aware that it was necessary to look to Australia or Europe for an expert who had the experience of handling large radio astronomy projects. DuBridge who was the president of Caltech then, hired British-Australian astronomer John Gatenby Bolton in 1955. Having years of experience working as a radio astronomer at the Commonwealth Scientific and Industrial Research Organization (CSIRO) in Australia, Bolton was the exact right candidate to take up the task. Bolton established the Owens Valley Radio Observatory (OVRO) and he started making plans of constructing a two-antenna interferometer in Owens Valley. Both antennas consisted of a 27 m diameter reflector and were polar-mounted. The mount had an axis parallel to the earth's pole and another at right angles. The antennas were affixed to L-shaped tracks and were, therefore, movable on the tracks (Cohen, 1994). Although the interferometer was initially planned to be operated at 400 MHz, higher frequencies was found to be desirable during the design and construction process. The first system installed into the interferometer operated at 750 MHz. But it was quickly replaced with 960 MHz (Cohen, 1994). In April 1959, the team started making observations using the antenna at the east-west arm. The telescope was regarded as state of its art at that time, though only one antenna had been completed then. The size of the antenna at the east-west arm was even larger than the Dwingeloo's telescope, and the only telescope which could provide better sensitivity and resolution than the Owens Valley's was the Lovell telescope (which was known as Mark I then). Various significant discoveries had been achieved using the single-antenna telescope. One of which was its success in identifying the 3C 295 galaxy to be eight times farther away than Cygnus A (Cohen, 1994). The entire interferometer was finally put in shape upon the completion of the north-south arm in late 1960. Being able to provide two dimensional measurements, the interferometer had been able to achieve even more excellent scientific contributions. The discovery that most extragalactic sources are double and determining the characteristics of the radio emitting cloud around Jupiter were some of the many breakthroughs (Cohen, 1994) achieved using the interferometer.

Parkes 64-m Telescope in 1961

The construction of the 64 m Parkes Telescope in Australia was initiated by Welsh physicist Edward George Bowen who was chief of the CSIRO's Radiophysics Laboratory then. Having fear that the Radiophysics Laboratory would lose its world leader status in radio astronomy when the Mark I telescope was completed, Bowen started to make plans to build a telescope which would be as good as, if not better than, the one in Jodrell Bank (Robertson, 2011). There was a dispute within the Radiophysics Laboratory about the types of radio telescope they were to build – whether it should be a large two dimensional cross array, known as Mills Super-Cross (which was obviously proposed by Mills), a large fully steerable dish, or a new solar radio telescope. At the end of the day, the fully steerable dish prevailed (Kellermann & Moran, 2001). In the early 1950s, Bowen successfully convinced two philanthropic organizations – the Carnegie Corporation and Rockefeller Foundation for financial sponsorships. The initial design of the structure was drawn out by English engineer Sir Barnes Neville Wallis (Bowen, 1981); whereas, Feeman Fox and Partners in London were chosen for the detailed design (Robertson, 2011). The final design of the telescope was completed in April 1959. In July 1959, German firm Maschinenfabrik Augsburg Nurnberg (MAN) was chosen to be the prime contractor for the construction work (Robertson, 2011). The size of the reflector dish was initially planned to be 70 m. When the construction process started to over-run, however, it was reduced to 64 m to contain costs (Bowen, 1981). In January, 1960, Australian government agreed to cover the remaining shortfall (Robertson, 2011). By August 1961, the construction work was completed and the Parkes Telescope had its official opening on 31st October, 1961 (Robertson, 2011). The telescope was built on the altitude/azimuth mount. One unique feature of it is that it consists of a master equatorial – a small optical telescope placed at the intersection of the altitude and azimuth axes of rotation. The direction of the reflector dish is driven by the master equatorial using a servo control mechanism, allowing the telescope to achieve high pointing accuracy (Robertson, 2011). Bolton who helped to build the Owens Velley interferometer returned to Australia and became the inaugural director of Parkes telescope (Robertson, 2011). Figure 8 shows the appearance of the telescope today (Kerton, 1989). For almost six decades, the Parkes telescope has been the driving force for the growth of radio astronomy in Australia.

Green Bank 90-m Telescope in 1962

On 24th August, 1956, the National Radio Astronomy Observatory (NRAO) was established to manage the national facility for research in radio astronomy in the United States (US) (Leverington, 2017). It is financially supported by the National Science Foundation (NSF) and managed by Associated Universities Incorporated (AUI) – a consortium of formed by Columbia, Cornell, Harvard, John Hopkins, Massachusetts Institute of Technology (MIT), Pennsylvania, Princeton, Rochester, and Yale Universities (Leverington, 2017). In June 1958, NRAO signed a contract to build a 42.67 m radio telescope at Green Bank. Having learned that the completion of the telescope would be significantly delayed, a few of the

Figure 8. The 64-m Parkes radio telescope (Kerton, 1989)

senior members of NRAO, including John Wilson Findlay, David Sutphin Heeschen, and Frank Donald Drake, proposed to build a 90 m telescope as quickly and as inexpensively as possible (Leverington, 2017). In 1959, the NRAO staff came out with the preliminary design of the telescope. Since the telescope was meant to be "inexpensive", the targeted lifetime of it was only five years (Leverington, 2017). In January 1960, the budget for building the telescope was approved and construction subsequently started in 1961. The duration of the entire construction process took only about 23 months. The 90 m radio telescope was erected in September 1962 (Leverington, 2017). The 90-m reflector was mounted on two towers and the surface of it comprised 16 mm Squarex aluminum (Leverington, 2017). The feed of the prime focus configuration was about 39 m above the reflector. The telescope had been an important instrument in radio astronomy. Among some of its significant findings include the discovery of numerous pulsars and the observation of the neutral hydrogen lines in the Milky Way. Very unfortunate though, the 90 m Green Bank telescope collapsed under its own weight on 15 November, 1988 (Leverington, 2017). The collapse of the entire structure should come in no surprise at all, since it was only initially meant to be functional over a period of five years. The extension of its lifetime for more than 20 years could be attributed to the regular inspection and fortification performed on the structure (Leverington, 2017).

Arecibo 300-m Telescope in 1963

The Arecibo radio telescope was the brainchild of American astronomer William Edwin Gordon. When working as a professor in Cornell University, Gordon was keenly devoted towards the study of Ionosphere using radar backscattering technique (Altschuler, 2002). When radar signal is scattered towards the ionosphere layer of the sky, most of the wave would penetrate the layer, with only a very small fraction of its energy reflected back to earth. In order to detect the weak reflection, Gordon realized that a very sensitive antenna is required. The idea prompted him to build a large parabolic reflector antenna. His persistent effort led to the construction of the Arecibo observatory in 1960. Three years later, on November 1, 1963, the Arecibo Ionospheric Observatory (AIO) held its inaugural opening ceremony. Located at Barrio Esperanza, Arecibo in Puerto Rico, the telescope is of the Gregorian configuration type – it constitutes a secondary ellipsoidal reflector placed behind the focal point of the primary parabolic reflector. The site was chosen primarily because it fulfills the criteria of (i) having the shape of a hemispherical valley the size of the primary reflector, which saved a considerable amount of excavation costs, (ii) the climate is reasonable (i.e. not too harsh), and that (iii) the country is politically stable (Altschuler, 2002). The primary reflector has a diameter of 305 m, about 51 m deep and covers an area of 20 acres. A 900 ton platform which supports the antennas and receivers is suspended about 137 m above the primary reflector by 18 cables. The cables are strung from three concrete towers which top at the same elevation (Altschuler, 2002). An azimuth arm lies beneath the platform. The receiver room is housed at one side of the arm, and a Gregorian dome at the other (Altschuler, 2002). Unlike typical Gregorian configuration which has only two reflectors, the Arecibo antenna consists of three – a tertiary reflector is employed to scatter the signal from the secondary reflector towards the feed of the receiver. Both the secondary and tertiary reflectors are mounted in the dome. Vegetation is grown beneath the primary reflector to prevent erosion (Altschuler, 2002). Although the initial idea of the antenna was for ionosphere study, the antenna, upon its completion, was also widely used for planetary studies and radio astronomy. Some of the many important findings made from the Arecibo telescope were the discovery of the binary pulsars and the exact rotation rate of Mercury (Altschuler, 2002). After having been around for almost 40 years, the Arecibo remains as the largest telescope on earth today.

Ryle's One-mile Telescope in 1964

In 1964, Ryle built the one-mile radio telescope at the Mullard Radio Astronomy Observatory (MRAO), Cambridge. The telescope constituted three 18-m reflector antennas arranged in line – two of which were affixed steadfast and were about 750 m apart from each other; whereas the third arm was moveable on 800 m rails (Leverington, 2017). This gave the telescopes a maximum separation of 1.55 km (Leverington, 2017). By allowing the moveable arm to collect signals at different locations spaced 23.5 m apart on each 64 successive days, an effective aperture equivalent to the size of a 1.6 km (i.e. one mile) diameter antenna could be synthesized (Kellermann & Moran, 2001). The telescope underwent its first tests and observation in 1964. The first tests of the telescope were conducted at 408 MHz; while the first observation at 1407 MHz (Leverington, 2017). The one-mile telescope was designed to analyze very faint distant cosmic objects and, in fact, it was successful in determining the properties of individual sources (Kellermann & Moran, 2001). The observations obtained from the telescope resulted in the publication of the 5C catalogue which recorded radio sources as faint as 2 mJy (1 Jy = 10^{-26} $Wm^{-2}Hz^{-1}$) (Kellermann & Moran, 2001). Upon the successful operation of the one-mile telescope, a half-mile telescope was built on the same rails as the one-mile. The half-mile telescope comprised four 9-m reflector antennas and was used specifically to study the 21 cm neutral hydrogen lines (Leverington, 2017).

Very Long Baseline Interferometry in 1967

In the early 1950s, British astronomer Robert Hanbury Brown conceived a novel type of interferometers – one in which the intensity of the radiation, instead of its electric field, collected at different locations to be cross-correlated (Kellermann & Moran, 2001). This type of interferometer allowed the local oscillators in the mixer circuits of different receivers to be independent among each other. Hence, the baselines (i.e. the farthest distance between two separate antennas) would not be limited by geographical factors. The intensity interferometer, however, imposed two limitations. Firstly, the phase was incoherent and, secondly, its sensitivity was relatively lower than the conventional interferometer which is phase-coherent. As a result of this, forming images using the intensity interferometer was rather difficult (Kellermann & Moran, 2001). In January 1967, nevertheless, an astronomical team in University of Florida was successful in using the intensity interferometer to detect the fringes from the burst radiation of Jupiter at 18 MHz (Moran, 1998).

By mid 1960s, astronomers realized that a telescope which could provide very high resolutions (i.e. angular sized less than 0.01 arcsec) was necessary to observe very small cosmic sources such as quasars, active galactic nuclei (AGN), and masers. In other words, interferometers with baselines of thousands of kilometers are much needed to collect signals from these cosmic sources. During this time, high-speed digital tape recorders which enabled the implementation of broadband analog systems and the availability of relatively stable atomic frequency standards became available. These two factors became the catalyst for the development of coherent interferometers with independent oscillators, over practically unlimited baselines (Kellermann & Moran, 2001). The term Very Long Baseline Interferometry (VLBI) has since been coined for this type of interferometers. In VLBI, several radio telescopes located at different parts of the earth are used to collect the intensity of the radio waves. The distance among the telescopes are computed based on the arrival time of the signals. On 8th May, 1967, the joint astronomy team between Cornell University and NRAO demonstrated the workability of the first VLBI. By collecting signals between the Naval Research Laboratory's 26 m antenna at Maryland Point and the 42.67 m Green Bank antenna, the team was successful in obtaining fringes over a baseline of 220 km, at a resolution of 0.5 arcsec (Kellermann & Moran, 2001). In response to the success that they achieved, the Cornell-NRAO team collaborated with the MIT-Haystack astronomy team to form a VLBI of 845 km baseline, between the 36.58 m Haystack antenna and the 42.67 m Green Bank antenna. The MIT team studied OH masers; while the Cornell-NRAO team studied quasars and AGN using the VLBI with a resolution of 0.04 arcsec (Kellermann & Moran, 2001). Having achieved successive accomplishments, the baseline of the VLBI continued to be extended. In 1984, 18 telescopes were used to form 153 baselines for VLBI observations (Kellermann & Moran, 2001). Images in the range of milliarcsec were obtained for the 3C 120 AGN, M 87, the 3C 236 giant galaxy, and the 3C 48 quasar (Kellermann & Moran, 2001).

Owens Valley 40-m Telescope in 1968

In 1962, Caltech submitted a financial proposal to the NSF to add four 38 m antennas to the Owens Valley interferometer and extending its track 390 m to the south (Leverington, 2017). In order to study the feasibility of the idea, the NSF funded the construction of a prototype 40 m antenna in 1964 first. In

1966, Caltech submitted another proposal again. This time, the university requested to build an additional eight 40 m telescopes along the 3 km east-west and 5 km north-south tracks. At this time, however, the NSF was considering the proposal from a few other radio telescope projects, including one which was later named the Very Large Array (VLA) upon its completion years later (Leverington, 2017). Since the NSF was under financial burden then, the plan proposed by Caltech was eventually turned down in 1971. Two years before the NSF made its decision, though, the Westinghouse Electric Corporation completed the construction of the 40 m prototype telescope in 1968. In April 1969, the 40 m telescope was included as part of the Very Long Baseline Interferometer (VLBI) experiment (Leverington, 2017).

Westerbork Synthesis Radio Telescope in 1970

The Westerbork Synthesis Radio Telescope (see Figure 9) was initially a collaboration project between the Netherlands and Belgium governments. Finding it hard to be managed by two nations and realizing the limited involvement it could have on the project, the Belgium government withdrew itself in mid-1967 (Leverington, 2017). The design of the telescope was based on the east-west linear array proposed by Swedish astronomer Jan Arvid Högbom (Leverington, 2017). However, it was subsequently simplified due to insufficient financial support. The modified design consisted of an array of ten 25-m reflector antennas fixed on a 1.6 km east-west baseline, with an additional 25 m antenna moveable along a 300 rail. A second moveable antenna was later added to the interferometer, extending the baseline to 3 km (Leverington, 2017). The telescope was erected in 1970. The telescope originally operated at 21 cm with an angular resolution of 23 arcsec, but was later improved to 2 arcsec at its shortest wavelength of 3.6 cm (Kellermann & Moran, 2001). In the early days of its operation, the telescope experienced considerable noise interference from the North Atlantic Treaty Organization (NATO) transmitter in a military base. The problem was avoided when the military agreed to change the frequency of the device in 1970 (Leverington, 2017). The telescope has been used to observe neutral hydrogen for faint cosmic sources, including the giant galaxies, jets, and clusters of galaxies (Kellermann & Moran, 2001).

The 5-km Ryle Telescope in 1970

Building on the achievements of the one-mile telescope in the 1960s, Ryle proceeded further to build a 5 km radio telescope. The telescope, which was later named as the Ryle telescope in the late 1980s, consisted of eight arms, with eight 13-m reflector antennas at each arm. Four of the arms were fixed at an interval of 1.2 km; whereas the remaining four were moveable on 1.2 km rails (Leverington, 2017). This resulted in a maximum east-west baseline of 4.8 km (Leverington, 2017). With its high sensitivity and resolution, the telescope was used to detect the faintest cosmic sources. The telescope was successful in detecting the Sunyaev-Zel'dovich effect – a phenomenon which causes distortion to the CMB when the radiation encounters high energy electrons in galaxy clusters.

Effelsberg 100-m Telescope in 1971

After World War II, the development of radio wave research in Germany was being restricted (Wielebinski, Junkes, & Grahl, 2011). Because of this reason, radio astronomy in Germany started much later than the US, the United Kingdom (UK), and many other European countries. With the restrictions lifted in the 1950s, the German government funded the construction of a 25-m fully steerable telescope on

Figure 9. The Westerbok Synthesis Radio Telescooe (WSRT)

the Stockert Mountain and a 36-m transit telescope in Berlin-Adlershof (Wielebinski, Junkes, & Grahl, 2011). The construction of the Berlin wall in 1961 prompted the director of Heinrich-Hertz-Institut (HHI), Otto Hachenberg to join the Astronomical Institute of the University of Bonn. Soon after Hachenberg joined the institute, he proposed to build an 80-m radio telescope, which could operate up to as high as 23 GHz (Wielebinski, Junkes, & Grahl, 2011). Owing to his vast experience, particularly in the design of the Dwingeloo Telescope, Hooghoudt had been hired to be the main consulting engineer of the project. Having received more funding from the non-profit organization, Volkswagenstiftung, the size of the reflector antenna was extended to 90 m, and subsequently 100 m. The site of the telescope was decided to be at the north-south valley close to the village of Bad Münstereifel-Effelsberg (Wielebinski, Junkes, & Grahl, 2011). The Effelsberg telescope was of the altitude/azimuth design and it was mounted on a 64-m diameter ring track. One unique feature of the antenna was that a series of reflecting mesh sections was installed to the periphery of the primary reflector, in order to reduce noise interference (Wielebinski, Junkes, & Grahl, 2011). Like the Arecibo telescope, the configuration of the antenna belonged to the Gregorian type. The secondary ellipsoidal reflector was 6.5 m in diameter. The telescope had its first observation on 23rd April, 1971 at 2.7 GHz. On 12th May, 1971, the inaugural opening ceremony of the telescope was held. Observation for the 408 MHz radio continuum survey was performed during its official opening (Wielebinski, Junkes, & Grahl, 2011). The Max-Planck-Institut für Radioastronomie (MPIfR) was founded to manage the operation of the telescope. Among some of the important discoveries made using the Effeslberg telescope were the production of the complete map of the Andromeda Nebula, which revealed its spiral structure, the discovery of halos around clusters of galaxies, the observation of the Zeeman effect, the first identification of the ammonia and water molecules in extragalactic sources, and the discovery of the millisecond pulsars (Wielebinski, Junkes, & Grahl, 2011). These discoveries were of significant values as they showed that the single antenna radio telescope could be as effective as the aperture synthesis array.

In 1973, the Effelsberg telescope was involved in the test of the transatlantic baselines of the VLBI. In 1979, the telescope became an associate member of the US VLBI network (Wielebinski, Junkes, & Grahl, 2011). In 1980, the European VLBI Network was founded (EVN) and Effelsberg telescope became one of its founding members. The telescope also participated in the Joint Institute for VLBI (JIVE) in Dwingeloo. In 1997, the Japanese Space Institute (ISAS) launched the HALCA VLBI satellite and this led to the involvement of the Effelsberg telescope in space VLBI projects (Wielebinski, Junkes, & Grahl, 2011). The telescope also plays an active role in the Megamaser Cosmology Project, in which the Hubble Constant is to be measured to 3% of accuracy and the flatness of the universe and the state of the dark energy are to be studied (Wielebinski, Junkes, & Grahl, 2011). For almost 30 years, the Effelsberg radio telescope had been the largest fully steerable single antenna telescope on earth. Today, the telescope is still actively used for radio astronomy observations. A recent image of the Effelsberg telescope is shown in Figure 10. Since 2007, the telescope has been collaborating with the European Low Frequency Array (LOFAR) telescope for making observations in the metre wavelengths (Wielebinski, Junkes, & Grahl, 2011).

Very Large Array in 1980

In January 1967, NRAO submitted a proposal to build the Very Large Array (VLA) to NSF. The proposal called for 36 25-m reflector antennas rail mounted on three moveable arms arranged in a Y shape. Each arm was 21 km long and the rail tracks of the arms could be propelled by transporters (Leverington, 2017). Due to financial constraints, however, NSF only approved the funding of the project when NRAO agreed to reduce the number of antennas to 27. Since radio wave energy is susceptible to water vapour, the site of the VLA project had to be high in altitude, which usually had low precipitable water in the atmosphere. The array was eventually selected to be on the Plains of San Augustin near Soccoro, New Mexico, which is 2120 m above sea level (Leverington, 2017). The construction of the VLA was approved in 1972. It was quite unfortunate that the US currency depreciated a year later. NRAO had to make adjustments to its original plan – the number of transporters had to be reduced from three to two and the idea of operating the telescope concurrently at two frequencies had to be abandoned (Leverington, 2017). The 25-m altitude/azimuth mounted antennas were based on the Cassgrain configuration where a secondary hyperboloid reflector is used to focus the signal from the primary reflector to the feeds. Four feed horns were placed off-axis beneath the primary reflector. Each feed had two outputs orthogonal to each other, so as to detect both co- and cross-polarization signals (Leverington, 2017). The first antenna was completed in 1975 and the first measurement using two antennas separated by a baseline of 1.2 km was obtained in February 1975 (Leverington, 2017). The first six antennas were operational in 1977 and the entire VLA project was erected in early 1981 (Leverington, 2017). The VLA initially operated at 1.43, 5, 15, and 23 GHz with angular resolutions ranging from 2.1 to 0.13 arcsec (Leverington, 2017). Since the interferometer commences its operation, it has been used by more than 2000 scientists around the world for more over 10,000 observations (Kellermann & Moran, 2001).

Figure 10. The Effelscberg 100-m radio telescope

IRAM 30-m Telescope and Interferometer in 1980s

In 1973, a Scientific Advisory Group for Millimeter Astronomy (SAGMA) was set up to investigate the development of millimeter-wave radio astronomy. A 30-m radio telescope and an interferometer which consisted of four 10-m antennas, operating at frequencies as high as 170 GHz, were proposed (Leverington, 2017). The 30-m Millimeter Radio Telescope (MRT) and interferometer projects were initially a joint collaboration among the UK, France and Germany. As the UK government was facing financial difficulties then, it decided to withdraw from the project. In 1978, France Centre National de la Recherche Scientifique (CNRS) and Germany Max-Planck-Gesellschaft (MPG) established Institut de Radioastronomie Millimetrique (IRAM) to coordinate the development of the project. Being interested in millimeter radio astronomy as well, Spain decided to fund 6% of the project and it became a full member of IRAM in 1990 (Leverington, 2017). The MRT and interferometer were decided to be constructed, respectively, in Pico Veleta, Spain and Plateau de Bure, France. In 1984, the MRT had its first observation at 80 GHz. In 1986, it started performing observations at 230 GHz. Subsequent improvement on it allowed the MRT to operate at 350 GHz. The IRAM interferometer started its construction in 1983 and it was planned to operate at 375 GHz. Three 15-m reflector fully steerable antennas were altitude/ azimuth mounted on a T-shaped rail with transporters used to propel the antennas (Leverington, 2017). The track stretched 160 m across north-south and 288 m across east-west (Leverington, 2017). Two of the antennas of the interferometer at Bure peak are depicted in Figure 11.

James Clerk Maxwell Telescope in 1987

After the UK retracted itself from the IRAM projects (i.e. the 30-m MRT and the interferometer), the UK Science Research Council (SRC) worked together with the Netherlands to build another Millimeter/Submillimeter radio telescope (Encrenaz, Gomez-Gonzalez, Laqueux, & Orchiston, 2011). The SRC planned for a 15-m telescope which could operate at 400 GHz and the site was chosen to be near the peak of Mauna Kea in Hawaii, which is more than 4000 m above sea level. Construction works commenced in 1983 and the telescope had its first observation in 1987. Being named the James Clerk Maxwell Telescope or JCMT in short, the telescope adopted the Cassegrain configuration and was altitude/azimuth mounted.

Australia Telescope Compact Array in 1988

In the mid-1970s, the Australian scientific community realized that the radio astronomical development in the country was gradually falling behind its competitors. In 1981, radio astronomers in Australia proposed to build a new aperture synthesis telescope which could perform observations from 300 MHz to 100 GHz (Leverington, 2017). More popularly referred to as the Australia Telescope Compact Array

Figure 11. Two of the antennas of the IRAM interferometer at Plateau de Bure, France

(ATCA), the interferometer provided a total baseline of 300 km which included the Compact Array at Culgoora, a new antenna at Mopra, and the Parkes 64-m antenna (Leverington, 2017). The ATCA comprised of six 22-m reflector antennas on a 6-km east-west baseline (Leverington, 2017). In order to have it inaugurated in conjunction with the bicentenary of the British settlement, the project was completed in 1988 and had is official opening on 2nd September the same year. However, the telescope was only fully functional in 1990 (Leverington, 2017). From time to time, the ATCA was connected to the VLBI to form a baseline of 3000 km. In later years, the telescope became part of the Australia Long Baseline Array (LBA) (Leverington, 2017). Figure 12 shows five of the 22-m antennas of ATCA (Masterson, n.d.).

Very Long Baseline Array in 1993

After its successful achievement on the VLBI, the NRAO proceeded further to build an array of telescopes distributed across the US in1975 (Leverington, 2017). Being known as the Very Long Baseline Array (VLBA), the interferometry array comprised ten 25-m antennas widely distributed in different parts

Figure 12. Five of the antennas of the Australia Telescope Compact Array (Masterson, n.d.)

of the country. The purpose of VLBA is to produce an effective aperture diameter of several thousand kilometres and angular resolutions in the range of milliarcsecond (Leverington, 2017). The construction project started in 1985 and upon its completion in 1993, the VLBA was able to extend 8600 km south-west and 2900 north-south (Leverington, 2017). The interferometer operated from 330 MHz to 43 GHz and was able to achieve resolutions from 22 to 0.17 milliarcsec (Leverington, 2017). The VLA, as well as the, Arecibo, Green Bank and Effelsberg telescopes subsequently worked together with the VLBA to form the High Sensitivity Array (Leverington, 2017).

The 100 by 110 m Green Bank Telescope in 2001

The collapse of the 90-m Green Bank telescope in late 1988 called for an urgent need to build a new one. In June 1989, NRAO submitted a proposal to the NSF to request for financial assistance for the development of a new radio telescope in Green Bank. The funding was approved in December 1990 and construction work was carried out immediately in 1991 (Prestage et al., 2009). Although the telescope was still amid construction, it had its first observation on 22nd August, 2000. The project was completed in the early fall of 2001 (Prestage et al., 2009). The new Green Bank Telescope or, more affectionately known as the GBT, adopted the altitude/azimuth mount, Gregorian offset configuration. The primary reflector is a 100-m section of a 208-m paraboloid, while the rim is elliptical 110 by 100 m and it provides a focal length of 60 m (Prestage et al., 2009). The offset ellipsoidal secondary reflector is 7.6 by 8.0 m, with a 190 m focal length (Prestage et al., 2009). The receiver placed at the secondary focus consists of eight feeds. Gregorian, rather than Cassegrain, was chosen because it exhibits much lower ground noise for an offset configuration. Without the central blockage, the offset reflectors also produce lower sidelobe levels and internal reflections (Prestage et al., 2009). The GBT (see Figure 13) operates from 100 MHz to 115 GHz. The size of the GBT surpasses the Effelsberg's, rendering it as the largest steerable radio telescope on earth.

Low Frequency Array in 2011

In 1990s, the ASTRON (which was formerly known as the NFRA) and a few other Dutch universities started to look into making observations at the lower end of the radio wave spectrum. Plans to build a Low Frequency Array (LOFAR) telescope were proposed. In November 2003, the Netherlands government agreed to provide strong financial support to the project. In response to the Netherlands' support, a supercomputer was installed at Groningen, Netherlands for processing data collected from LOFAR in 2004 (Leverington, 2017). In 2006, 96 dual-dipole antennas were set up in four clusters in Exloo, Netherlands. 48 of the antennas were at the central and 16 were in the remaining three clusters (Leverington, 2017). In 2007, the first LOFAR station was built next to the Effelsberg telescope in Germany. Since then, more stations were being built in the Netherlands and a few other European countries. LOFAR operates in the range of 120 to 240 MHz and 30 to 80 MHz (Leverington, 2017). The antennas are distributed at an area of 100 km in diameter in the Netherlands and over 1500 km throughout Europe (ASTRON, n.d.). LOFAR celebrated its inaugural opening in June 2010 and started observations in 2011.

Figure 13. The Green Bank Telescope

Jansky Very Large Array in 2012

From 2001 to 2012, the VLA underwent a major upgrade. The project was initially coined the Expanded VLA (EVLA). However, upon its completion in 2012, it was renamed to Jansky VLA (Leverington, 2017). The sensitivity of the new interferometry array has enhanced by 10 times and it can produce angular resolution from 24 to 0.04 arcsec, at a frequency range of 75 MHz to 43 GHz (Leverington, 2017).

Atacama Large Millimeter/Submillieter Array in 2013

The plan to construct the Atacama Large Millimeter/Submillimeter Array or ALMA in short can be traced back to the mid-1980s. Back then, three astronomical teams were looking into the possibility of studying radio waves in the millimeter and submillimeter regions. NRAO proposed to build the Millimeter Array (MMA) which performed observations at millimeter wavelengths from 30 to 350 GHz. The construction of 40 8-m reflector antennas was planned for the MMA. At about the same time, the European Southern Observatory (ESO) was planning to start the Large Southern Array (LSA) project, which also operated in the millimeter wavelengths below 350 GHz. 50 16-m antennas were considered for LSA. The National Astronomical Observatory of Japan (NAOJ) was slightly more ambitious. It was looking into the construction of the Large Millimeter/Submillimeter Array (LMSA) which included frequencies in both the millimeter and submillimeter wavelengths. 50 10-m antennas were proposed for the LMSA project (ALMA, n.d.). Having realized that the projects were simply too tough to be handled alone, the three teams decided to consolidate their projects together. In 1999, a resolution was signed among Europe, North America and Japan to construct ALMA (ALMA, n.d.). Today, the ALMA international consortium has expanded to a much larger group. It includes involvement from the East Asia which consist of Japan, Taiwan, and South Korea; North America which consists of the US and Canada; ESO which represents 16 European countries; and Chile (Hiramatsu, 2017).

The ALMA telescope comprises 66 reflector antennas, distributed widely across an area of 16 km in diameter, allowing it to produce an angular resolution as fine as 0.01 arcsec (Hiramatsu, 2017). The antennas can be divided into two systems, i.e. the 12-m Array which is managed by the US and European teams and the Atacama Compact Array (ACA) which is managed by the Japanese team. The 12-m Array constitutes 50 of the 12-m antennas; whereas, the ACA comprises four 12-m antennas and 12 7-m antennas (Hiramatsu, 2017). The ACA system is also alternatively known as the Morita Array, in honour of the Japanese astronomer Koh-Ichiro Morita of NAOJ, who passed away in Chile amid his work on ALMA.

Sited on the Chajnantor plateau which is 5000 m above sea level in Northern Chile, ALMA is currently the world's most powerful millimeter/submillimeter interferometric radio telescope, operating from 31 to 950 GHz (Yeap & Tham, 2018). ALMA began its observations in September 2011 when it was still partially finish. The telescope was officially commissioned in March 2013 (Yeap & Tham, 2018).

Square Kilometer Array in 2020s

In September 1993, the International Union of Radio Science (URSI) established the Large Telescope Group to chart out plans for a next generation radio observatory (SKA, n.d.). On 10th August, 2000, representatives from 11 countries – China, Australia, Germany, India, Italy, Canada, the Netherlands, US and UK – came together to establish the International Square Kilometer Array Steering Committee (ISSC) (SKA, n.d.). In December 2011, the SKA Organization was formed to manage the development of the Square Kilometer Array (SKA) project. Much similar to the case of the ALMA telescope, although the SKA project started out with the involvement of only 11 countries, the members has now expanded to more than 20 countries.

The Karoo region in South Africa has been chosen to host the antennas for the mid and high frequencies observations – the eventual plan is to extend the antennas all over the African continent though; the Murchison Shire in Western Australia, on the other hand, has been chosen to host the antennas which operate in the low-frequencies (SKA, n.d.). The ultimate plan of the project is to set up a million of low-frequency antennas in Australia and about 2000 mid- and high-frequency antennas in the African continent (SKA, n.d.). The design and development stage of the project started in 2012 and observations with a partial array were deemed to start in the mid-2020s (SKA, n.d.).

Table 1. Nobel laureates in radio astronomy (Nobel Media AB, 2020)

Year of Award	Name	Contributions
1909	Guglielmo Marconi and Karl Ferdinand Braun	in recognition of their contributions to the development of wireless telegraphy
1974	Sir Martin Ryle and Antony Hewish	for their pioneering research in radio astrophysics: Ryle for his observations and inventions, in particular of the aperture synthesis technique, and Hewish for his decisive role in the discovery of pulsars
1978	Arno Allan Penzias and Robert Woodrow Wilson	for their discovery of cosmic microwave background radiation
1993	Joseph Hooton Taylor, Jr. and Russel Alan Hulse	for the discovery of a new type of pulsar, a discovery that has opened up new possibilities for the study of gravitation
2006	John Cromwell Mather and George Fitzgerald Smoot III	for their discovery of the blackbody form and anisotropy of the cosmic microwave background radiation
2011	Saul Perlmutter, Brian P. Schmidt and Adam G. Riess	for the discovery of the accelerating expansion of the Universe through observations of distant supernovae
2019	James Peebles	for his theoretical discoveries on the evolution of the universe
2019	Michel Mayor and Didier Queloz	For their discovery of an exoplanet orbiting a sun-like star

NOBEL PRIZE IN PHYSICS

Throughout the past century, Nobel Prizes in Physics have been awarded to scientists who had made significant breakthroughs in the field of radio astronomy. A list of these Nobel laureates and a summary of their contributions are tabulated in Table 1.

REFERENCES

ALMA. (n.d.). Origins. Retrieved from https://www.almaobservatory.org/en/about-alma-at-first-glance/origins/

Altschuler, D. R. (2002). The national astronomy and ionosphere's center's (NAIC) Arecibo observatory in Puerto Rico. *Single-Dish Radio Astronomy: Techniques and Applications*, *278*, 1–24.

ASTRON. (n.d.). *About LOFAR*. Retrieved from http://www.lofar.org/about-lofar/about-lofar

Baars, J. W. M., & Karcher, H. J. (2017). *Radio telescope reflectors: Historical development of design and construction*. Switzerland: Springer.

Bennett, A. S., & Simth, F. G. (1962). The preparation of the revised 3C catalogue of radio sources. *Monthly Notices of the Royal Astronomical Society*, *125*(1), 75–86. doi:10.1093/mnras/125.1.75

Bowen, E. G. (1981). History of Australian astronomy. *Publications of the Astronomical Society of Australia*, *4*(2), 267–273. doi:10.1017/S132335800001660X

Cohen, M. H. (1994). The Owens Valley radio observatory: Early years. *Engineering and Science, 57*, 8–23.

Edge, D. O., Shakeshaft, J. R., McAdam, W. B., Baldwin, J. E., & Archer, S. (1959). A survey of radio sources at frequency of 159 Mc/s. *Memoirs of the Royal Astronomical Society, 68*, 37–60.

Encrenaz, P., Gomez-Gonzalez, J., Lequeux, J., & Orchiston, W. (2011). Highlighting the history of French radio astronomy. 7: The genesis of the institute of radioastronomy at millimeter wavelengths (IRAM). *Journal of Astronomical History and Heritage, 14*, 83–92.

Feain, I., Cornwell, T., Ekers, R., Morganti, R., & Junkes, N. (n.d.). The Centaurus A Galaxy. CSIRO. Retrieved from https://www.scienceimage.csiro.au/image/10850/the-centaurus-a-galaxy

Frater, R. H., Goss, W. M., & Wendt, H. W. (2017). *Four Pillars of Radio Astronomy: Mills, Christiansen, Wild, Bracewell*. Switzerland: Springer. doi:10.1007/978-3-319-65599-4

Halpern, P., & Tomasello, N. (2016). Size of the observable universe. *Advances in Astrophysics, 1*(3), 135–137. doi:10.22606/adap.2016.13001

Hiramatsu, M. (2017). ALMA and its observational achievements: Unveiling the mysteries of the dark universe. *Fujitsu Scientific and Technical Journal, 53*, 9–14.

Kellermann, K. I., & Moran, J. M. (2001). The development of high-resolution imaging in radio astronomy. *Annual Review of Astronomy and Astrophysics, 39*(1), 457–509. doi:10.1146/annurev.astro.39.1.457

Kerton, R. (1989). Parkes radio telescope. CSIRO. Retrieved from https://www.scienceimage.csiro.au/image/2720/parkes-radio-telescope/

Kerton, R. (2009). The 70 metre antenna (DSS 43) at the Canberra Deep Space Communication Complex. CSIRO. Retrieved from https://www.scienceimage.csiro.au/image/10808/the-70-metre-antenna-dss-43-at-the-canberra-deep-space-communication-complex/

Leverington, D. (2017). *Observatories and Telescopes of Modern Times: Ground-based Optical and Radio Astronomy Facilities since 1945*. UK: Cambridge University Press. doi:10.1017/9781139051507

Masterson, J. (n.d.). Australia Telescope Compact Array. CSIRO. Retrieved from https://www.scienceimage.csiro.au/image/3881/five-antennas-at-narrabri/

Mather, J. C., & Hinshaw, G. F. (2008). Cosmic background explorer. *Scholarpedia, 3*(3), 4732. doi:10.4249cholarpedia.4732

Moran, J. M. (1998). Thirty years of VLBI: Early days, successes, and future. *International Astronomical Union Colloquium*, 164, 1–10.

Nobel Media AB. (2020). All Nobel prizes in physics. Retrieved from https://www.nobelprize.org/prizes/lists/all-nobel-prizes-in-physics/

Oort, J. H., Kerr, F. T., & Westerhout, G. (1958). The galactic system as a spiral nebula. *Monthly Notices of the Royal Astronomical Society, 118*(4), 379–389. doi:10.1093/mnras/118.4.379

Planelles, S., & Biffi, V. (2017). A mysterious universe: Revealing the bright and dark sides of the cosmos. *Metode Science Studies Journal, 7*, 153–161.

Prestage, R. M., Constantikes, K. T., Hunter, T. R., King, L. J., Lacasse, R. J., Lockman, F. J., & Norrod, R. D. (2009). The Green Bank telescope. *Proceedings of the IEEE, 97*(8), 1382–1390. doi:10.1109/JPROC.2009.2015467

Robertson, P. (2011). An Australian icon: Planning and construction of the Parkes telescope. Science with Parks @ 50 Years Young, Australia.

Rowe, J. (n.d.). An artist's impression of magnetar XTE J1810-197 showing the radio emissions and the magnetic field. CSIRO. Retrieved from https://www.scienceimage.csiro.au/image/3210/an-artist-s-impression-of-magnetar-xte-j1810-197-showing-the-radio-emissions-and-the-magnetic-field-/

Ryle, M., Smith, F. G., & Elsmore, B. (1950). A preliminary survey of the radio stars in the northern hemisphere. Monthly Notices of the Royal Astronomical Society, 110, 508–523. doi:10.1093/mnras/110.6.508

Ryle, M., & Vonberg, D. D. (1948). An investigation of radio-frequency radiation from the sun. *Proceedings of the Royal Society of London. Series A, Mathematical and Physical Sciences, 193*(1032), 98–120. doi:10.1098/rspa.1948.0036 PMID:18865816

Shakeshaft, J. R., Ryle, M., Baldwin, J. E., Elsmore, B., & Thomson, J. H. (1955). A survey of radio sources between declinations -38° and +83°. *Memoirs of the Royal Astronomical Society, 67*, 106–154.

SKA. (n.d.). *The SKA project.* Retrieved from https://www.skatelescope.org/

Strom, R. G. (2013). How was atomic HI (λ = 21 cm line) in space discovered? *International Journal of Modern Physics: Conference Series, 23*, 474–477.

Sullivan, W. T., III. (1991). Some highlights of interferometry in early radio astronomy. *International Astronomical Colloquium*, 131, 132–149.

Wendt, H., Orchiston, W., & Slee, B. (2008). W. N. Christiansen and the development of the solar grating array. *Journal of Astronomical History and Heritage, 11*, 173–184.

Wielebinski, R., Junkes, N., & Grahl, B. H. (2011). The Effelsberg 100-m radio telescope: Construction and forty years of radio astronomy. *Journal of Astronomical History and Heritage, 14*, 3–21.

Woerden, H. V., & Strom, R. G. (2007). Dwingeloo – the golden radio telescope. *Astronomische Nachrichten: Astronomical Notes, 328*(5), 376–387. doi:10.1002/asna.200710767

Yeap, K. H., & Hirasawa, K. (2019). Introductory Chapter: Electromagnetism. In K. H. Yeap, & K. Hirasawa (Ed.), Electromagnetic fields and waves (pp. 1–12). UK: InTechOpen.

Yeap, K. H., & Tham, C. Y. (2018). Optimization of an offset receiver optics for radio telescopes. *Journal of Infrared, Millimeter, and Terahertz Waves, 39*(1), 64–76. doi:10.100710762-017-0449-z

Chapter 2
Fundamentals of a Radio Telescope

Kim Ho Yeap
https://orcid.org/0000-0001-7043-649X
Universiti Tunku Abdul Rahman, Malaysia

Kazuhiro Hirasawa
https://orcid.org/0000-0002-1510-1141
Universiti of Tsukuba, Japan

ABSTRACT

In radio astronomy, radio telescopes are used to collect radio waves emanated from cosmic sources. By analyzing these signals, the properties of the sources could be unraveled. A telescope typically consists of the following astronomical instruments: a primary and a secondary reflector, receiver optics which usually includes a lens or a pair of mirrors and a pair of feed horns (one for each orthogonal polarization [or simply a corrugated horn with an orthomode transducer OMT]), waveguides, a mixer circuit, a local oscillator, amplifiers, a detector circuit, and a data processing unit. This chapter provides a concise but complete overview of the working principle of the astronomical instruments involved in the construction of a radio telescope. The underlying physics of the components in a radio telescope, ranging from the antenna to the front-end and back-end systems, are illustrated.

INTRODUCTION

In radio astronomy, radio telescopes are built to observe naturally occurring signal emission from celestial objects and phenomena, such as stars, galaxies, planets, quasars, pulsars, active galactic nuclei (AGN), etc. Extraterrestrial signals within the radio wave regime are rich with spectral and spatial information which is important in the field of astrophysics. By analyzing these signals, knowledge about the universe, as a whole, and a specific cosmic source, in particular, could be unraveled. The signal emission detected from the dust clouds in the interstellar medium (ISM), for instance, allows astrophysicists to study the physical and chemical properties during the formation of a star. Similarly, the Cosmic Microwave Back-

DOI: 10.4018/978-1-7998-2381-0.ch002

ground (CMB) radiation provides information of the early universe when it was approximately 375,000 years old. It also reveals the age and composition of the universe.

The latest generation of radio telescopes, such as those being employed in the Atacama Large Millimeter/submillimeter Aray (ALMA) and Square Kilometer Array (SKA) projects, are designed for multiple frequency bands with receivers for each individual band offset from the antenna axis. These telescopes typically consists of the following astronomical instruments: a primary and a secondary reflector, receiver optics which usually includes a lens or a pair of mirrors and a pair of feed horns (one for each orthogonal polarization) (Yeap & Tham, 2018), waveguides, a mixer circuit, a local oscillator, amplifiers, a detector circuit, and a data processing unit. These components are summarized in the functional block diagram depicted in Figure 1. To understand the process how a radio telescope detects and interpret the signal emanated from a celestial object, one can imagine the electromagnetic energy propagates from a far-field radio source. When the beam, in the form of a collimated bundle of radio waves, is incident upon the primary reflector of the dish antenna, it is scattered to the secondary reflector. The secondary reflector, in turn, illuminates the first and then the second mirrors (assuming that the focusing elements in the receiver optics are a pair of mirrors). The radio frequency (RF) signal is eventually focused to a feed antenna, which is usually a horn. At the feed, the RF signal is directed, via waveguides, to a receiver circuit. The mixer circuit at the receiver modulates the RF signal with another signal generated by a local oscillator. The process of modulation down converts the signal to an intermediate frequency (IF) signal, while maintaining its original phase. This IF signal is then fed to a detector circuit after going through amplification. The purpose of the detector is to perform wave rectification, allowing it to be processed by a data processing unit. In general, the input signal from the reflector antennas undergoes two stages, namely, the frond-end and the back-end. The components from the receiver optics to the mixer are considered as part of the front-end system; whereas, those from the amplifier to the data processing unit (which includes both the computer and display) belong to the back-end system. In this chapter, the astronomical instruments involved in the construction of the radio telescope, ranging from the parabolic dish antenna to the front-end and back-end components will be illustrated.

REFLECTOR ANTENNAS

Due to its large collecting areas and relatively high angular resolution over a wide frequency range, circular parabolic reflector antennas are commonly employed in radio telescopes. According to the theory of reciprocity, when a beam of collimated rays is incident upon a parabolic reflector, it will converge to a focal point; and, likewise, when a point source is placed at the focal point, the rays scattered from the reflector will emerge as a collimated beam (Balanis, 2005). Although the purpose of the antenna is to collect radio signals from distant celestial sources, the opposite approach is usually adopted when designing a radio telescope. Since the location among different sources may vary, it is, therefore, more convenient to assume that the antenna is transmitting rather than receiving during its design stage.

Various geometrical configurations exist in the design of reflector antennas. The most common types are the Cassegrain, Gregorian, and prime focus designs. Figures 2 to 6 show the optical arrangement for these three types of design. The design of the prime focus (which is also known as the front-fed) is the simplest. It comprises only a single primary reflector, with a feed antenna mounted directly at the focal point. The aperture of the feed is facing downwards towards the reflector dish. The Cassegrain and Gregorian antennas, on the other hand, are multiple-reflector antennas, i.e. they consist of at least two

Figure 1. A simplified functional block diagram of a radio telescope

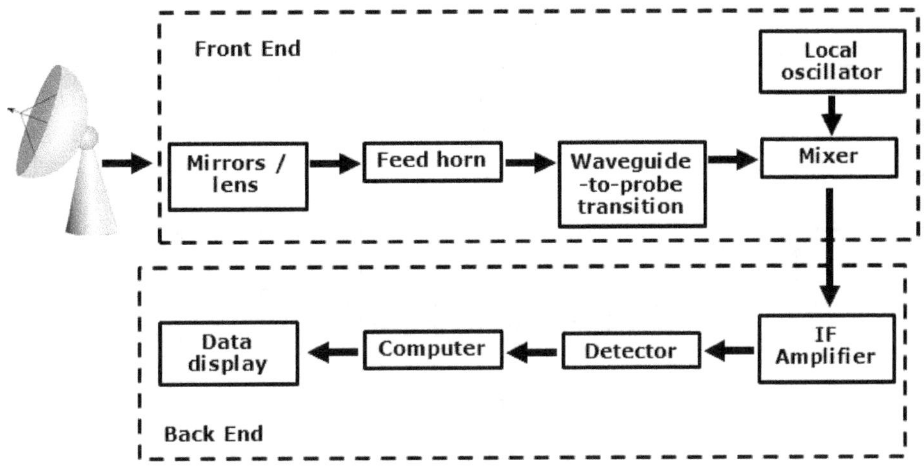

reflectors. Both these antennas were named, respectively, after their inventors, Laurent Cassegrain and James Gregory. Unlike the prime focus, the feed of the Cassesgrain and Gregorian is placed below the primary reflector with its aperture facing upward. Multiple-reflector types are usually preferred over the single reflector prime focus for several reasons. Firstly, the magnifying secondary reflector can increase the ratio of the effective focal length f to reflector diameter D, i.e. f/D ratio. This produces a larger focal plane which allows multiple feed horns to be installed onto it. Secondly, the feed is directed towards the cold sky, instead of the warm ground. This arrangement minimizes the risk of the antenna from picking up thermal noise interference. Thirdly, loss from transmission lines can be reduced. Since the feed of the prime focus is installed above the dish, long transmission lines are required to channel the energy intercepted by the antenna to the receiver. Owing to the skin effect, transmission lines turn out to be very lossy at very high radio frequencies. It is to be noted that most emissions from distant celestial sources are usually very faint upon its detection by telescopes – the signals are measured in Jansky (Jy), which is in the order of 10^{-26} W/m^2/Hz. The magnitude of the signal would therefore be severely affected when propagating through a lossy guiding structure. This problem can, however, be minimized if both the

Figure 2. A prime focus configuration

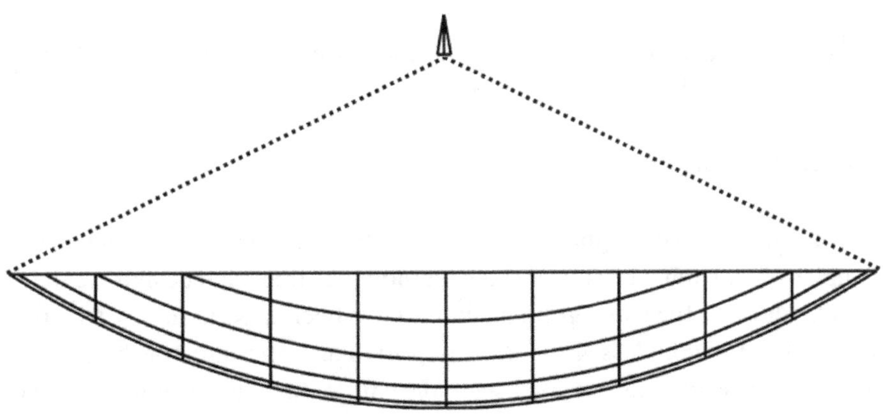

Figure 3. An on-axis Cassegrain configuration

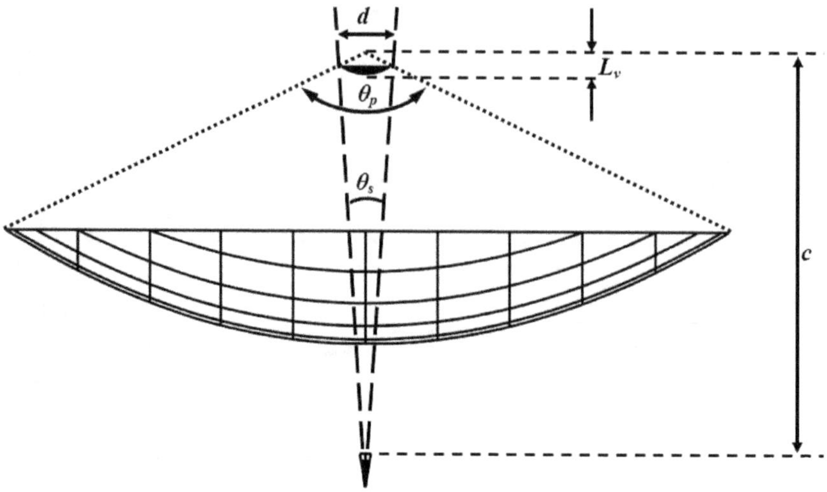

Figure 4. An on-axis Gregorian configuration

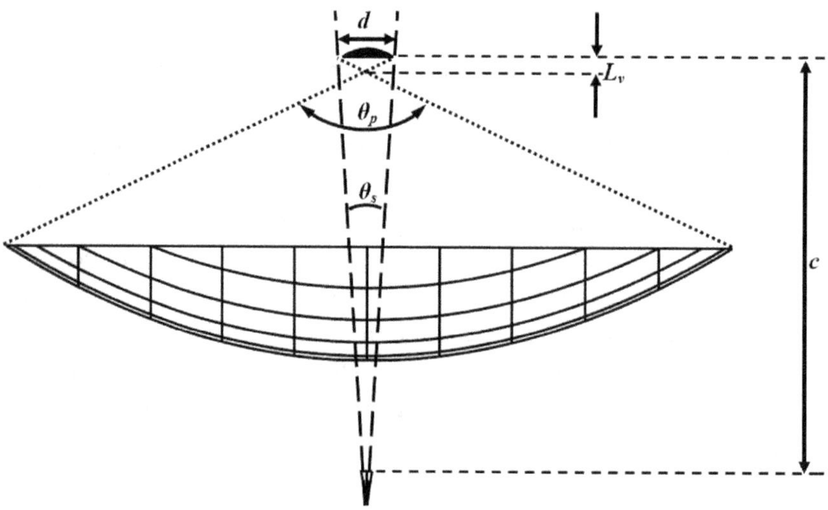

feed and the receiver are placed close together, near the vertex of the primary reflector, instead of its primary focal point (Yeap et al., 2016).

Multiple-Reflector Designs

Both Cassegrain and Gregorian configurations have been very well received in the construction of radio telescopes. Examples of existing telescopes which employ the Gregorian design are the Green Bank and Arecibo telescopes; whereas, the Cassegrain antennas have been used in the ALMA project. As can be observed in Figures 3 to 6, both Cassegrain and Gregorian antennas are constructed using two reflectors – a larger primary reflector and a smaller secondary reflector. The design for each of the reflectors

Figure 5. An offset Cassegrain configuration

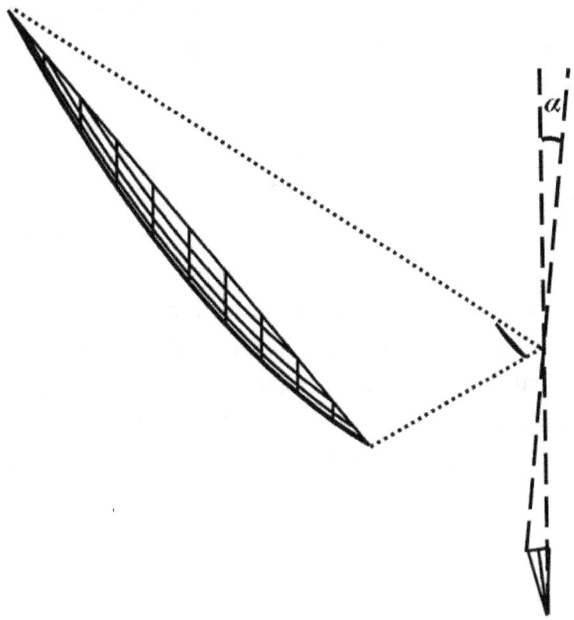

Figure 6. An offset Gregorian configuration

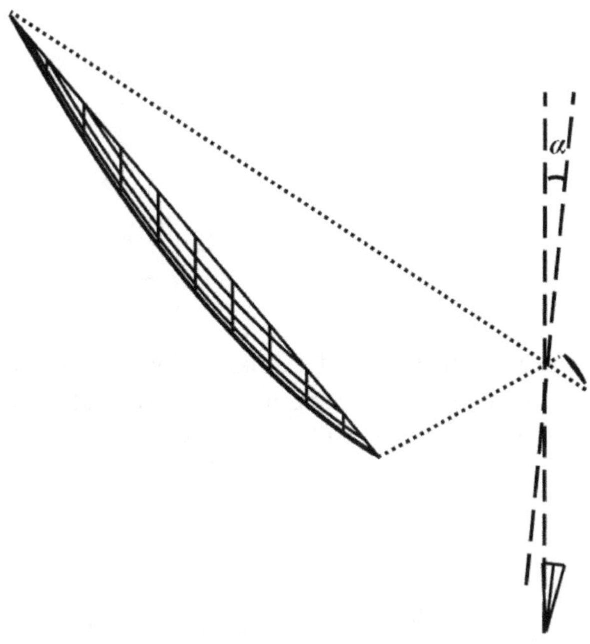

has a separate impact on the field scattered at the aperture plane. It has been shown by Green (1963) and Balanis (2005) that over the aperture plane of a two-reflector design with high magnification, the amplitude and phase distributions are influenced separately by the curvatures of the secondary and primary reflectors, respectively.

To focus a bundle of parallel rays upward to a focal point and vice-versa, the primary reflector has to be in the form of a paraboloid. The secondary reflector is used to intercept the energy scattered from the primary reflector and focus it downwards to a feed. In order to obtain good radiation characteristics, the diameter of the secondary reflector must be at least a few wavelengths (Balanis, 2005). For the Cassegrain design, the secondary reflector is in the form of a hyperbola and it has to be mounted beneath the focal point of the primary reflector. The Gregorian secondary reflector, on the other hand, is ellipsoid in shape and it is mounted above the primary focal point. Since the only difference between both designs lies on the geometrical structure of the secondary reflector, changing from the Cassegrain to the Gregorian design and vice versa requires re-calculating the secondary eccentricity e, the diameter of the secondary reflector d, as well as the distance between the secondary vertex and the primary focal point L_v. The equations for these three parameters are given in (1) to (3) below (Hannan, 1960; Lamb, 1999)

$$e = \frac{M+1}{M-1}, \tag{1}$$

$$\frac{1}{\tan\theta_p} + \frac{1}{\tan\theta_s} = \frac{2c}{d}, \tag{2}$$

$$1 - \frac{\sin\left(\frac{\theta_p - \theta_s}{4}\right)}{\sin\left(\frac{\theta_p + \theta_s}{4}\right)} = \frac{2L_v}{c}. \tag{3}$$

where c is the distance between the primary and secondary focal points, M the magnification, and θ_p and θ_s are, respectively, the angle subtended by the primary and secondary reflectors. For the Gregorian configuration, a negative sign must be employed for M and θ_p.

Offset Designs

The Cassegrain and Gregorian designs shown in Figures 3 and 4 have the feed mounted along the same axis of the primary reflector. Since the feed and the secondary reflector are aligned along the same axis, this type of on-axis configurations has the disadvantage of producing a shadowy region in the aperture. In other words, certain fraction of the wave energy fails to be intercepted by the antenna and as a consequence of this; the spillover efficiency of the antenna deteriorates. To reduce aperture blocking and voltage-standing-wave-ratio (VSWR), offset reflector configurations are employed. As can be seen in Figures 5 and 6, the secondary reflector and feed horn of the offset Cassegrain and Gregorian antennas are placed at one side of the primary reflector. Unlike its on-axis counterpart in which the primary focal point is affixed along the central axis, the focal point of the offset antenna is tilted to the side of the primary reflector. This type of optical arrangement eliminates the spillover effect due to the central blockage. Besides improving the spillover efficiency, the offset design also allows larger *f/d* ratio, exhibits lower sidelobes, and is more effective in suppressing cross-polarized fields (Balanis, 2005). The *f/d* ratio dictates the subtended angle of the primary reflector. By adjusting the *f/d* ratio, the feed pattern can therefore be improved. The principle polarization intended to be received by the antenna is typically referred to as the co-polarization; whereas, the field component perpendicular to it represents the unwanted cross-polarization. In practice, the cross-polarized fields are to be suppressed so as to optimize the energy carried by the co-polarized fields. The condition to cancel cross polarization in offset designs can be determined by (4) below (Mizugutch, Akagawa, & Yokoi, 1976)

$$M = \frac{|1-e^2|}{(1+e^2)-2e\cos\alpha} \tag{4}$$

where α is the angle subtended between the axis of the primary reflector and that of the sub-reflector. Very often, the offset Gregorian configuration is preferred over its Cassegrain counterpart. This is because the Gregorian offset reflector picks up lower ground noise and, with its arm at the top, allows easier access to the primary focus (Prestage et al., 2009).

Resolution and Sensitivity

The two most critical factors which dictates the performance of a radio telescope is its angular resolution and sensitivity. In simple terms, the resolution of a telescope is the degree of detail produced in an image. The higher is the resolution of a telescope, the finer is the visible image it produces and the easier it is in resolving very close objects. As shown in (5) below, the angular separation θ_r (measured in radians) is in direct proportion to the wavelength of the signal λ, and inversely proportional to the diameter of the primary reflector (or the baseline of an interferometer) *D*.

$$\theta_r = 1.22\frac{\lambda}{D}. \tag{5}$$

The angular separation θ_r defines how close two objects can be resolved. This is to say that, the smaller is the value of θ_r, the better is the ability of the telescope in distinguishing objects placed in very close proximity.

The sensitivity of a telescope is its ability in collecting energy from a radio source. An antenna with a large diameter is able to collect more electromagnetic energy from the sky than a small one. In order to detect very faint celestial objects, a highly sensitive radio telescope with sufficiently large primary reflector is required. The sensitivity of a radio telescope is closely related to the aperture efficiency η_a of the antenna. The aperture efficiency η_a can be defined as the ratio of the maximum effective area of the antenna to its physical area (Balanis, 2005). For a reflector antenna, η_a is a function of the subtended angle and the feed pattern (Balanis, 2005). In general, η_a can be further broken down into the product of the (Balanis, 2005).

1. Fraction of the energy intercepted by the reflector, i.e. spillover efficiency η_s.
2. Uniformity of the amplitude distribution of the radiation over the reflector surface, i.e. taper efficiency η_t.
3. Uniformity of the signal phase over the aperture plane, i.e. phase efficiency η_p.
4. Uniformity of the wave polarization over the aperture plane, i.e. polarization efficiency η_x.
5. Fraction of the energy being blocked from reaching the reflector, i.e. blockage efficiency η_b.
6. Surface irregularities of the reflector, i.e. random error efficiency η_r.

By carefully designing and fabricating the feed and reflector, however, the effect imposed by the non-uniformity of the signal amplitude and phase, as well as the presence of a blockage and surface irregularities can be considered to be negligible (i.e. $\eta_p = \eta_x = \eta_b = \eta_r$» 1). Hence, the remaining factors which influence the aperture efficiency are the spillover and taper efficiencies. In other words, the aperture efficiency η_a of a very carefully designed reflector antenna depends solely on the product of the spillover η_s and taper efficiency η_t, i.e.

$$\eta_a = \eta_s \times \eta_t \,. \tag{6}$$

Figure 7 depicts the relationship of the spillover, taper, and aperture efficiencies as a function of the relative power density at the edge of the aperture (i.e. the edge taper T_e). It is apparent from the figure that both the curves of the spillover and taper efficiencies contradict each other – the spillover efficiency increases along with T_e; while, the taper efficiency gradually decreases. In order to achieve the optimum aperture efficiency η_a, a compromise between the spillover efficiency η_s and the taper efficiency η_t is therefore necessary (Balanis, 2005; Yeap & Tham, 2018). For an unblocked reflector illuminated by an ideal Gaussian beam, the maximum aperture efficiency that can be obtained is 81.45% (Yeap & Tham, 2018).

RECEIVER OPTICS

In order to accommodate multiple frequency bands, the receiver for each band is offset from the primary antenna axis. As can be observed in Figure 8, the receiver optics typically consists of two focusing elements, in the form of ellipsoidal mirrors along the optical path between the feed horn and the antenna focus. In order to ensure that the beam from the secondary reflector could be coupled effectively to the feed horn, it is important to optimize the geometries and optical arrangement of the mirrors. Physical optics (PO) has been known as a common technique for optics design. Despite its high accuracy, however, this technique requires large computational time to search for the optimum parameters for a frequency independent design. It, therefore, renders infeasibility and impracticality to adopt PO for the optimization process. Since Gaussian quasi-optics technique includes the effects of diffraction within a reasonable limit and it provides sufficiently accurate results, it has been adopted to search for the optimum parameters instead (Yeap & Tham, 2018). PO, in this case, is used at the final stage, for the verification of the system performance. In this section, the description of beam propagation in the receiver optics is modeled using the quasi-optics technique.

Figure 7. The aperture efficiency (solid line), spillover efficiency (dashed line), and taper efficiency (dotted line) for an unblocked Cassegrain aperture as a function of edge taper

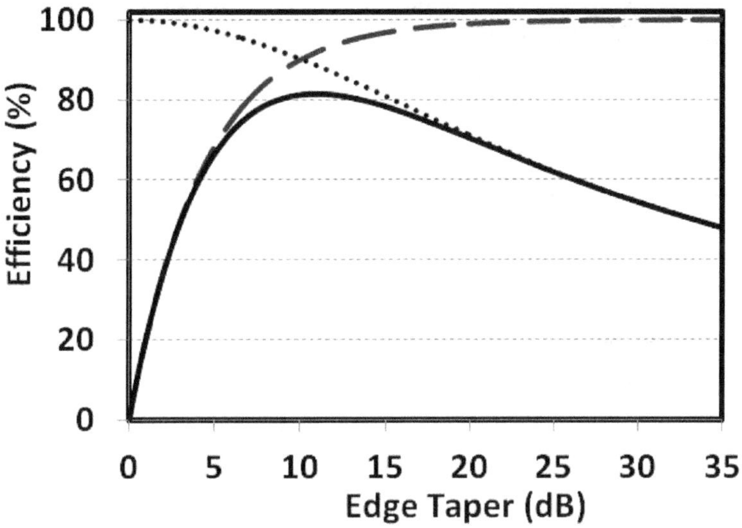

Figure 8. The optical arrangement of a receiver optics

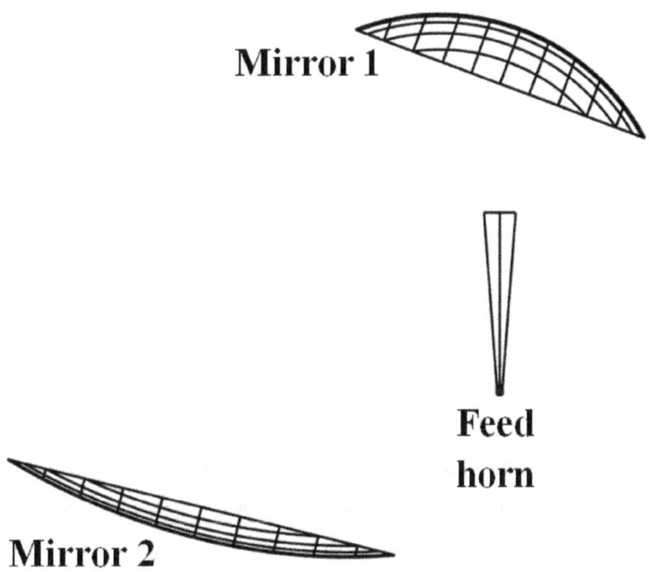

In a quasioptical system, the radius of curvature R, beam radius w and phase slippage ϕ_0 completely characterize the beam at each location along the propagation path. The beam parameters R, w and ϕ_0 are first calculated at the aperture of the feed horn. The optimum beam radius that couples to the fundamental Gaussian mode is given as $w = 0.644a$ for a corrugated horn and $0.76a$ for a conical horn (Goldsmith, 1998) where a is the aperture radius. The radius of curvature R is equivalent to the slant length of the horn R_h and the reference phase slippage at the aperture is $\phi_0 = 0$. The beam is propagated using thin lenses to approximate the ellipsoidal mirrors. The Gaussian beam quasi-optics equivalent of the optical system in Figure 8 is shown in Figure 9. The beam transformation properties of this quasi-optical system can be found using the ABCD law (Halbach, 1964). The beam parameters R and w at free space propagation are obtained as (Yeap & Tham, 2018)

$$R_{out} = \frac{R_{in}^{2}\left[\left(1+\frac{L}{R_{in}}\right)^{2}\left(\pi w_{in}^{2}\right)^{2}+\left(L\lambda\right)^{2}\right]}{\left(\pi w_{in}^{2}\right)^{2}\left(R_{in}+L\right)+L\left(\lambda R_{in}\right)^{2}} \tag{6}$$

$$w_{out} = \sqrt{w_{in}^{2}\left[\left(1+\frac{L}{R_{in}}\right)^{2}+\left(\frac{L\lambda}{\pi w_{in}^{2}}\right)^{2}\right]} \tag{7}$$

where the subscripts *in* and *out* denote respectively the input and output of the optical system, and L is the propagation distance. The phase slippage is given as

$$\Delta\varphi_{out} = \tan^{-1}\left(\frac{\pi w_{out}^{2}}{\lambda R_{out}}\right)-\tan^{-1}\left(\frac{\pi w_{in}^{2}}{\lambda R_{in}}\right) \tag{8}$$

Similarly, the beam parameters R and w at the lenses in Figure 9 (i.e. mirror 1 and mirror 2) are obtained as

$$R_{out} = \frac{fR_{in}}{f-R_{in}} \tag{9}$$

$$w_{out} = w_{in} \tag{10}$$

where f is the focal length of the lens. The value of the radius of curvature R is used to locate the position of the secondary reflector. When the antenna is at focus, the equiphase of R is identical with the surface of the reflector.

FEED HORN

The field structure coupled to the feed can be expressed as a spectrum of hybrid modes of circular waveguides (Balanis, 2005). Horn antennas are commonly used as the feed in the receiver optics, although dipole antennas are also used in some occasions. Generally, the horn is simply a hollow pipe tapered to a larger opening (Balanis, 2005). Horns come in various geometrical structures – some of which are

Figure 9. Gaussian beam quasioptics equivalent of the receiver optics in Figure 8

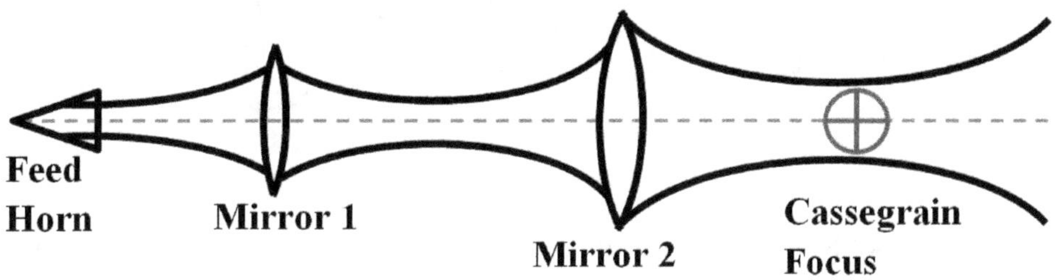

the corrugated, conical (smooth-walled), rectangular, pyramidal, and diagonal feed horns. The basic working principles of the horns are the same – they are used to couple the incident wave and transmit them to the receiver circuits. Here, the two most commonly used structures in radio telescopes, i.e. the conical and corrugated horns will be discussed.

Conical Horn

Due to its simplicity and ease of fabrication, the smoothed-wall conical horn has been one of the most popular options for the feed in receiver optics. As can be observed from Figure 10, the horn has a circular aperture and the metallic boundary is tapered in the form of a cone. The wave propagating in the horn is the dominant TE11 mode of a circular waveguide (Goldsmith, 1998). Hence, the electric field is transverse, but is not azimuthally symmetric. As compared to its corrugated counterpart, the smooth-walled conical feed horn produces relatively poorer co-polarization, higher level of sidelobes, poorer coupling to the fundamental Gaussian mode, and lacks the pattern symmetry (Goldsmith, 1998). Despite its complicated structure, the corrugated horn is considered a better candidate for wave coupling.

Corrugated Horn

As shown in Figure 11, the corrugated horn consists of uniform parallel grooves within the surface of the horn. The wall of the horn presents the same boundary conditions to all polarizations, effectively getting rid of spurious diffractions at the edges of the aperture (Goldsmith, 1998). The corrugated grooves modify the electric field distribution which is uniform at the waveguide-to-horn interface to

Figure 10. The structure of a conical (smooth-walled) horn

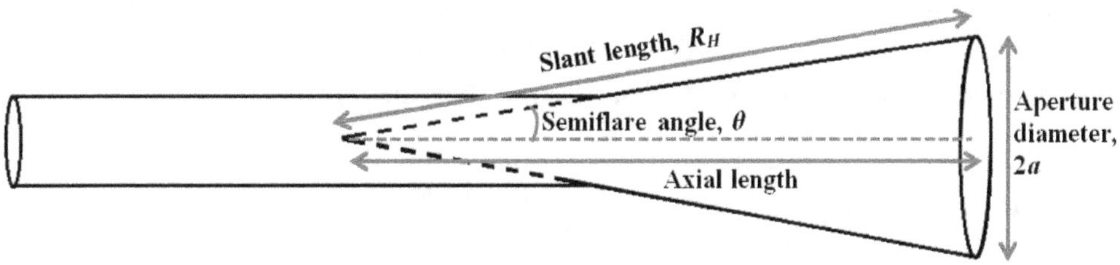

Figure 11. The structure of a corrugated horn

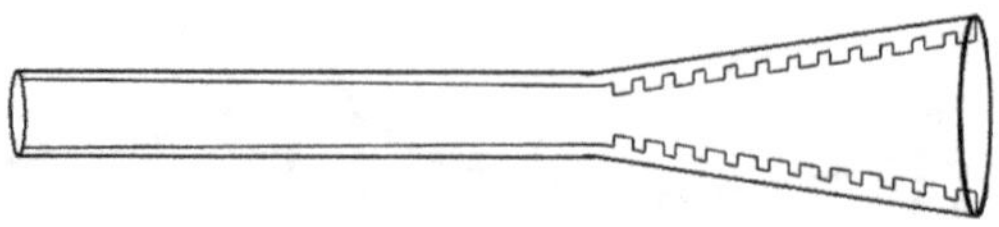

cosine at its aperture (Goldsmith, 1998). Unlike the conical horn, the field pattern of the corrugated horn is azimuthally symmetric and it is, therefore, an excellent candidate for illuminating symmetric reflector antennas (Goldsmith, 1998). Owing to its angle independent field distribution, the corrugated horn is also commonly referred to as the scalar feed horn. The fundamental Gaussian mode resembles very closely the main lobe of the radiation field from the feed horn (Goldsmith, 1998).

WAVEGUIDES

Circular and rectangular waveguides are widely used in radio receiver systems to channel and couple signals to the mixer circuits. The RF signal coupled to the horn is first channeled to a circular waveguide. The structure subsequently undergoes a circular-to-rectangular waveguide transition before allowing the signal to be coupled to the mixer circuit via a microstrip probe. Rectangular waveguides are usually preferred over circular ones since they can operate in single mode at larger bandwidth. Figures 12 and 13 show, respectively, the structure of hollow circular and rectangular waveguides. The front-end receiver noise temperature T_R is determined by a number of factors, such as the mixer noise temperature T_M, the conversion loss C_{Loss}, the noise temperature of the first IF amplifier T_{IF}, and the coupling efficiency between the IF port of the junction and the input port of the first IF amplifier η_{IF}. A comparison of the performance of different SIS waveguide receivers is listed in Table 1 (Walker et al., 1992; Yeap, Tham, Yassin, & Yeong, 2011). It is apparent that the value of T_R for the 230 GHz system is a factor of 3 to 4 times less than that of the 492 GHz system. Upon close inspection on the table, it could be observed that the decrease in system performance at 492 GHz is due to the increase of C_{Loss} and T_M by a factor of approximately 3. Since extraterrestrial signals are usually extremely faint, it is therefore of primary importance to minimize the conversion loss C_{Loss} of the mixer circuit. One way is to ensure that the energy of the LO and, in particular, the RF signals is channeled and coupled from the waveguides to the mixer circuit in a highly efficient manner. This is to say that, energy loss in the waveguide has to be minimized. To do so, the surface wall of the waveguides has to be made of highly conducting material, such as copper or brass. When designing the waveguides, accurate formalisms are necessary to determine the attenuation α in waveguides.

Assuming that the wave is propagating along the z-direction of the circular waveguide shown in Figure 14, the wave loss can be computed based on the transcendental equation in (11) below (Yeap, Tham, Yeong, & Woo, 2010; Yeap, Wong, Nisar, Lai, & Ng; 2017)

$$\left[jk_r^{\,2}\sqrt{\frac{\mu_w}{\varepsilon_w}}+\omega\mu_d k_r\frac{J_n'(u)}{J_n(u)}\right]\left[jk_r^{\,2}\sqrt{\frac{\varepsilon_w}{\mu_w}}+\omega\varepsilon_d k_r\frac{J_n'(u)}{J_n(u)}\right]=\left[\frac{nk_z}{a_r}\right]^2 \tag{11}$$

Figure 12. Hollow circular waveguides

Figure 13. Hollow rectangular waveguides

Table 1. Comparison of SIS receiver performance (Walker et al., 1992)

SIS Junction	Nb	Pb	Nb
Center Frequency (GHz)	230	345	492
T_R **(K)**	48	159	176
T_M **(K)**	34	129	123
C_{Loss} **(dB)**	3.1	8.1	8.9
T_{IF} **(K)**	7.0	4.2	6.8

where $k_r = \sqrt{k_d^2 - k_z^2}$, $J_n(u)$ and $J_n'(u)$ represent the Bessel function of the first kind and its derivative, ω the angular frequency, n the order of the Bessel function, k_d, μ_d and ε_d are respectively the wavenumber, permeability and permittivity of the dielectric core material (which is air for hollow waveguides), μ_w and ε_w are respectively the permeability and permittivity of the conducting wall, and a_r is the inner radius of the waveguide. The argument u is given as

$$u = \sqrt{a_r^2 \left(k_d^2 - k_z^2\right)} \tag{12}$$

Solving numerically for the root of (11), the propagation constant k_z can then be found. The propagation constant is a complex variable comprising the phase constant β_z and attenuation constant α, as shown in (13)

$$k_z = \beta_z - j\alpha_z \tag{13}$$

By extracting the imaginary part of (13), the loss in the circular waveguide can therefore be determined.

The process of finding the loss in a rectangular waveguide is lengthier compared to its circular counterpart. Here, we consider a rectangular waveguide with its width a and height b lying, respectively, along the x- and y-directions and the wave propagating in the z-direction, as shown in Figure 15. The solution involves the concurrent search for the roots of the pair of transcendental equations in (14) (Yeap, Tham, Yassin, & Yeong, 2011; Yeap, Wong, Nisar, Lai, & Ng; 2017)

$$\left[\frac{j\omega\mu_d k_y \tan\left(u_y\right)}{k_x^2 + k_y^2} + \sqrt{\frac{\mu_w}{\varepsilon_w}}\right]\left[\frac{j\omega\varepsilon_d k_y \cot\left(u_y\right)}{k_x^2 + k_y^2} - \sqrt{\frac{\varepsilon_w}{\mu_w}}\right] = \left[\frac{k_z k_x}{k_x^2 + k_y^2}\right]^2 \tag{14a}$$

Figure 14. Cross section of a circular waveguide

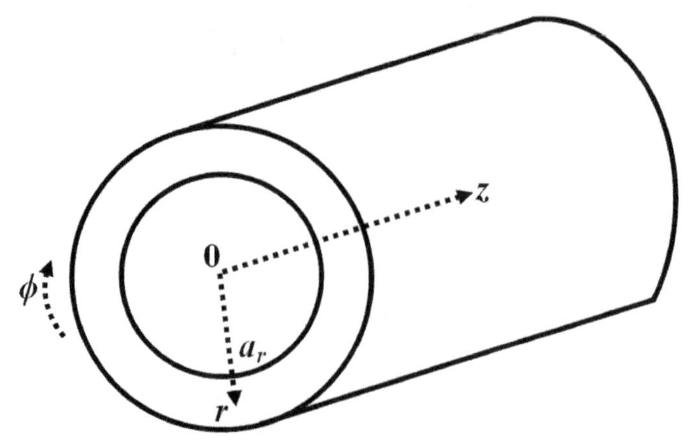

$$\left[\frac{j\omega\mu_d k_x \tan(u_x)}{k_x^{\ 2}+k_y^{\ 2}}+\sqrt{\frac{\mu_w}{\varepsilon_w}}\right]\left[\frac{j\omega\varepsilon_d k_x \cot(u_x)}{k_x^{\ 2}+k_y^{\ 2}}-\sqrt{\frac{\varepsilon_w}{\mu_w}}\right]=\left[\frac{k_z k_y}{k_x^{\ 2}+k_y^{\ 2}}\right]^2 \tag{14b}$$

where k_x and k_y are the transverse wavenumbers in the x and y directions, respectively. The arguments u_x and u_y are given as

$$u_x = \frac{k_x a + p\pi}{2} \tag{15a}$$

$$u_y = \frac{k_y b + q\pi}{2} \tag{15b}$$

where p and q represent the number of half wave cycle in the x and y directions, respectively. Like the case of the circular waveguide, the transverse wavenumbers k_x and k_y can be found by solving (14) numerically. The loss in the waveguide α can then be computed by substituting k_x and k_y into the dispersion relation of the rectangular waveguide in (16) and extracting the imaginary values out from k_z.

$$k_z = \sqrt{k_d^{\ 2} - k_x^{\ 2} - k_y^{\ 2}} \tag{16}$$

The dimensions of the waveguides can be estimated based on the cutoff frequency f_c of the lossless propagating modes. For a perfectly conducting circular waveguide, the cutoff frequency for TE (i.e. f_{cTE}) and TM modes (i.e. f_{cTM}) can be, respectively, expressed as (Balanis, 1989)

$$f_{cTE} = \frac{\chi'_{mn}}{2\pi a_r \sqrt{\mu_d \varepsilon_d}} \tag{17a}$$

$$f_{cTM} = \frac{\chi_{mn}}{2\pi a_r \sqrt{\mu_d \varepsilon_d}} \tag{17b}$$

where χ'_{mn} and χ_{mn} denote, respectively, the n th zero of $J_n'(u)$ and $J_n(u)$ of order m. For a perfectly conducting rectangular waveguide, the cutoff frequency for TE and TM modes is the same, as shown in (18) below

$$f_c = \frac{1}{2\sqrt{\mu_d \varepsilon_d}}\sqrt{\left(\frac{p}{a}\right)^2 + \left(\frac{q}{b}\right)^2} \tag{18}$$

Figure 15. Cross section of a rectangular waveguide

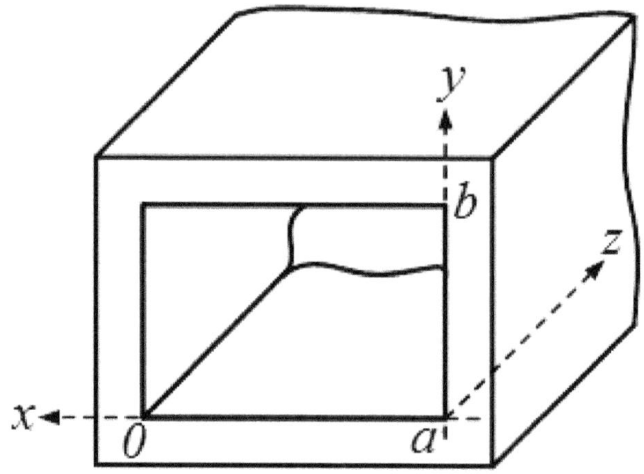

RECEIVER CIRCUIT

To couple the wave at the feed to the electronics, receivers are implemented. In general, there are two types of receivers, i.e. the coherent and incoherent receivers. Coherent receivers – which are more commonly referred to as the superheterodyne or heterodyne receivers – are widely used to observe signal in the low and high spectral resolution. Incoherent receivers, on the other hand, are more common in detecting signal at low spectral resolution. The most commonly used incoherent receivers in radio astronomy are bolometers. While the incoherent receivers could only detect the intensity of the incident RF signal, the coherent receivers account for both the intensity and phase.

The noise added to the input signal in the receivers follows closely Gaussian distribution (Wilson, 2013). The noise in heterodyne receivers is measured in Kelvins (K); whereas, that in bolometers is based on the Noise Equivalent Power or NEP (measured in $WHz^{-0.5}$). NEP is the input power level which produces twice of the output power (Wilson, 2013).

Coherent Receiver

In practice, it is often easier to build high gain, stable amplifiers and other related electronics for low frequency signals. Because of this reason, it is desirable to down convert the frequency of the incoming RF signal before feeding the signal for further amplification and analysis. The down conversion process is performed in the mixer circuit of a heterodyne receiver. In order to produce output power which is linear with the input power, non-linear square law devices are used as mixers. For signal above the sub-millimeter frequencies, such as infrared and visible light, the non-linear mixing device is a photo-detector; for signal in the range of radio waves, a diode is commonly adopted, instead. The superconductor-insulator-superconductor SIS device is the most efficient diode used for mixing signals in the millimeter and sub-millimeter regimes.

At the front-end of the radio telescope, the received RF signal at the aperture of the feed horn is channeled through a circular and subsequently a rectangular waveguide to a microstrip probe. The signal impinged on the probe is fed to the mixer circuit, built on the same substrate as the microstrip. At the same time, a signal which oscillates at a slightly lower frequency than the RF signal is generated by a local oscillator (LO). The LO signal is fed to the mixer circuit via waveguide-to-microstrip transition. The RF and LO signals are eventually added together at the beam splitter and the resultant is sent to the mixer. Heterodyning takes place at the mixer circuit – the RF signal is down converted to an intermediate IF signal. The IF signal oscillates at a frequency f_{IF} which is the difference between the frequencies of the RF and LO signals. When mixing both signals, the amplitude and phase of the LO signal is held constant. By doing so, the amplitude and phase of the input RF signal can be traced. Hence, the IF signal consists of the amplitude and phase of the RF signal (Wengler, 1992). In linear mixing, the amplitude of the LO signal is higher than the RF signal (Wengler, 1992). As a result of this, the intensity of the mixed IF signal is amplified. The amplification factor is referred to as the gain of the mixer.

SIS Mixer

Owing to its low noise temperatures and LO power requirement in comparison to the conventional Schottky diode mixer, the superconductor-insulator-superconductor SIS tunnel junction diode has been a popular mixer device in radio telescopes. The SIS mixer is built with a very thin sheet of insulator sandwiched between two layers of superconductors. Since these layers are deposited via lithography on a substrate, the SIS mixer also has the advantage of being planar in structure (Wilson, 2013). The working principle of a superconductor can be explained based on the Bardeen–Cooper–Schrieffer (BCS) theory (Bardeen *et al.*, 1957). When an electron passes by positively charged ions in the lattice of a superconductor, the positively charged atoms would be attracted towards the electron. This forms a trough of positive charges around the electron. The strong force exerted by the positive charges predominate that of the negatively charged electron and is therefore capable of curbing the repulsion effect of electrons approaching the trough. Before the lattice returns to its original position, another electron is drawn into the trough. This results in two electrons which appear to be paired up together. The coupled electrons are known as Cooper pairs (Cooper, 1956). This phenomenon overcomes the collisions of electrons with the lattice structure, rendering literally infinitesimal resistance to a superconductor. A material would only exhibit the properties of superconductivity when the operating temperature is below the critical temperature T_c. Figure 16 depicts the energy band diagram of a superconductor (Wengler, 1992). As can be observed from the band diagram, an energy gap $E_g = 2\Delta$ exists in the Cooper pairs. The energy gap represents the binding energy of the loosely bound Cooper pairs. Hence, the electronic states below the gap is filled with Cooper pairs; whereas, above the gap is an empty continuum of states according to BCS theory (Wengler, 1992). Although the BCS theory predicts that there is no energy state within the band gap, recent findings actually suggest that intragap states may exist (Yeap, Ong, Nisar, Lai, & Ng, 2016; Yeap, Teh, Nisar, & Hirasawa, 2015; Mitrovic & Rozema, 2008; Noguchi, Suzuki, Endo, & Tamura, 2009; Noguchi, Suzuki, & Tamura, 2011; Noguchi, Naruse, & Sekimoto, 2012; Noguchi, Naruse, & Sekimoto, 2013). The band diagram is somewhat analogous to that of an intrinsic semiconductor, where the lower valence band is filled with electrons and the upper conduction band is empty. Thermal energy or radiation higher than or equivalent to 2Δ can break Cooper pairs. The electrons which break free from the pairs are known as quasiparticles (Yeap, Yeong, Tham, & Nisar, 2016). In many ways, quasiparticles behave like normal electrons. Since the energy gap of Cooper pairs is 2Δ, the field frequency of the radiation absorbed by the carriers must be at least $2\Delta/h$ in order to release the quasiparticles, where h is the Planck's constant. The threshold frequency of $2\Delta/h$ is known as the gap frequency f_g. The voltage which corresponds to the energy gap is known as the gap voltage V_g which is equivalent to $2\Delta/e$ where e

is the charge of an electron (Wengler, 1992). For the superconductors used in SIS mixers, V_g lies in the range of mV. Some of the superconducting materials commonly applied in SIS tunnel junction are lead alloys (V_g » 2.5 mV), lead-bismuth alloys (V_g » 3 mV), niobium (V_g » 3 mV), and niobium-nitride alloys (V_g » 4 to 5 mV) (Wengler, 1992).

Figure 17 shows the band diagram of the SIS mixer when it is zero biased. Since the energy bands of the superconductors at both sides are aligned together, the Cooper pairs remain at the so-called "valence band". Hence, no quasiparticles can tunnel across the insulator layer. When the DC voltage supply is increased beyond V_g, however, the filled valence states on the left is facing directly a high density of empty conduction states at the right (Wengler, 1992). This allows the quasiparticles to break free and subsequently tunnel across to the conduction states, as graphically shown in Figure 18. The tunnelling effect produces a sharp rise of current when the voltage is merely slightly above V_g. As the voltage is progressively increased, the current starts to rise linearly with the voltage.

When radiation with frequency above f_g is incident upon the SIS tunnel junction, the photons absorbed by the quasiparticles at the left side of the superconductor raise the energy of the carriers, such that they could be promoted to the empty states at the right side (see Figure 19). This phenomenon is referred to as the photon-assisted tunnelling and it is applicable for the condition of the voltage supply when it is either below or above V_g. This is to say that, current would still be induced in the SIS device even if the voltage supply is below the threshold V_g, given that the frequency is above the gap frequency f_g. If the voltage is above V_g, additional current would be superimposed onto the DC *I-V* curve forming current steps on the curve. Since the radiation excites additional quasiparticles to the conduction states according to the same rate as photons absorption, the device is capable of detecting individual quanta. In a detector of perfect quantum efficiency, the detector current includes an electron for every photon absorbed (Wengler, 1992).

Figure 16. Energy band diagram of a superconductor

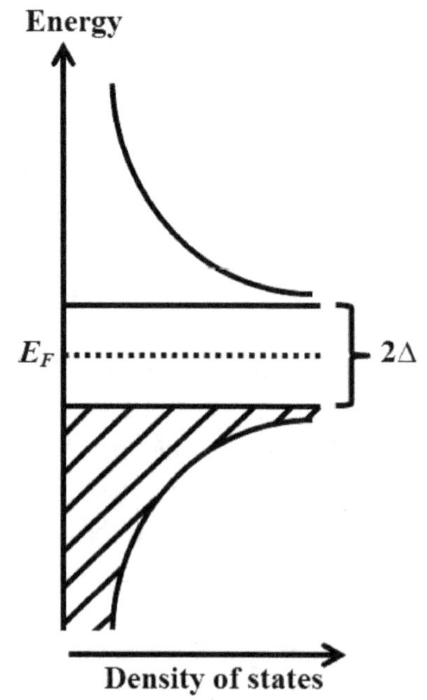

Figure 17. Energy band diagram of a zero-biased SIS mixer

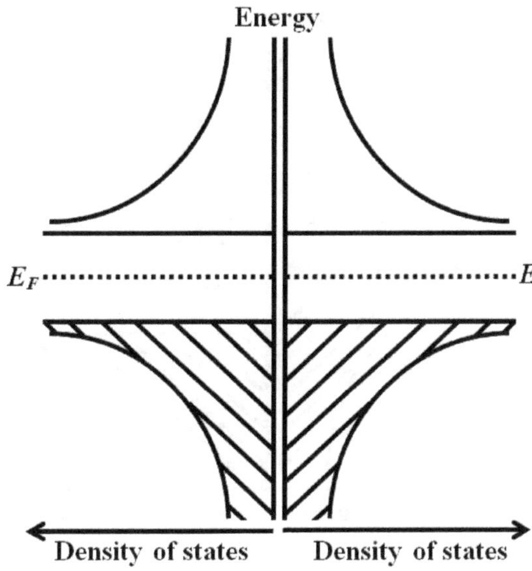

Figure 18. Energy band diagram of an SIS mixer when the DC bias voltage is larger than the gap voltage

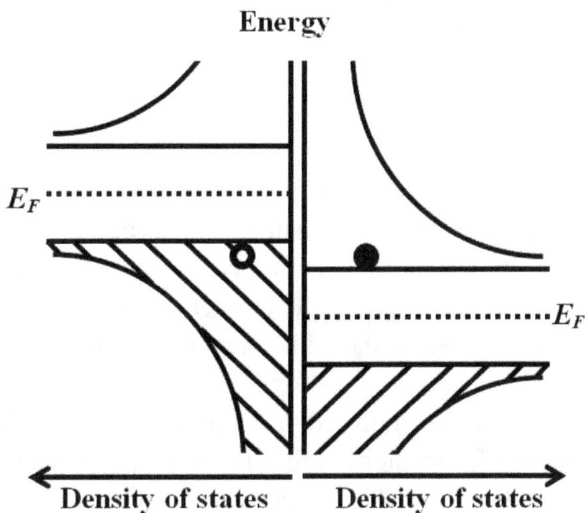

In the presence of the RF and LO signals, the voltage *V*, current *I*, and power *P* across the mixer are given as (Wengler, 1992)

$$V_{IF}(t) = V_{LO}\cos(2\pi f_{LO}t) + V_{RF}\cos(2\pi f_{RF}t) \tag{19}$$

$$P_{IF}(t) = P_{LO} + P_{RF} + 2\sqrt{P_{LO}P_{RF}}\cos(2\pi f_{IF}t) \tag{20}$$

Figure 19. Photon assisted tunneling takes place in an SIS mixer

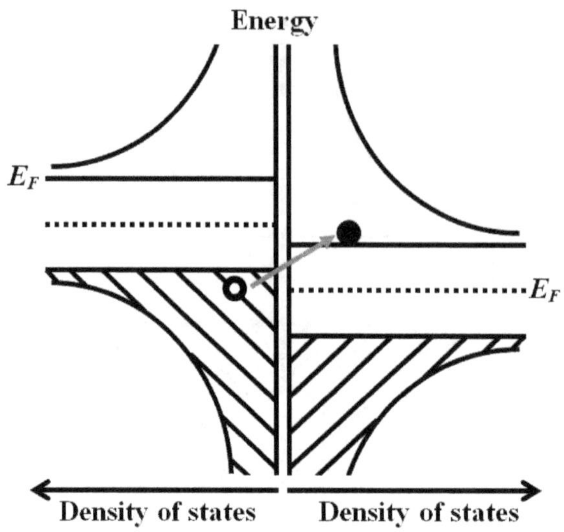

$$I_{IF}(t) = R_t 2\sqrt{P_{LO}P_{RF}}\cos(2\pi f_{IF}t) \tag{21}$$

where R_t is the current responsivity of the device and the subscripts RF and LO denote, respectively, the components of the RF and LO signals.

Incoherent Receiver

Another popular mixing device used in radio telescopes is the bolometer. To put it in simple terms, a bolometer or bolometric sensor is nothing more than just a thermal or power detector. Bolometers are widely implemented in the millimeter and sub-millimeter range to perform photometry. In order to ensure high sensitivity, it is cooled to the cryogenic temperatures (in the range of 50 to 300 mK). It is able to detect radiation over a wide bandwidth and is, thus, widely used to observe cold dust clouds and the core of celestial objects.

As depicted in the block diagram in Figure 20, a bolometer consists of three main components – a layer of absorber (typically made of Tungsten), a thermometer, and a heat sink. The working principle of a bolometer is similar to a calorimeter. The incident RF radiation is used to heat up the absorber and, as a result of the rise in temperature, the resistance of the absorber increases proportionately with the absorbed power. For a given fixed voltage, the current in the bolometer will in turn drops. In other words, the output power changes accordingly with the temperature of the absorber. A highly sensitive bolometer should constitute a narrow transition, low thermal conductance (to ensure that it is sufficiently sensitive) and low heat capacity (so as to produce fast response time). The thermometer and absorber is connected via a weak thermal link to a heat sink. The purpose of the heat sink is to cool the absorber back to thermal equilibrium with its ambient temperature after the photon energy has been absorbed. The process of cooling is, however, very slow even for very sensitive bolometers. Since the device is dictated by thermal effect, it is independent of frequency, phase, and polarization of the incoming radiation. The bolometer is therefore a broadband device (Wilson, 2013).

Figure 20. Model of a bolometer

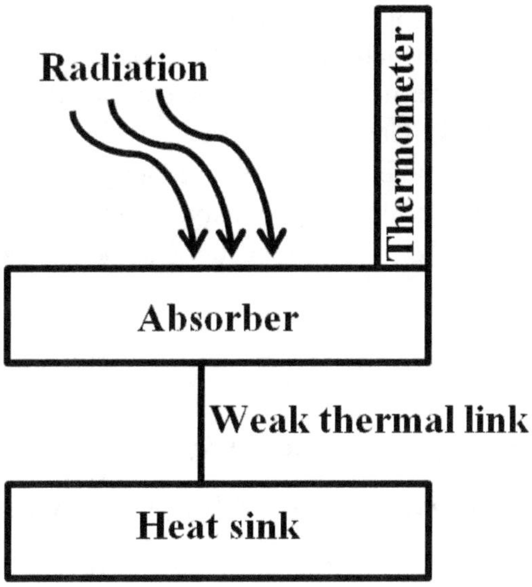

The Transition Edge Sensor (TES) is considered the most efficient bolometer. Being made of superconducting materials, the TES bolometer is able to perform two to three times better than its semiconductor counterpart, when it is used in ground-based telescopes (Wilson, 2013). The TES bolometer is also capable of processing higher number of pixels (Wilson, 2013).

BACK END

Due to the presence of white noise, the signal from the receiver output is an AC voltage, i.e. it oscillates in the positive and negative phases. If a DC voltmeter is used to observe the reading, it will show an average value of zero. In order to demodulate the modulating signal from its carrier, i.e the white noise, DC rectification is to be performed at the back end. A non-linear square-law detector such as a diode is used for the process of demodulation. Since the DC output voltage of the detector circuit is the squared of its AC input voltage, the output power is linear with its input power. The output of the detector is subsequently sent to a low-pass filter, allowing the intensity which varies at slower frequencies to be analysed.

To observe the spectral lines, Fourier Transform (FT) is performed on the signal fed from the filter. The voltage from FT is squared so as to obtain the Power Spectral Density in an autocorrelator (Wilson, 2013). Correlation is the process of time-averaging the vector product of signals (Fisher, 2002). When the signals from two orthogonal feed polarization outputs are correlated, information of the polarized radio source can then be obtained (Fisher, 2002). To acquire the spectral and phase information of the celestial sources, the signal has to be sampled. This is done using an analogue-to-digital converter (ADC) (Fisher, 2002). In order to prevent the spectral information from aliasing, the sampling rate has to abide by Nyquist's sampling theorem, i.e. the sampling rate must be at least equivalent to twice the signal bandwidth. Practically speaking, however, the sampling rate should be greater than twice the bandwidth so that clear spectral lines could be reconstructed.

Very often, the IF band from the receiver output is segmented via bandpass filters into various frequency intervals. Signals at each interval are then rectified by a separate detector circuit. The output spectrum from each detector circuits can be selectively viewed via a multiplexer. The electronics involved in constructing the filter bank is laborious, particularly, when the number of interval segments is large. Another way of observing spectrum with a wide bandwidth (in the range of GHz) is using an Acoustic Optical Spectrometer (AOS). The AOS works based on the principle of light diffraction caused by ultrasonic waves (Wilson, 2013). When the IF signal propagates into the crystal medium, it will be modulated. The modulated signal is detected by a charge coupled device (CCD). A typical AOS can provide a bandwidth as large as 2 GHz and 2000 spectral channels (Wilson, 2013).

REFERENCES

Balanis, C. A. (1989). *Advanced Engineering Electromagnetics*. US: John Wiley & Sons.

Balanis, C. A. (2005). *Antenna Theory: Analysis and Design*. US: Wiley-Interscience.

Bardeen, J., Cooper, L. N., & Schrieffer, J. R. (1957). Theory of superconductivity. *Physical Review*, *108*(5), 1175–1204. doi:10.1103/PhysRev.108.1175

Cooper, L. N. (1956). Bound electrons pairs in a degenerate Fermi gas. *IEEE Transactions on Microwave Theory and Techniques*, *104*, 1189–1190.

Fisher, J. R. (2002). Back-ends. In Single-Dish Radio Astronomy: Techniques and Applications, 278, 113 – 122.

Goldsmith, P. F. (1998). *Quasioptical systems; Gaussian beam, quasioptical propagation and applications*. New York: IEEE Press.

Green, K. A. (1963). Modified Cassegrain antenna for arbitrary aperture illumination. *IEEE Transactions on Antennas and Propagation*, *AP-11*(5), 589–590. doi:10.1109/TAP.1963.1138084

Halbach, K. (1964). Matrix representation of Gaussian optics. *American Journal of Physics*, *32*(2), 90–108. doi:10.1119/1.1970159

Hannan, P. W. (1961). Microwave antennas derived from the Cassegrain telescope. *IRE Transactions on Antennas and Propagation*, *9*(2), 140–153. doi:10.1109/TAP.1961.1144976

Lamb, J. W. (1999). Optimized optical layout for MMA 12-m antennas. *MMA Memo* 246.

Mitrovic, B., & Rozema, L. A. (2008). On the correct formula for the lifetime broadened superconducting density of states. *Journal of Physics Condensed Matter*, *20*(1). doi:10.1088/0953-8984/20/01/015215

Mizugutch, Y., Akagawa, M., & Yokoi, H. (1976). Offset dual reflector antenna, *Int. Symp. of Antennas and Propagation Society*, 14, 2 – 5.

Noguchi, T., Naruse, M., & Sekimoto, Y. (2012). RF conductivity and surface impedance of a superconductor taking into account the complex superconducting gap energy. *Physics Procedia*, *36*, 318–323. doi:10.1016/j.phpro.2012.06.166

Noguchi, T., Naruse, M., & Sekimoto, Y. (2013). Contribution of quasiparticles in the subgap states to the surface impedance of superconductors. *IEEE Transactions on Applied Superconductivity*, *23*(3). doi:10.1109/TASC.2013.2240761

Noguchi, T., Suzuki, T., Endo, A., & Tamura, T. (2009). Contribution of the imaginary part of the superconducting gap energy on the SIS tunneling current. *Physica. C, Superconductivity*, *469*(15-20), 1585–1588. doi:10.1016/j.physc.2009.05.218

Noguchi, T., Suzuki, T., & Tamura, T. (2011). Subgap tunneling current at low temperature in Nb/Al-AlN/Nb SIS junctions. *IEEE Transactions on Applied Superconductivity*, *21*(3), 756–759. doi:10.1109/TASC.2010.2089033

Prestage, R. M., Constantikes, K. T., Hunter, T. R., King, L. J., Lacasse, R. J., Lockman, F. J., & Norrod, R. D. (2009). The Green Bank telescope. *Proceedings of the IEEE*, *97*(8), 1382–1390. doi:10.1109/JPROC.2009.2015467

Walker, C. K., Kooi, J. W., Chan, M., LeDuc, H. G., Schaffer, P. L., Carlstrom, J. E., & Phillips, T. G. (1992). A low-noise 492 GHz SIS waveguide receiver. *International Journal of Infrared and Millimeter Waves*, *13*(6), 785–798. doi:10.1007/BF01011595

Wengler, M. J. (1992). Submillimeter-wave detection with superconducting tunnel diodes. *Proceedings of the IEEE*, *80*(11), 1810–1825. doi:10.1109/5.175257

Wilson, T. L. (2013). Techniques of Radio Astronomy. In T. D. Oswalt, & H. E. Bonds (Eds.), *Planets, stars, and stellar systems* (pp. 283–323). The Netherlands: Springer. doi:10.1007/978-94-007-5618-2_6

Yeap, K. H., Loh, M. C., Tham, C. Y., Yiam, C. Y., Yeong, K. C., & Lai, K. C. (2016). Analysis of reflector antennas in radio telescopes. *Advanced Electromagnetics*, *5*(3), 32–38. doi:10.7716/aem.v5i3.402

Yeap, K. H., Ong, S. S., Nisar, H., Lai, K. C., & Ng, C. A. (2016). Attenuation in superconducting circular waveguides. *Advanced Electromagnetics*, *5*(2), 34–38. doi:10.7716/aem.v5i2.388

Yeap, K. H., Teh, J. S. M., Nisar, H., Yeong, K. C., & Hirasawa, K. (2015). Attenuation in superconducting rectangular waveguides. *Frequenz Journal of RF-Engineering and Telecommunications*, *69*, 111–117.

Yeap, K. H., & Tham, C. Y. (2018). Optimization of an offset receiver optics for radio telescopes. *Journal of Infrared, Millimeter, and Terahertz Waves*, *39*(1), 64–76. doi:10.100710762-017-0449-z

Yeap, K. H., Tham, C. Y., Yassin, G., & Yeong, K. C. (2011). Attenuation in rectangular waveguides with finite conductivity walls. *Radioengineering Journal*, *20*, 472–478.

Yeap, K. H., Wong, E. V. S., Nisar, H., Lai, K. C., & Ng, C. A. (2017). Attenuation in circular and rectangular waveguides. *Electromagnetics*, *37*(3), 171–184. doi:10.1080/02726343.2017.1301198

Yeap, K. H., Yeong, K. C., Tham, C. Y., & Nisar, H. (2016). Analysis of Energy Loss in Superconducting Waveguides. In V. Ponnusamy, N. Zaman, T. J. Low, & A. H. M. Amin (Eds.), *Biologically-Inspired Energy Harvesting through Wireless Sensor Technologies* (pp. 71–83). US: IGI Global. doi:10.4018/978-1-4666-9792-8.ch004

Chapter 3
Development of a Simple Home-Brew Radio Telescope

Kim Ho Yeap
https://orcid.org/0000-0001-7043-649X
Universiti Tunku Abdul Rahman, Malaysia

Wei Gie Lim
Universiti Tunku Abdul Rahman, Malaysia

Tee Tang
Queensland University of Technology, Australia

Veerendra Dakulagi
https://orcid.org/0000-0003-1056-0004
Lincoln University College, Malaysia

ABSTRACT

In this chapter, the design and construction of a simple home-brew radio telescope are illustrated. The radio telescope is built essentially from commercial off-the-shelf components. The main components include an offset prime-focus Ku-band reflector antenna, RF detector circuit, Arduino microcontroller, and a computer. To demonstrate the viability of the telescope, a drift-scan of the sun was performed. From the measurement, it is observed that the radiation signal from the sun has a peak power of –34.3 dBm, half power of –35.4 dBm, and a beamwidth of 3.13°. The impact of rain on the signal from the sky was also investigated. The result shows that there was a noticeable degradation of the power received. A significant amount of energy is absorbed by the dense water vapour in the atmosphere.

INTRODUCTION

When scientists discovered that additional information of the sky could be obtained by analysing the signal at different wavelengths other than that visible to the eyes (such as microwaves, radio waves, ultraviolet, infrared, x-rays, and gamma rays), they started to look for ways to view the sky. The full range

DOI: 10.4018/978-1-7998-2381-0.ch003

of electromagnetic wave is called the electromagnetic spectrum. The invention of radio telescopes was one of the early successes in detecting signals which are not visible to human eyes (Wood, 2015). Radio telescopes create a picture of the sky, not in visible light, but in radio waves (Yeap et al., 2013; Yeap et al., 2016a; Yeap et al., 2016b). The invention of the radio telescope is very useful to mankind, because it is able to detect the invisible and hidden activities across the universe. By analysing the information obtained from the radio telescope, some of the greatest mysteries of the cosmos can therefore be unravelled. It is the advent of radio telescopes which has given birth to the field of radio astronomy in physics. To put it in simple terms, radio astronomy is nothing more than the study of cosmic sources which emit radio waves, such as stars, quasars, pulsars, planets, etc.

Constructing a radio telescope presents exceptional difficulties. This is mainly due to the specificity of this discipline and the limited availability of dedicated instrumentation. The main difficulty faced by amateur astronomers in building a radio telescope is the detection of the extremely weak signal radiated from cosmic sources – the magnitude of some of these signals may be comparable to that of thermal noise. Although the underlying operational concept of a radio telescope is pretty much similar to that of a typical commercial radio found in most households, the antenna of the radio telescope is significantly much larger in size. Since signals from the cosmic sources are usually extremely weak, the sensitivity and resolution of the radio telescope have to be high in order to detect these signals. As a result, the size of the parabolic dish antenna has to be sufficiently large, i.e. the diameter of a typical dish is usually larger than 10 m, and the electronics components have to sustain low loss and high signal-to-noise ratio – some of which may have to operate in cryogenic temperature so as to minimize thermal noise (Yeap & Tham, 2018). Hence, building a radio telescope is extremely costly. This is the main reason why courses related to radio astronomy are not commonly offered in secondary and, not to mention, tertiary education. To introduce radio astronomy as a common course, it may be necessary to simplify the process of building a radio telescope, so much so that, it can be easily built in schools and tertiary institutions.

The purpose of this chapter is therefore to introduce an effective approach to design and develop an affordable and relatively simple radio telescope. Since signals from the sun has the highest intensity and can easily be detected, it shall be used to demonstrate the operability of the telescope. At the end of this chapter, the readers should be able to learn how to:

1. Construct a radio telescope using off-the-shelf components.
2. Perform simple measurements using the telescope.
3. Interpret and analyze the obtained data.

A KU-BAND RADIO TELESCOPE

The Ku-band radio telescope can be separated into the outdoor and indoor units. The outdoor unit comprises mainly a parabolic reflector antenna and a low noise block (LNB). The reflector antenna focuses incoming electromagnetic waves towards the LNB. Satellite dishes which operate in the frequency range of 10.7 GHz to 12.75 GHz are often used as the reflector antenna for the detection of the signals in the Ku-band. The LNB consists of the mixer and amplifier units. It is used to down convert the frequencies of the signals from the Ku-band to an intermediate frequency and to amplify the magnitude of the signals (Furse & Bhatia, 2006). The indoor unit consists of a satellite receiver, rotator controller and computer controller. The most important and difficult task in building the indoor unit is the selection of the most suitable and useful receiver device.

RADIO FREQUENCY INTERFERENCE

Electromagnetic or, to be more specific, radio frequency interference (RFI) is a major concern in radio astronomy. Owing to the highly non-ergodic nature of signals, identifying unintentional radio frequency interference signals (for example, from equipment operating in the neighbourhood of a radio telescope) is a rather challenging task (Czech, Mishra & Inggs, 2016).

The technological advancement in the field of telecommunication is the main culprit of man-made RFI in developed cities and it poses a threat to the development of radio astronomy. In bands below 2 GHz, the interferences mainly come from broadcasting services, communication data, satellite communication, aeronautical satellites, meteorological satellite and radio navigation satellite. Although filters could be used, they are usually not adequate in minimizing the impact of RFI.

It is also worthwhile noting that, radio telescopes can be easily affected by signals emitted in the nearby bands, since the received signals of interest are very weak. The unwanted signals may couple to the antenna and propagate to the receiver system. Because of the high gain of radio-telescope antennas and the fact that celestial signals are generally a quantity in comparison, the primary path for interference is through the antennas (Adnan, 2010). Table 1 shows the radio frequency interference sources that can disturb the radio astronomical observation in Malaysia (Abidin, Ibrahim, Adnan, & Annuar, 2008).

DEVELOPMENT OF THE HOME-BREW TELESCOPE

Figure 1 depicts a simplified block diagram of a radio telescope. As can be seen from the figure, a simple radio telescope consists of four main components, i.e. a parabolic dish antenna, a radio frequency (RF) detector, an analog-to-digital converter ADC and a computer.

The dish antenna is supported by a telescope mount. Telescope mounts are designed to allow for accurate pointing of the instrument. It therefore allows the motion of the stars to be tracked as the Earth rotates. For practical purposes, equatorial mount is often preferred to azimuth-elevation mount. For simplicity, an off-the-shelf Ku-band dish is used to collect radio signals from space and focuses them onto the antenna. The Ku-band dish is selected in this case because it can be easily salvaged from a television (TV) broadcast unit. The LNB is the receiving device mounted onto the satellite dish. The LNB down converts the radio waves collected by the dish which ranges from 10.7 to 12.75 GHz to the range of 950 to 2150 MHz. The process of down conversion is necessary since it allows the signal to propagate in transmission lines (such as an RG-6 cable) with less attenuation (Kirkman-Bey & Xie, 2014). An external DC 12 V is required to supply power to the LNB.

An RF detector is used to monitor the output of the LNB and develop an output voltage which is directly proportional to the logarithmic received power (dBm). Figure 2 shows the circuit schematic of the RF detector used which operates from 950 MHz to 2150 MHz. The LT5534 chip shown in the figure is a log detector. A 3.3 V voltage source is supplied to the chip. The voltage source is fed from Arduino – a microcontroller-based board used particularly for digital interactive project. To ensure that the RF detector is operating accurately, it has to be calibrated first before it is used. Figure 3 depicts the

Table 1. RFI sources in Malaysia

No	Signal Sources	Frequency (MHz)
1	Radio Broadcasting- Traxx FM	80 - 108
2	Aeronautical Mobile	125 - 150
3	Broadcasting Mobile (Tv-Channel 5)	175 - 217
4	Deuterium (DI), Fixed and Mobile	327
5	Mobile Satellite (intermittent)	150
6	Broadcasting (Tv-channel 38)	574 - 700
7	Mobile Phone(Celcom, Maxis, Digi)	890 - 933
8	Mobile Phone (GSM) (Celcom, Maxis, Digi)	1735 - 1880
9	Telekom Malaysia	1962

measurement setup for calibrating the RF detector. An RF signal source which generates signal at different frequencies and power is connected to the RF detector. The output voltage of the detector is then measured. The purpose of calibrating the detector is to determine the relationship between the input power P_{in} and the output DC voltage V_{out}. The calibration process is performed at the centre frequency (i.e. 1550 MHz) of the detector. Figure 4 illustrates the measurement obtained during the calibration. Upon close inspection on the figure, it can be observed that the output voltage literally varies linearly with the input power source. By finding the gradient of the straight line and extrapolating the line so that it crosses the axis where $V_{out} = 0$, the relationship between P_{in} and V_{out} can then be expressed as

$$P_{out} = 28.57V_{out} - 66.99\,. \tag{1}$$

The output of the RF detector V_0, which consists of a low frequency signal (i.e. less than 10 kHz), is subsequently connected to Arduino so as to undergo analog-to-digital conversion, as shown in Figures 2 and 5. The converter has 10 bits resolution, returning integers from 0 to 1023. The output digital data is then processed and saved in a computer. Figure 6 depicts the complete picture of the entire set up.

MEASUREMENT SETUP

For effective measurement of the cosmic signal, the antenna is to be placed at an open area which is free from electromagnetic source interference. In order to place the dish to face the correct direction, a magnetic compass is first used to measure and align the direction of the mount polar-axis scope towards earth's magnetic north pole, which is close to earth's north celestial pole (NCP). Extreme care has to be taken when using the compass since it can be easily deflected by metallic objects or surfaces at the vicinity.

As can be seen in Figure 7, the satellite dish used is an off-axis prime-focus antenna. This is to say that, the antenna is not in perpendicular direction with the RF signal from the sky (which is the signal of the sun, in this case). This makes it difficult to align the antenna to the main lobe of the RF signal. Since the dish scatters not only radio waves, but visible light towards the antenna feed as well, a shiny object (such as a coin) can be attached to the centre of the dish to allow easier visualization, as shown

Figure 1. Block diagram of a simple radio telescope

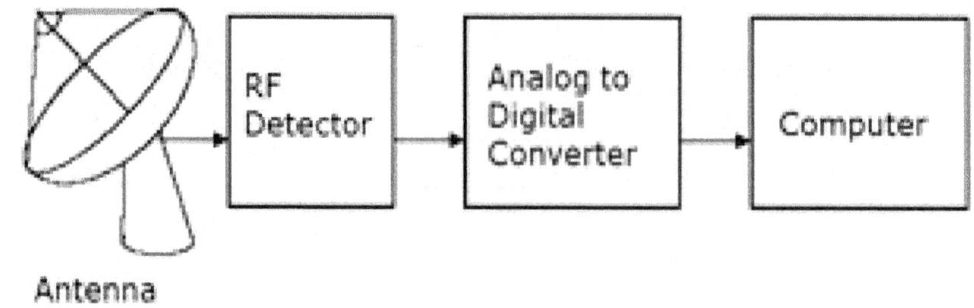

Figure 2. RF detector schematic

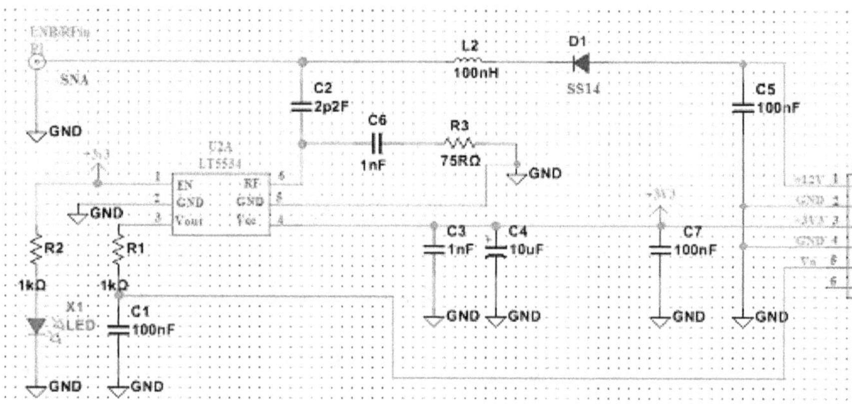

Figure 3. Measurement setup for calibrating the RF detector

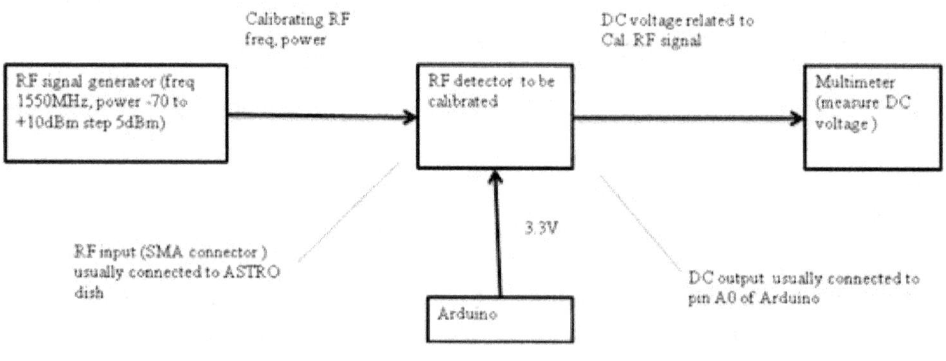

Figure 4. Power versus DC voltage at frequency = 1550 MHz

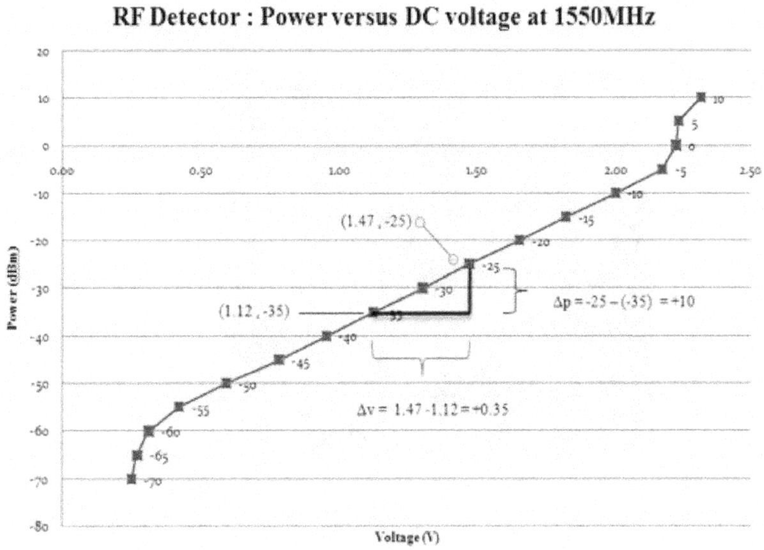

Figure 5. The screenshot of the RF detector connecting to the Arduino kit

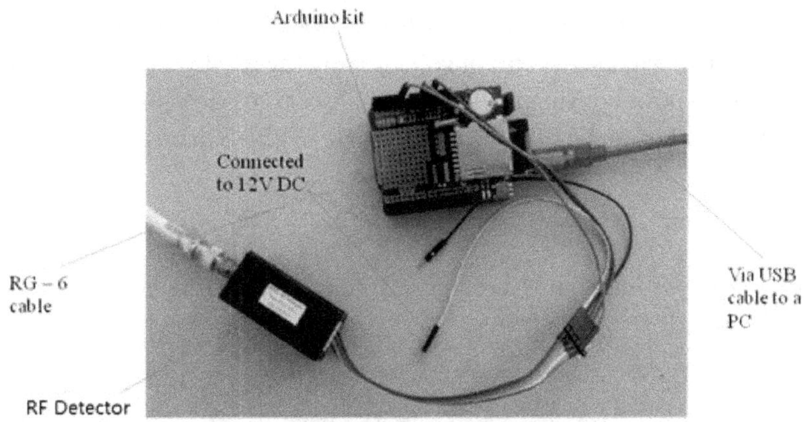

in Figure 8. This is because the glitter caused by the reflection of the coin will create a bright spot at the feed when the light is focused onto it.

SUN DRIFT MEASUREMENT

The RF beamwidth φ of the dish antenna is typically around a few degrees angle. The beamwidth is determined by the half-power point of the RF main-lobe as shown in Figure 9. The parameters $D = 1.392 \times 10^6$ km and $d = 149{,}598{,}261$ km are, respectively, the diameter of the sun and its distance from Earth (as graphically shown in Figure 10), the angular diameter of the sun δ can be found trigonometrically as

Figure 6. The antenna is first connected to an RF detector and an arduino board, before placing it at an open area to perform drift-scan

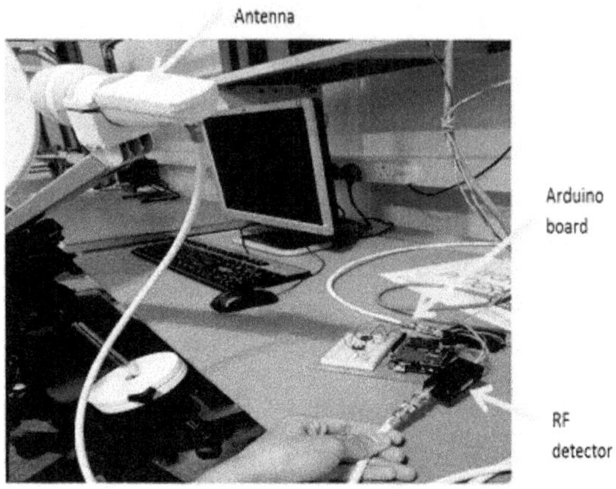

$$\delta = 2\tan^{-1}\left[\frac{D}{2d}\right] = 0.5^{\circ}. \tag{2}$$

When the sun drifts with respect to the stationery dish over time, the radio wave from the sun (which is focused by the parabolic dish to the receiver) also drifts across the antenna feed, as graphically depicted in Figure 11. By using (1) to correlate the output voltage from the RF detector V_{out} with the RF power P_{out}, the variation of the received power during the drift-scan process was recorded. As can be observed from Figure 12, the radiation signal from the sun has a peak power of –34.3 dBm and half power of –35.4 dBm.

Figure 7. The off-axis reflector antenna is supported by a tripod telescope mount

Figure 8. The reflection of the coin can be visualized at the feed

The time duration from the first half power level to the second is approximately 12.5 mins. Since it takes 24 hours for earth to rotate 360°, 12.5 mins corresponds to a Half Power Beamwidth (HPBW) of 3.13°.

THE EFFECT OF RAIN

Radio waves are highly susceptible to water vapour, rendering it exceptionally difficult to detect cosmic signal at places imbued with high humidity. In order to investigate the impact of water vapour on the detection of cosmic signal, an experimental measurement was performed during which the sky changed from clear to cloudy to raining. The reflector antenna was pointed towards the direction of a TV satellite. The signal from a TV satellite was used for detection partly because it possessed relatively high intensity; partly, it is also because the satellite is fixed at a position in the sky with respect to Earth. This type of satellite is called geostationary and rotates together with the earth so that it appears stationary when viewed from the earth. Figure 13 shows the variation of the power received by the telescope as the

Figure 9. Measurement of the signal from the sun when it drifts across the reflector antenna

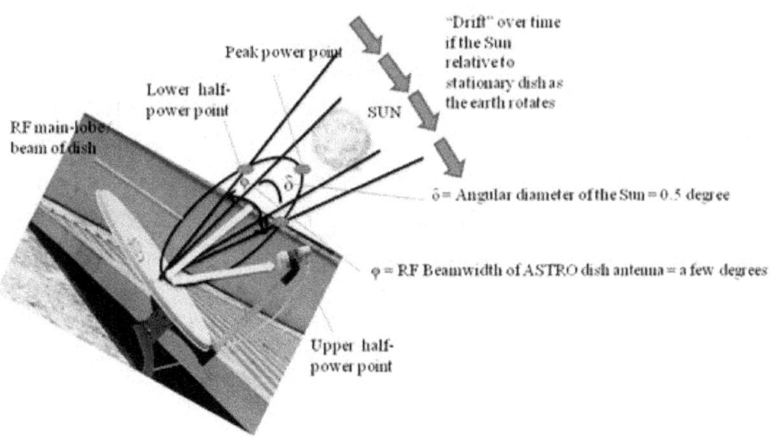

Figure 10. Graphical representation of the sun seen from Earth

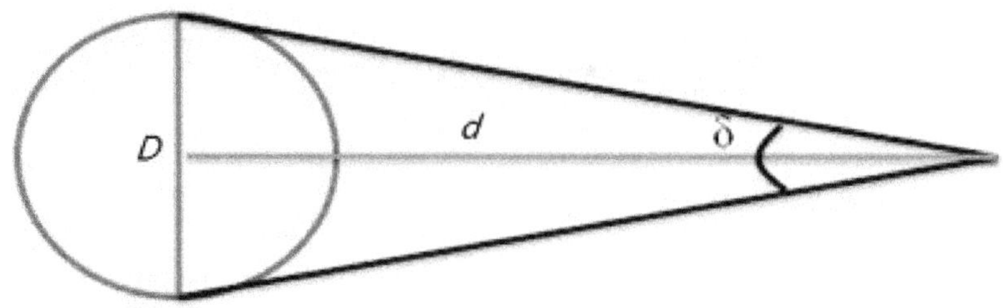

Figure 11. Sun beam drifts across the antenna feed

weather changed. As can be observed from the figure, there is a noticeable degradation in the received power when rain started to fall. It is apparent that the energy from the source has been strongly absorbed by the dense water droplets in the atmosphere.

CONCLUSION

This chapter illustrates the process of constructing a simple home-brew radio telescope using essentially off-the-shelf components. The telescope consists of a prime-focus reflector antenna which is used to detect signals in the Ku-band, an RF-detector, an Arduino-based ADC system, and a computer. The radio telescope is capable of detecting electromagnetic signals from 10.7 GHz to 12.75 GHz. The function of data logging is incorporated into the system, allowing it to record measured data at 1 second interval for up to several days. The telescope has been used to carry out a few scientific experiments including measuring the signal radiated from the sun and observing the effect of rain on satellite signals.

Figure 12. The power received by the telescope when drift-scanning the sun

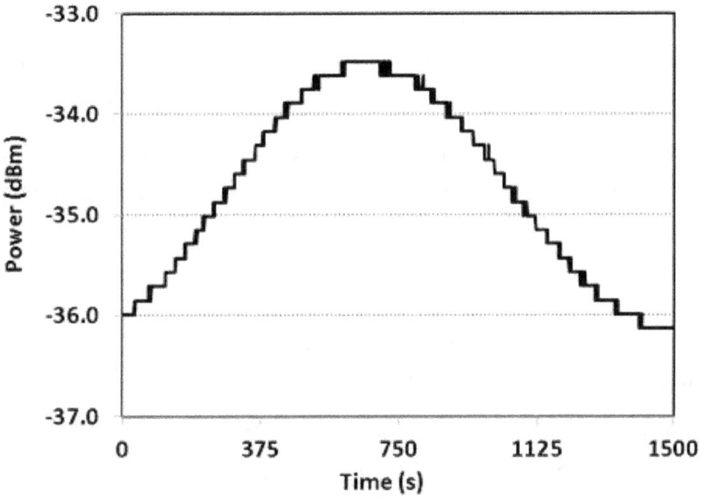

Figure 13. The power received by the telescope when the sky changed from clear to raining

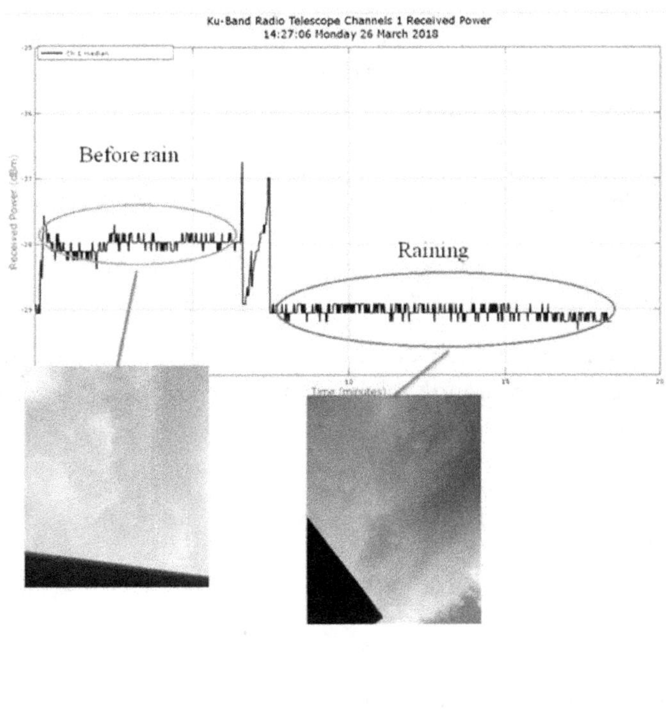

REFERENCES

Abidin, Z. Z., Ibrahim, Z. A., Adnan, S. B. R. S., & Anuar, N. K. (2009). Investigation of radio astronomical windows between 1 MHz – 2060 MHz in University Malaya. *New Astronomy, 14*(6), 579–583. doi:10.1016/j.newast.2009.02.003

Adnan, S. B. R. B. S. (2010). The study of radio frequency interference (RFI) for radio astronomy in some remote locations in peninsular Malaysia. Master Dissertation. University of Malaya.

Czech, D., Mishra, A. K., & Inggs, M. (2016). *Identifying radio frequency interference with hidden Markov model. Radio Frequency Interference* (pp. 17–20). Socorro, NM: RFI.

Furse, C., & Bhatia, R. S. (2006). A Simple Radio Telescope Operating at Ku Band for Educational Purposes. *IEEE Antennas & Propagation Magazine, 48*(5), 144–152. doi:10.1109/MAP.2006.277116

Kirkman-Bey, M., Cintron-Tirado, P., Sebsebie, R., Dogan, N., & Xie, Z. (2014). Design and simulation of Ku-band low-noise block down-converter in 0.18 micrometer CMOS technology. In *Proceedings of the IEEE SoutheastCon*, 13-16. Lexington, KY. 10.1109/SECON.2014.6950648

Wood, D. (2015). Radio Telescope Definition, Parts & Facts. [online] Available at https://study.com/academy/lesson/radio-telescope-definition-parts-facts.html [Accessed 1 February 2018]

Yeap, K. H., Law, Y. H., Rizman, Z. I., Cheong, Y. K., Ong, C. E., & Chong, K. H. (2013). Performance analysis of paraboloidal reflector antennas in radio telescopes. *International Journal of Electronics, Computer, and Communications Technologies, 4*(1), 21–25.

Yeap, K. H., Loh, M. C., Tham, C. Y., Yiam, C. Y., Yeong, K. C., & Lai, K. C. (2016a). Analysis of reflector antennas in radio telescopes. *Advanced Electromagnetics, 5*(3), 32–38. doi:10.7716/aem.v5i3.402

Yeap, K. H., & Tham, C. Y. (2018). Optimization of an offset receiver optics for radio telescopes. *Journal of Infrared, Millimeter, and Terahertz Waves, 39*(1), 64–76. doi:10.100710762-017-0449-z

Yeap, K. H., Yiam, C. Y., Lai, K. C., Loh, M. C., Lim, S. K., & Rizman, Z. I. (2016b). Analysis of offset antennas in radio telescopes. *International Journal on Advanced Science, Engineering. Información Tecnológica, 6*(6), 997–1004.

KEY TERMS AND DEFINITIONS

ADC: Analog-to-Digital Converter.

DC: Direct Current.

Half Power Beamwidth: Half Power Beamwidth (HPBW) is the aperture angle in which the magnitude of the radiation reduces by half (i.e. -3 dB) from the amplitude of the main lobe.

Ku-band: According to the IEEE radar bands, the Ku-band spans from 12 GHz to 18 GHz of the electromagnetic spectrum. This range of frequencies is primarily used for satellite communication.

LNB: Low Noise Block.

RF: Radio Frequency.

RFI: Radio Frequency Interference.

Section 2
Electromagnetism

Chapter 4
Basic Electromagnetic Theory

Magdalene Wan Ching Goh
https://orcid.org/0000-0002-5554-1635
Wawasan Open University, Malaysia

ABSTRACT

Electromagnetic theory covers the basic principles of electromagnetism. This chapter explores relationships between electric and magnetic fields. The chapter describes the behaviour of electromagnetic wave. The four sets of Maxwell's equations which underpin the principles of electromagnetism are briefly explained. An illustration on wave polarization and propagation is presented. The author describes the classification of waves according to their wavelengths (i.e. the electromagnetic spectrum).

THE ELECTROMAGNETIC WAVES

Light waves or visible lights are the most common electromagnetic (EM) waves. An EM wave does not need a medium to propagate, unlike mechanical waves (like sound waves or water waves) which need a medium to propagate. It can travel through air, solid materials or the vacuum of space. EM waves cannot exist in the absence of a magnetic waves. Vibration of electric charge must occur for an EM wave to form. This leads to the formation of the magnetic and the electric components in an EM wave. When the two get charged, they move together in the direction of the wave and are perpendicular to each other, but maintaining the direction of the formed EM wave. This can be illustrated in Figure 1.

WAVES IN GENERAL

Waves are disturbances that relate to any function that moves, be it the ocean waves, sound waves, or mechanical waves (like vibrations on a guitar string) and EM fields. The disturbances are called travelling waves if they are moving away from the source that created them. Standing waves are formed whenever two waves of identical frequency interfere with one another while traveling in the opposite directions along the same medium. There are two types of wave motions, namely the transverse wave and the longitudinal wave. A transverse wave is a travelling wave in which a disturbance is created in a

DOI: 10.4018/978-1-7998-2381-0.ch004

Figure 1. Electromagnetic wave propagation

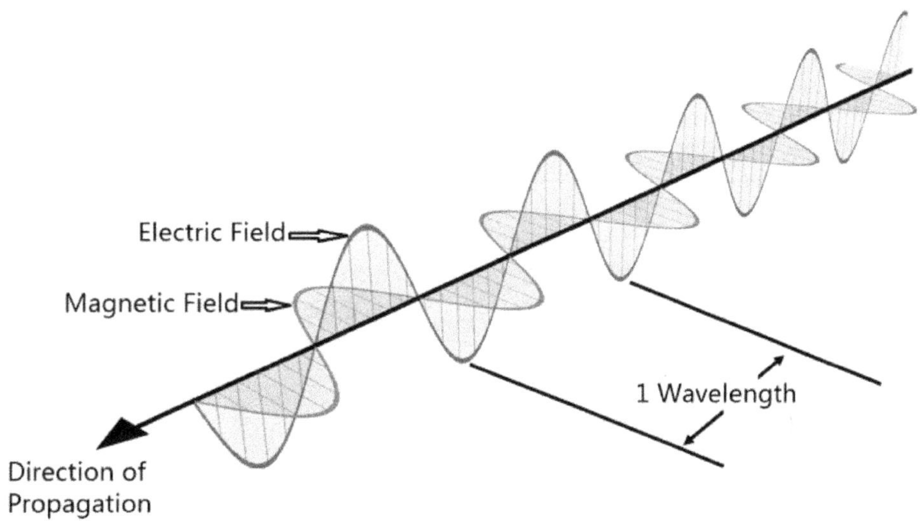

direction perpendicular to the direction in which the wave is travelling. One example of a transverse wave is an EM wave travelling in free space. A longitudinal wave is a wave that vibrates back and forth along or parallel to the direction in which the wave is travelling. It is also sometimes referred to as compression wave. Sound wave is an example of longitudinal wave. The highest point of a wave is known as "peak", the lowest "trough", as illustrated in Figure 2. The distance between two consecutive "peaks" is called the wavelength while the distance between the "peak" and "trough" is called the amplitude. EM waves can be split into a range of frequencies. This is known as the electromagnetic spectrum. Some of the applications of EM waves include the transmission of long and short radio waves, frequency modulation (FM) and amplitude modulation (AM) waves, as well as, television (TV), telephone, and wireless signals or energies. They are also responsible for transmitting energy in the form of microwaves, infrared radiation (IR), visible lights (VIS), ultraviolet light (UV), X-rays, and gamma rays. EM waves travel in the speed of light, which is a constant velocity of $c = 3\times10^{8}\,ms^{-1}$ in vacuum. The speed will be slightly lower when they travel in a medium, where the speed will be determined by the type of medium, they are travelling in. They are neither deflected by the electric field, nor by the magnetic field. However, they are capable of showing interference or diffraction. As mentioned earlier, EM waves are 'transverse' waves. This means that they are measured by their amplitude and wavelength, as illustrated in Figure 2.

The frequency, wavelength and period of an EM wave travelling in free space can be calculated using the following formulas:

Let $u = f\lambda$, where u is the speed at which the EM wave is travelling, f is the frequency of the EM wave and λ is the wavelength. The wavelength of the EM wave would be (Sadiku, 2007):

$$\lambda = \frac{u}{f} \tag{1}$$

Figure 2. Illustration of "peak", "trough", wavelength and amplitude

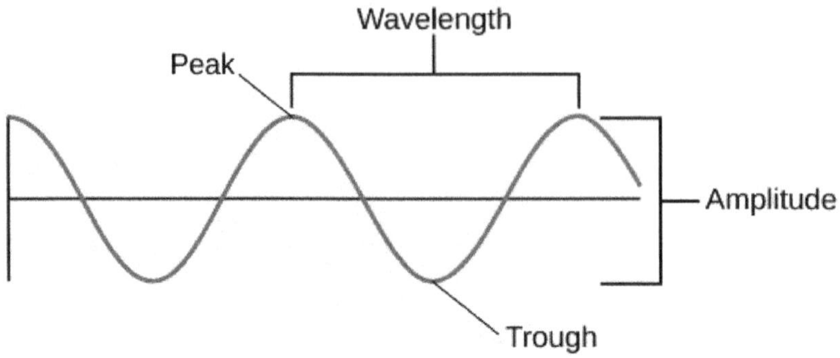

The period, T of the EM wave would be:

$$T = \frac{1}{f} \tag{2}$$

and, the frequency of the EM wave would be:

$$f = \frac{u}{\lambda} \tag{3}$$

MAXWELL'S EQUATIONS

Maxwell's equations unify the electrical and magnetic phenomena. They basically summarise the fundamentals of electricity and magnetism. Maxwell's equations form a significant portion of modern physics. The understanding of the Maxwell's equations has led to many scientific discoveries and applications in modern technology. Our understanding of the EM waves come primarily from Maxwell's equations. Optical fibre, lasers, radio and TV transmissions and wireless communications are among many innovations that came about in no small part because of our understanding of Maxwell's equations. Maxwell's equations comprise four differential equations shown in modern notation in Table 1 (Yeap & Hirasawa, 2019).

Below is a brief discussion of each law:

- **Gauss' Law for Electricity:** Electric charges produce an electric field. The electric flux across a closed surface is proportional to the charge enclosed.
- **Gauss' Law for Magnetism:** There are no magnetic monopoles. The magnetic flux across a closed surface is zero.
- **Faraday's Law:** Time-varying magnetic fields produce an electric field.
- **Ampère's Law:** Steady currents and time-varying electric fields produce a magnetic field.

The electric force fields are described by the quantities, E, called the electric field and $D = \varepsilon E$, which is the electric displacement. The latter includes how the electrical charges in a material become polarized in an electric field. The magnetic force fields are described by H, which is the magnetic field, and $B = \mu H$ which is the magnetic flux density. The latter accounts for the magnetization of a material.

The equations can be considered in two pairs. The first pair consists of Gauss' law for electricity and Gauss' law for magnetism. Gauss' law for electricity describes the electric force field surrounding a distribution of electric charge ρ. It shows that the electric field lines diverge from areas of positive charge and converge into areas of negative charge, as illustrated in Figure 3. Gauss' law for magnetism shows that magnetic field lines curl to form closed loops, as illustrated in Figure 4 and with the implication that every north pole of a magnet is accompanied by a south pole. The second pair, Ampère's law and Faraday's law, describes how electric and magnetic fields are related. Faraday's law describes how a time-varying magnetic field will cause an electric field to curl around it, while Ampère's law describes how a magnetic field curls around a time-varying electric field, or an electric current flowing in a conductor.

These equations can explain how your hair stands on end when you comb your hair with a plastic comb, how the needle of a compass always points north, how electricity can be generated by a power station turbine, and how a loudspeaker can convert an electric current into sound. When combined, these equations also describe the transmission of radio waves and the propagation of light.

WAVE POLARIZATION

The polarization of an EM wave indicates the plane in which it is vibrating. As EM waves consist of an electric and a magnetic field vibrating perpendicular to each other, it is necessary to adopt a convention to determine the polarization of the signal. The plane of the electric field is used for this purpose.

Wave polarization is not an easy concept to be visualized, as it takes place in three dimensions and across time. The most straightforward forms are the vertical and horizontal polarizations. They fall into a category known as linear polarization (see Figure 5). In linear polarization, the wave can be thought

Table 1. Maxwell's equations

Differential Form	Integral Form	Related to
$\nabla \bullet \vec{D} = \rho_v$	$\oint_c \vec{D} \bullet d\vec{s} = -\int_v \rho_v \bullet d\vec{v}$	Gauss' law for electricity
$\nabla \bullet \vec{B} = 0$	$\oint_c \vec{B} \bullet d\vec{s} = 0$	Gauss' law for magnetism
$\nabla \bullet \vec{H} = \vec{J} + \frac{\partial \vec{D}}{\partial t}$	$\oint_c \vec{H} \bullet d\vec{l} = -\int_s \vec{J} \bullet d\vec{s} + \int_s \frac{\partial \vec{B}}{\partial t} \bullet d\vec{s}$	Ampère's law
$\nabla \times \vec{E} = -\frac{\partial \vec{B}}{\partial t}$	$\oint_c \vec{E} \bullet d\vec{l} = -\int_s \frac{\partial \vec{B}}{\partial t} \bullet d\vec{s}$	Faraday's law

Figure 3. Lines of electric fields surrounding a positive and a negative charge

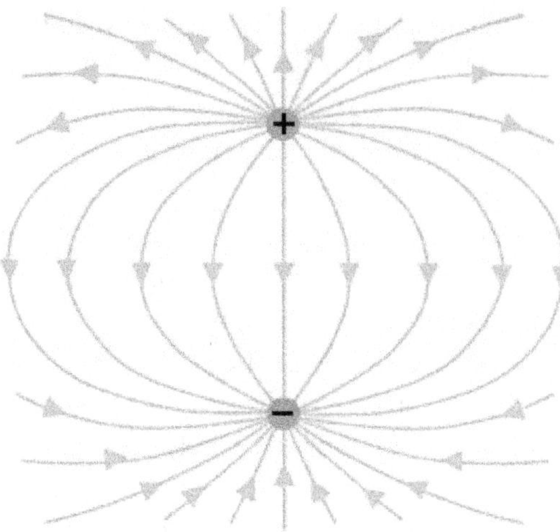

Figure 4. Lines of magnetic field around a bar of magnet and a current-carrying conductor

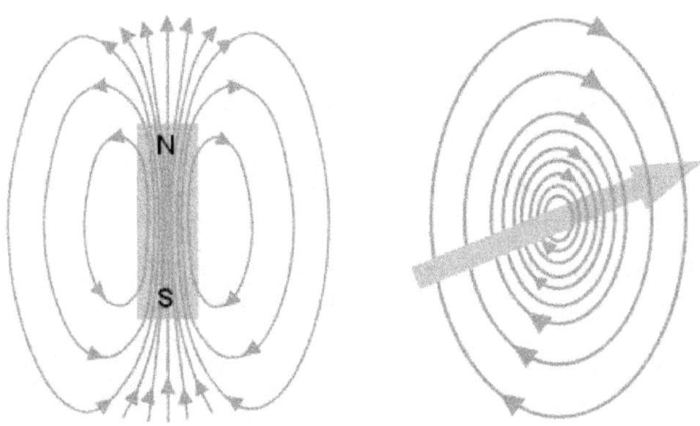

of as vibrating in one plane, i.e. up and down, or side to side. This form of polarization is the most commonly used.

It is also possible to generate waveforms that have circular polarization (see Figure 6). Circular polarization can be visualized by imagining a signal propagating from a rotating source, i.e. an antenna that is rotating. The tip of the electric field vector can be seen to trace out a helix or corkscrew as it travels away from the antenna. There are two types of circular polarization, namely the right-handed or left-handed polarization, dependent upon the direction of rotation as seen from the transmitting antenna. This is illustrated in Figure 7 (a) and (b).

Another form of wave polarization is called elliptical polarization, shown in Figure 8. This occurs when there is a combination of both linear and circular polarizations. Again, this can be visualized by imagining the tip of the electric field tracing out an elliptically shaped corkscrew. Radio frequency radiation coming from extraterrestrial sources in space may be linearly or circularly polarized, or anything in between, or unpolarized. The polarization of the waves provides additional information about the source.

Figure 5. Linear polarization

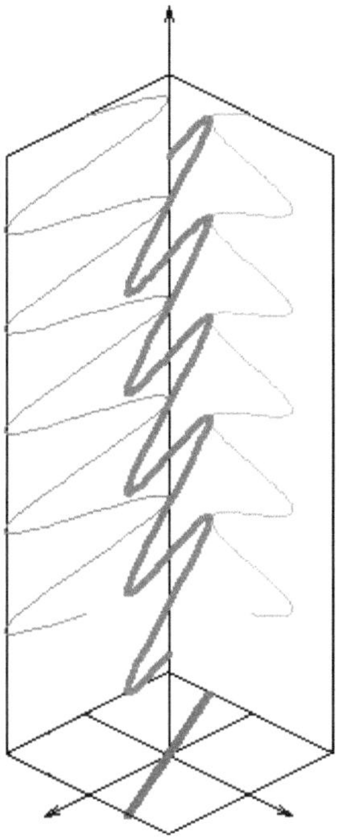

Figure 6. Circular polarization.

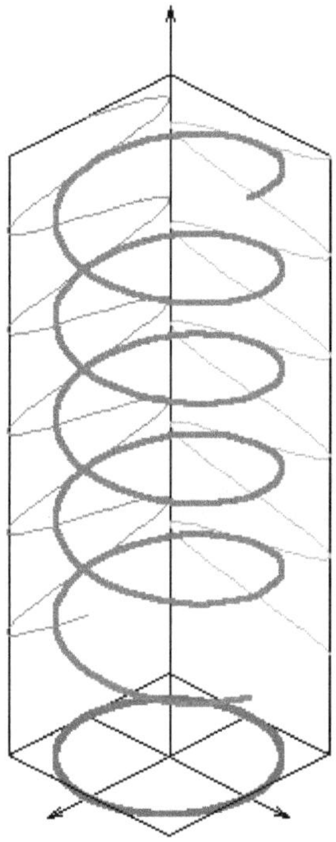

Figure 7a. Right-handed polarization

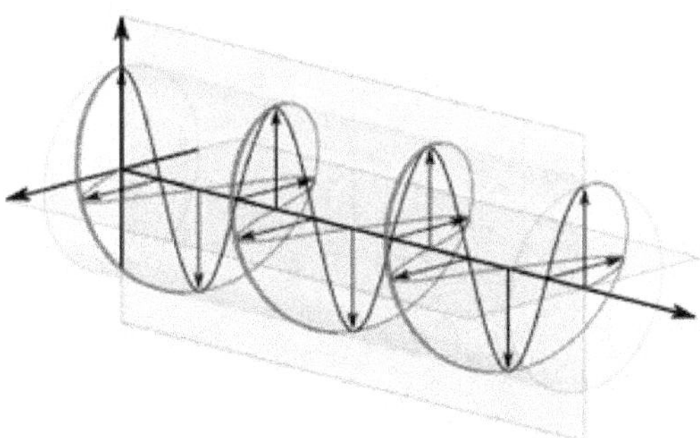

INVERSE SQUARE LAW

When an EM radiation leaves its source, it spreads out as it travels along straight lines, as if it is covering the surface of an ever-expanding sphere. As the radiation travels a distance R, its strength drops in proportion to $\frac{1}{R^2}$, as depicted in Figure 9 (Cheng, 1989). This is known as the inverse-square law of propagation, and it accounts for loss of signal strength over space, called space loss. For example, Saturn is approximately ten times farther from the sun than our planet Earth. Note that the distance between the sun and Earth is defined as one astronomical unit (AU). By the time the sun's radiation reaches Saturn, it is spread over 100 times the area it covers at one AU. As a result, Saturn only receives 1/100th of the solar energy flux that Earth receives. Solar energy flux is defined as energy per unit area.

Figure 7b. Left-handed polarization

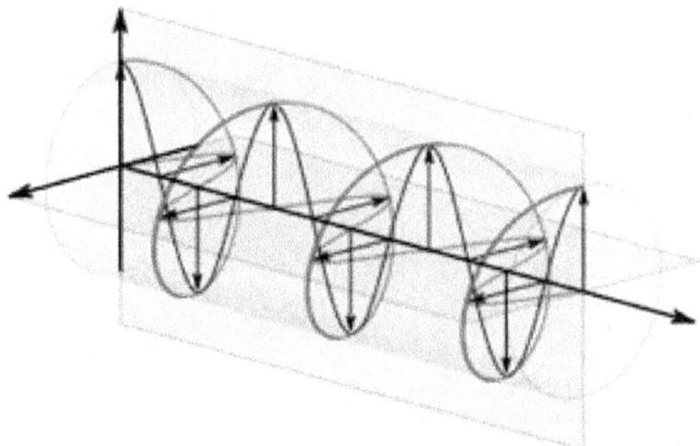

Figure 8. Elliptical polarization

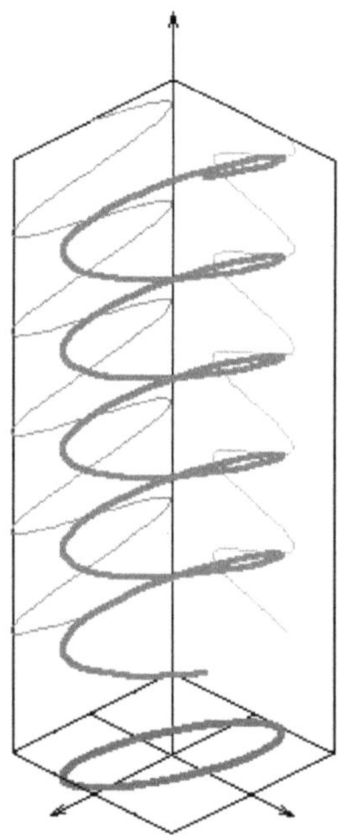

Understanding the inverse square law is important to the exploration of the universe. The concentration of EM radiation decreases very rapidly with increasing distance from the source of the EM radiation. Whether the source of the EM radiation is from the low power antenna of a space craft, an extremely powerful star or a radio galaxy, because of the great distances the EM radiation has to travel, it will deliver only a tiny amount of energy to a detector on Earth.

THE ELECTROMAGNETIC SPECTRUM

The electromagnetic (EM) spectrum describes all the wavelengths of light, both seen and unseen. The *electromagnetic spectrum* is the term used by scientists to describe the entire range of light that exists,

Figure 9. The inverse square law

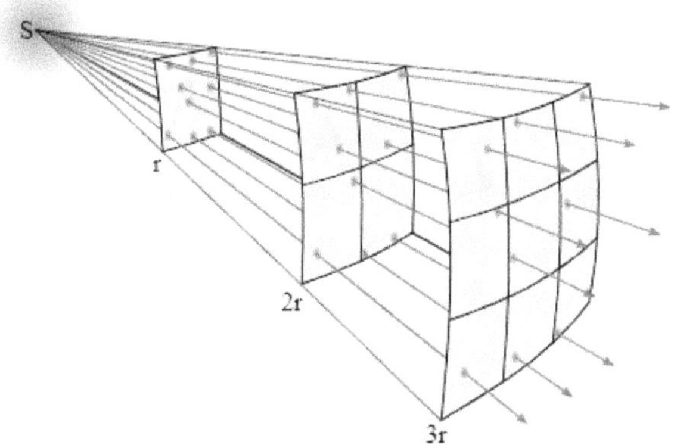

as illustrated in Figure 10. The visible part of the spectrum, shown in Figure 11, is just a very small portion of the entire EM spectrum. From radio waves to gamma rays, most of the light in the universe is, in fact, invisible to us! The radio waves, or radio frequencies (RF) cover the frequencies between 3 kHz to 300 GHz, and they are divided into bands as illustrated in Table 2.

Some of the popular transmission systems that work in the RF spectrum band includes analogue radio, aircraft navigation, marine radio, amateur radio, TV broadcasting, mobile networks and satellite systems. The RF spectrum is a broad spectrum and many of its characteristic has not been fully experimented yet. It has lots of possibilities in medical applications like magnetic resonance imaging (MRI) technology, seismography and oceanic studies. RF transceivers are significant components in interplanetary missions such as Mars exploration mission. Our future digital communication systems, like the 5G technology relies on high frequency bands of the RF spectrum since it can support higher bandwidth.

Various astronomical phenomena can only be observed via specific wavelengths different from visible light. By scanning the sky in the complete spectrum of electromagnetic radiation via optical telescopes,

Figure 10. Electromagnetic wave spectrum (Source NASA/Wikipedia)

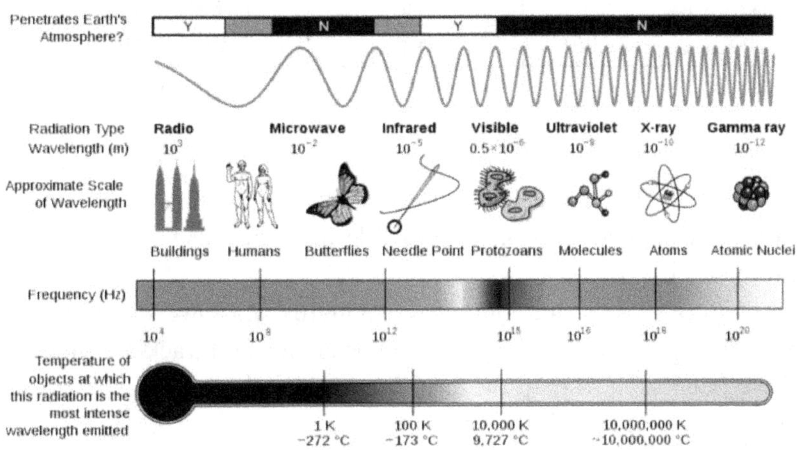

Figure 11. The electromagnetic spectrum with the visible lights showing

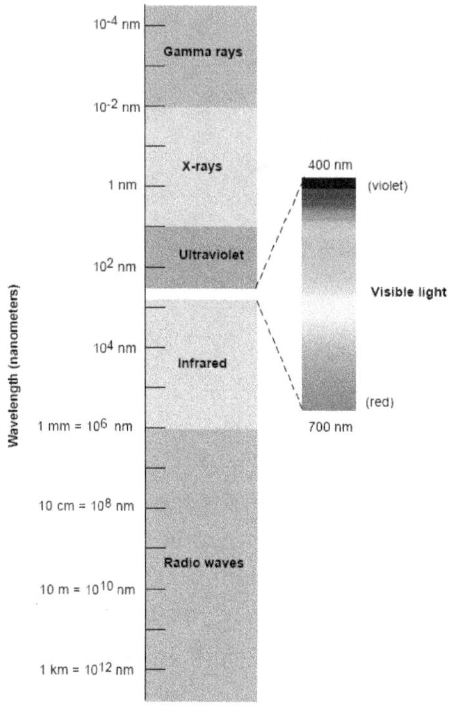

Table 2. Radio frequency spectrum

Designation	Abbreviation	Frequencies	Wavelength
Very Low Frequency	VLF	3 kHz – 30 kHz	100 km – 10 km
Low Frequency	LF	30 kHz – 300 kHz	10 km – 1 km
Medium Frequency	MF	300 kHz – 3 MHz	1 km – 100 m
High Frequency	HF	3 MHz – 30 MHz	100 m – 10 m
Very High Frequency	VHF	30 MHz – 300 MHz	10 m – 1 m
Ultra High Frequency	UHF	300 MHz – 3 GHz	1 m – 100 mm
Super High Frequency	SHF	3 GHz – 30 GHz	100 mm – 10 mm
Extremely High Frequency	EHF	30 GHz – 300 GHz	10 mm – 1 mm

X-ray telescopes, microwave telescopes and radio telescopes, astronomers gather information that would not be accessible if they were just observing via visible light, because visible light is only a very tiny fraction of the entire EM spectrum.

Radio waves and microwaves are the longest wavelengths and lowest energies of EM waves. They are used by astronomers to peer inside dense interstellar clouds and track the motion of cold, dark gas. Radio telescopes have been used to map the structure of our galaxy while microwave telescopes are sensitive to the remnant glow of the Big Bang.

Infrared telescopes are used for finding cool, dim stars, slicing through interstellar dust bands. They can also be used to measure the temperatures of planets in other solar systems. The wavelengths of infrared light are long enough to navigate through clouds that would otherwise block our view. By using large infrared telescopes, astronomers have been able to peer through the dust lanes of our Milky Way galaxy and into the core of our galaxy.

The majority of stars emit most of their EM energy as visible light. Because wavelength correlates with energy, the colour of a star tells us how hot it is. For example, red stars are the coolest and blue stars are the hottest. The coldest of stars emit hardly any visible light at all; they can only be seen with infrared telescopes.

The ultraviolet (UV) light has wavelengths shorter than violet. Astronomers use it to look for the most energetic form of stars and to identify regions of star birth or stellar nurseries. When viewing distant galaxies with UV telescopes, most of the stars and gas disappear, and all the stellar nurseries flare into view.

X-rays and gamma rays are the highest energies in the EM spectrum. Our atmosphere blocks this spectrum of energies, so astronomers must rely on telescopes in space to see the universe in X-ray and gamma ray. Sources of X-rays include exotic neutron stars, the vortex of superheated material spiralling around a black hole, or diffuse clouds of gas in galactic clusters that are heated to many millions of degrees. Meanwhile, gamma rays, the shortest wavelength of light unveil violent supernova explosions, cosmic radioactive decay, and even the destruction of antimatter. Gamma ray bursts, which is the brief flickering of gamma ray light from distant galaxies when a star explodes and creates a black hole, are among the most energetic singular events in the universe.

REFERENCES

Cheng, D. K. (1989). *Field and Wave Electromagnetics*. US: Addison-Wesley.

Sadiku, M. N. O. (2007). *Elements of Electromagnetics*. US: Oxford University Press.

Yeap, K. H., & Hirasawa, K. (2019). Introductory Chapter: Electromagnetism. In K. H. Yeap, & K. Hirasawa (Ed.), Electromagnetic fields and waves (pp. 1 – 12). UK: InTechOpen.

Chapter 5
Electromagnetic Wave Propagation

Magdalene Wan Ching Goh
https://orcid.org/0000-0002-5554-1635
Wawasan Open University, Malaysia

ABSTRACT

Electromagnetic wave propagation is an invisible phenomenon that cannot be detected by the human senses. To understand wave propagation, one must first learn what wave propagation is and the basic principles that affect wave propagation. This chapter introduces the atmospheric windows which allow electromagnetic radiation from bands to penetrate Earth. Helmholtz equations, i.e. the equations which govern wave propagation, and the properties of waves (such as propagation constant and characteristic impedance) are then briefly explained. When waves encounter different media during its propagation, they may be reflected, refracted, or diffracted. These phenomena are also covered. The last part of this chapter concisely explains the terminologies commonly used to describe electromagnetic radiation.

THE ATMOSPHERIC WINDOWS

The composition of the Earth's atmosphere is such that it allows certain wavelengths of electromagnetic (EM) waves to pass through and absorb some in certain wavelengths. The areas of the EM spectrum that are absorbed by the atmospheric gasses, such as water vapor, carbon dioxide and ozone, are known as the absorption bands. In contrast to the absorption bands, there are areas of the EM spectrum where the atmosphere is transparent to specific wavelengths. These wavelength bands are known as the *atmospheric windows* (see Figure 1).

The atmosphere absorbs most of the wavelengths shorter than ultraviolet, most of the wavelengths between infrared and microwaves, and most of the longest radio waves. Only the visible light, some ultraviolet, infrared and short wavelength radio waves are able to penetrate the atmosphere, and bring information about the universe to our instruments here on the ground. The main frequency ranges allowed to pass through the atmosphere are called the *radio window* and the *optical window*. The radio

DOI: 10.4018/978-1-7998-2381-0.ch005

window is the range of frequencies from about 5 MHz to 300 GHz (wavelengths of 100 m to 1 mm). The low frequency end of the window is limited by signal absorption in the ionosphere. The upper limit is determined by signal attenuation caused by water vapor, carbon dioxide and ozone in the atmosphere. In order to observe the universe in all the EM spectrum, some of the observation equipment are put in space.

The optical window (thus optical astronomy) can be severely limited by atmospheric conditions, such as an overcast sky, air pollution, and light pollution from cities or populated area, and blinding interference from the bright sun light. Radio astronomy is not hampered by most of the conditions mentioned here. Observation can be carried out using a radio telescope in broad daylight, which cannot be done with optical telescope. However, at the higher frequencies in the atmospheric radio window, clouds and rain can cause signal attenuation. For this reason, sub-millimetre wavelengths radio telescopes are built on the highest mountains, where the atmosphere has the least chance for attenuation, due to the low humidity level at higher altitude. One good example of such telescope is the Atacama Large Millimetre/ submillimetre Array (ALMA) (Yeap & Tham, 2018), where the array of telescopes was built on the Andes in Chile.

ABSORPTION AND EMISSION LINES

When the radiation from an object passes through a gas cloud, some of the electrons in the atoms and molecules of the gas absorb some of the energy of the EM radiation. The EM radiation emerging from the gas cloud will be missing those wavelengths that are absorbed. The spectrum will show dark absorption lines. The atoms or molecules in the gas will re-emit EM radiation at those same wavelengths. If we then observe this re-emitted EM radiation from the clouds of gas in the space between the stars, we will see bright emission lines at the exact frequencies of the absorption lines. These phenomena are known as *Kirchhoff's laws of spectral analysis* (see Figure 2).

Figure 1. The atmospheric windows

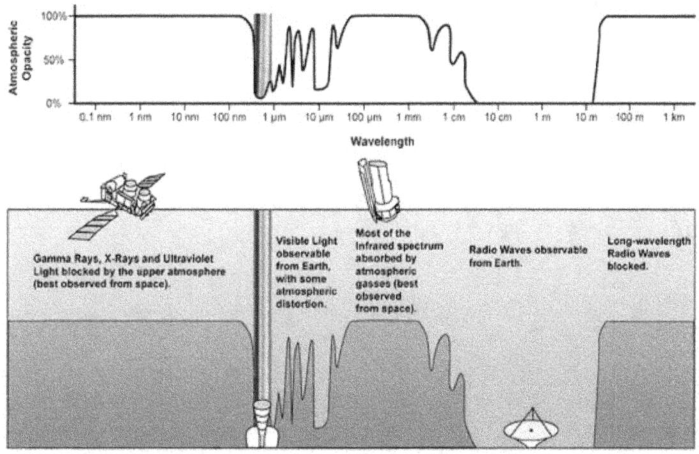

Figure 2. Kirchhoff's law of spectral analysis

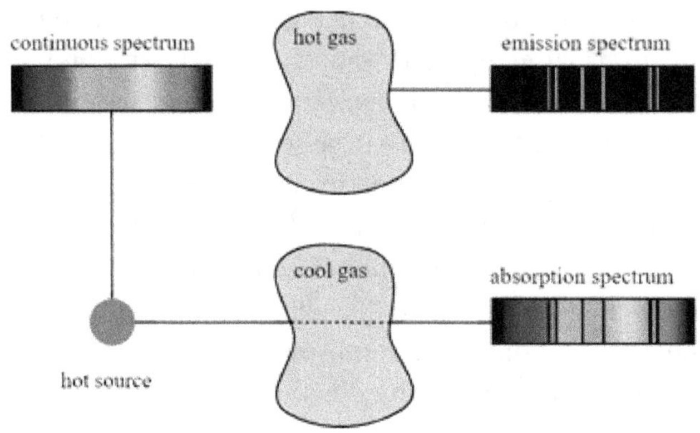

HELMHOLTZ EQUATIONS

The wave equations for a lossless medium are given as:

$$\nabla^2\vec{E} - \mu\varepsilon\frac{\partial^2\vec{E}}{\partial t^2} = 0 \text{ or } \frac{\partial^2\vec{E}}{\partial x^2} + \frac{\partial^2\vec{E}}{\partial y^2} + \frac{\partial^2\vec{E}}{\partial z^2} - \mu\varepsilon\frac{\partial^2\vec{E}}{\partial t^2} = 0 \tag{1}$$

$$\nabla^2\vec{H} - \mu\varepsilon\frac{\partial^2\vec{H}}{\partial t^2} = 0 \text{ or } \frac{\partial^2\vec{H}}{\partial x^2} + \frac{\partial^2\vec{H}}{\partial y^2} + \frac{\partial^2\vec{H}}{\partial z^2} - \mu\varepsilon\frac{\partial^2\vec{H}}{\partial t^2} = 0 \tag{2}$$

These equations are called the time-dependent Helmholtz equations. The absence of the first-order term signifies that the electromagnetic fields do not decay as they propagate in a lossless medium. In Cartesian coordinates, the three scalar equations of Equation 1 are:

$$\begin{aligned}
\frac{\partial^2 E_x}{\partial x^2} + \frac{\partial^2 E_x}{\partial y^2} + \frac{\partial^2 E_x}{\partial z^2} &= \mu\varepsilon\frac{\partial^2 E_x}{\partial t^2} \\
\frac{\partial^2 E_y}{\partial x^2} + \frac{\partial^2 E_y}{\partial y^2} + \frac{\partial^2 E_y}{\partial z^2} &= \mu\varepsilon\frac{\partial^2 E_y}{\partial t^2} \\
\frac{\partial^2 E_z}{\partial x^2} + \frac{\partial^2 E_z}{\partial y^2} + \frac{\partial^2 E_z}{\partial z^2} &= \mu\varepsilon\frac{\partial^2 E_z}{\partial t^2}
\end{aligned} \tag{3}$$

and for equation (2) are:

$$\frac{\partial^2 H_x}{\partial x^2}+\frac{\partial^2 H_x}{\partial y^2}+\frac{\partial^2 H_x}{\partial z^2}=\mu\varepsilon\frac{\partial^2 H_x}{\partial t^2}$$
$$\frac{\partial^2 H_y}{\partial x^2}+\frac{\partial^2 H_y}{\partial y^2}+\frac{\partial^2 H_y}{\partial z^2}=\mu\varepsilon\frac{\partial^2 H_y}{\partial t^2} \quad (4)$$
$$\frac{\partial^2 H_z}{\partial x^2}+\frac{\partial^2 H_z}{\partial y^2}+\frac{\partial^2 H_z}{\partial z^2}=\mu\varepsilon\frac{\partial^2 H_z}{\partial t^2}$$

Both sets of equations describe coupled space- and time-dependent fields, and both have the form of the differential wave equation.

TRANSMISSION LINE IN PHASOR FORM

Assume that the EM wave is harmonic time dependence, we have:

$$V(z,t)=R\mathrm{e}\left[V_s(z)e^{i\omega t}\right] \quad (5)$$

$$I(z,t)=R\mathrm{e}\left[I_s(z)e^{i\omega t}\right] \quad (6)$$

where $V_s(z)$ and $I_s(z)$ are the phasor forms of $V(z,t)$ and $I(z,t)$.

The equation can be rewritten as:

$$-\frac{dV_s}{dz}=(R+j\omega L)I_s \quad (7)$$

$$-\frac{dI_s}{dz}=(G+j\omega C)V_s \quad (8)$$

Wave Equation for Voltage and Current

If we decoupled the equations by taking the second derivative, we get:

$$\frac{d^2V_s}{dz^2}=(R+j\omega L)(G+j\omega C)V_s \quad (9)$$

$$\frac{d^2V_s}{dz^2}-\gamma^2V_s=0 \quad (10)$$

$$\frac{d^2 I_s}{dz^2} - \gamma^2 I_s = 0 \tag{11}$$

where where $\gamma = \alpha + j\beta = \sqrt{(R + j\omega L)(G + j\omega C)}$.

Note that the wave equations for voltage and current are similar to the wave equations of plane waves.

Solutions of Wave Equations

Before we go into further details, let us first get ourselves familiar with a few terms:

- **Propagation constant (γ):** This is a measure of changes in a sinusoidal EM wave, in terms of amplitude and phase, while the wave is propagating through a medium. The medium can be a transmission line or free space. Note that the propagation constant is a dimensionless quantity, and is given by $\gamma = \alpha + j\beta$.
- **Attenuation constant (α):** It causes the signal amplitude to decrease while propagating through a transmission line which is assumed lossy. It always has a positive value, and its SI unit is Np/m (although dB/m is often used as well).
- **Phase constant (β):** It is the imaginary component of the propagation constant. It gives us the phase of the signal along a transmission line, at a constant time. Its unit is radians/meter, but it is often converted into degrees/meter.

The wavelength of an EM wave is given as $\lambda = \frac{2\pi}{\beta}$ and the wave velocity is $u = \frac{\omega}{\beta} = f\lambda$. So, the solutions of the wave equations, i.e Equations (10) and (11), are:

$$V_s(z) = V_o^+ e^{-\gamma z} + V_o^- e^{\gamma z} \tag{12}$$

$$I_s(z) = I_o^+ e^{-\gamma z} + I_o^- e^{\gamma z} \tag{13}$$

where the superscripts + and – denote, respectively, the amplitude of the positive and negative traveling waves.

Characteristic Impedance

The characteristics impedance, Z_o of the line is the ratio of the positive traveling voltage wave to current wave at any point on the line. The characteristic impedance can be represented mathematically as:

$$\begin{aligned} Z_o &= \frac{V_o^+}{I_o^+} = -\frac{V_o^-}{I_o^-} = \frac{R + j\omega L}{\gamma} = \frac{\gamma}{G + j\omega C} \\ Z_o &= \sqrt{\frac{R + j\omega L}{G + j\omega C}} = R_o + jX_o \end{aligned} \tag{14}$$

where R_o and X_o are, respectively, the real and imaginary parts of Z_o.

REFLECTION, REFRACTION AND DIFFRACTION

Reflection: Radio frequency (RF) radiation generally travels in a straight line through space and can be reflected by certain substances. The angle at which a radio wave is reflected from a smooth metal surface is equivalent to the angle at which it approaches the surface. In other words, the angle of reflection of RF waves equals their angle of incidence. This is shown in Figure 3.

This principle of RF reflection is used in antenna design to focus transmitted EM waves into a narrow beam and to collect and concentrate received RF signals into a receiver. If a reflector is designed with the reflecting surface in the shape of a paraboloid, EM waves approaching parallel to the axis of the antenna will be reflected and focused above the surface of the reflector at the feed horn. This type of arrangement is called prime focus and it provides the large antenna surface area necessary to receive very weak signals. Note that the term *aperture* is normally used for parabolic antenna surface area.

A major problem however, with prime focus arrangements for large aperture antennas is that, the equipment required at the prime focus is heavy and the supporting structure tends to sag under the weight of the equipment, which can affect calibration. One solution is the Cassegrain focus arrangement. Cassegrain antennas employ a secondary reflecting surface to "fold" the EM waves back to a prime focus near the primary reflector (see Figure 4).

The reflective properties of EM waves have also been used to investigate the planets using a technique called planetary radar. With this technique, EM waves are transmitted to the planet, where they reflect off the surface of the planet and are received at one or more receiving stations on Earth. Sophisticated signal processing techniques are used, where the signal in terms of time, amplitude, phase and frequency are carefully analysed to develop detail images of the surface of the planets.

Refraction: Refraction is the bending or the deflection of EM waves when they pass from one kind of transparent medium into another. The index of refraction is the ratio of the speed of the EM energy in a vacuum to the speed of the EM energy in the observed transparent medium. The law of refraction

Figure 3. RF reflected

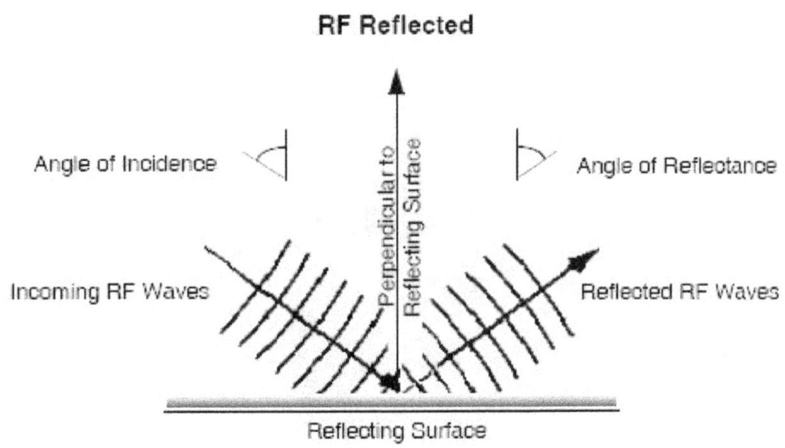

Figure 4. Prime focus and Cassegrain focus antennas

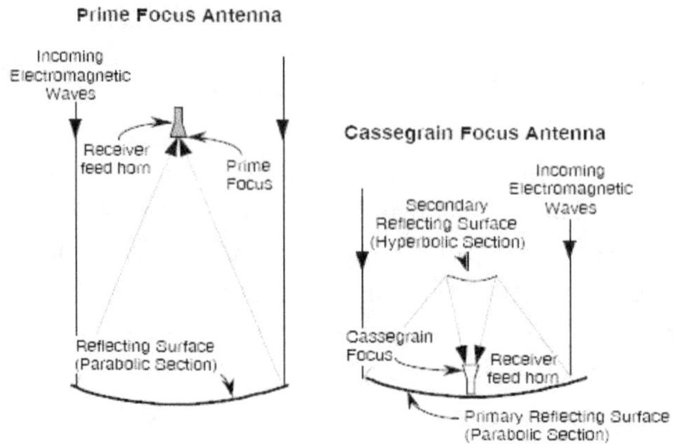

states that EM waves passing from one medium into another medium of a different refraction index, will be bent in their direction of travel.

Substances of higher densities would normally have higher indices of refraction. The index of refraction of a vacuum for example, is 1.0, by definition. The index of refraction of air is 1.00029, water is 1.3, glass 1.5 and diamond 2.4. Because air and glass have different indices of refraction, the path of EM waves travelling from air into glass at an angle will be bent towards the perpendicular plane as they travel into the glass. Similarly, the path will be bent to the same extent away from the perpendicular plane, when they exit the other side of the glass and into the air (see Figure 5).

In a similar manner, the EM waves entering Earth's atmosphere from space are slightly bent by refraction. Atmospheric refraction is the greatest for EM radiation coming from sources near to the horizon, ie, below 15° elevation, and causes the apparent altitude of the source to be higher than its actual height. As Earth rotates and the object gains attitude, the refraction effect decreases and becoming zero at zenith, which is directly overhead. Refraction's effect on sunlight adds about 5 minutes to the daylight at equatorial latitudes, since the Sun appears higher in the sky than it actually is (see Figure 6).

Figure 5. Air to glass to air refraction

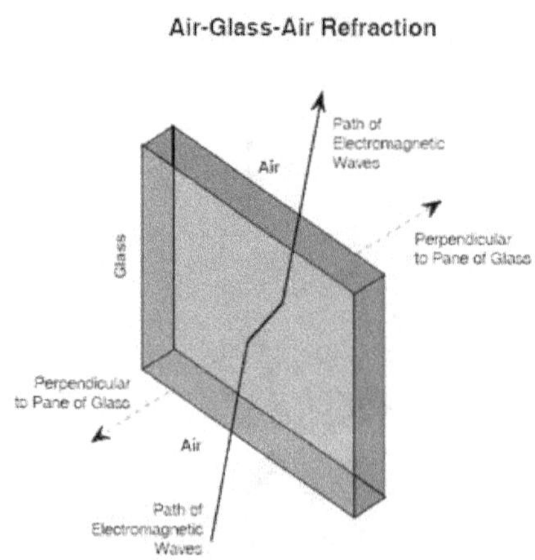

Figure 6. Diagram showing displacement of the Sun's image at sunrise and sunset due to atmospheric refraction

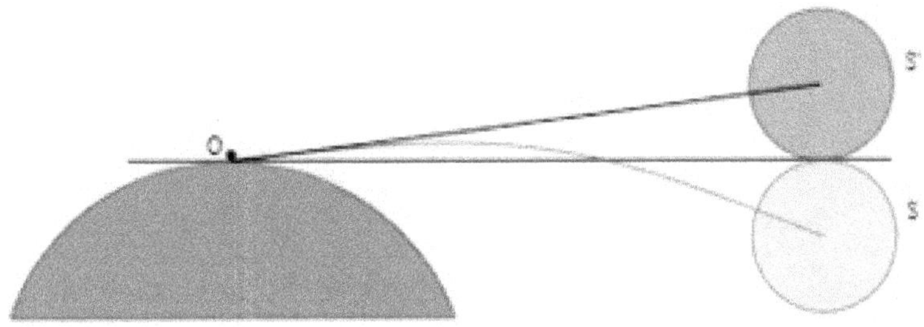

Diffraction: When an EM wave passes by an object in space, the wave is bent around the object. This phenomenon is known as diffraction. The effects of diffraction are usually very small and hence, almost unnoticeable. However, the effect of diffraction can be observed with a simple experiment using a source of light (like a light bulb). Hold two fingers about 10 cm in front of one eye. Keep the space between your fingers at about 1 mm. Look at the light source through the space between your fingers. By adjusting the spacing between your fingers, you will see a series of dark and light lines. They are caused by constructive and destructive interference of light diffracting around your fingers. Recall the inverse square law of propagation from the previous chapter. EM energy may be considered to propagate from a point source in plane waves. The inverse square law applies not only to the source of the energy, but also to any point on a plane wave. This means that, from any point on the plane wave, the EM energy propagates as if the point is the source of the EM energy. As a result, EM waves may be considered to be continuously created from every point on the plane and propagate in all direction (see Figure 7).

For an infinite plane wave, the sideways propagation from each point is balanced by the propagation from its neighbours. Thus, the EM wave continues on as a plane wave. However, when an EM wave encounters an object, the effect at the edges of the object is that the path of the EM wave is bent slightly, as depicted in Figure 8.

Suppose that now the EM wave (light wave for example) is then intercepted by a surface, say a screen, a short distance from the object. Compared to the parallel EM waves that have passed farther

Figure 7. Huygens' plane waves

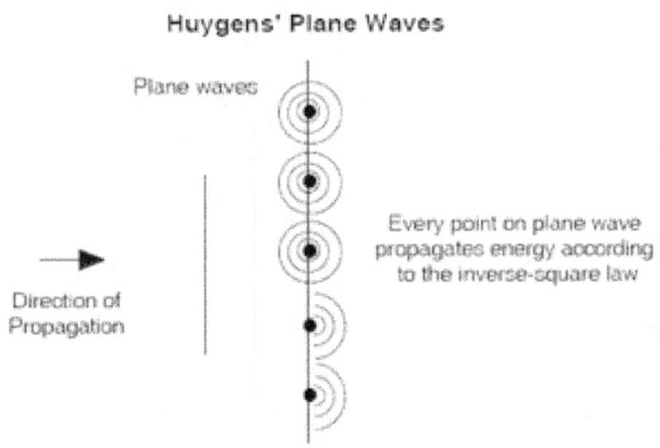

Figure 8. EM waves bend slightly at the edges of object they encountered

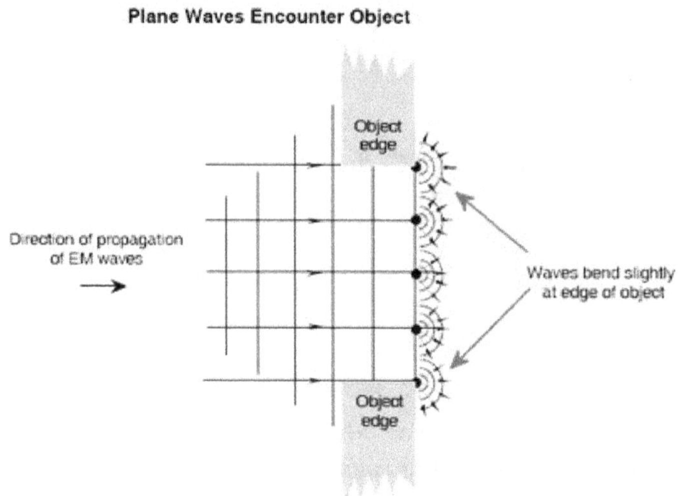

from the object's edge (for example "waves B, C and D" as depicted in Figure 9), the waves at the edges (for example "waves A and E" in Figure 9) will have travelled a shorter distance from the object to any given point on the screen.

The effect is that the light waves are out of phase when they arrive at any given point on the surface. If they are out of phase by 180°, they will cancel each other out (destructive interference) and produce a dark line. However, if they are in phase, they will add together (constructive interference) and produce a bright line.

Diffraction is most noticeable when an EM wave passing around an object or through an opening in an object, like the slit between your fingers, is monochromatic (the same frequency). Figure 10 shows a typical diffraction pattern seen when observing a star through a telescope with a lens that focuses the light to a point.

FLUX DENSITY, INTENSITY AND LUMINOSITY

This section will deal briefly on some of the basic concepts of EM radiation theory. Let us assume that some EM radiation is passing through a surface with an area of dA, as depicted in Figure 11. Some of the EM radiation will leave the surface within a beam of solid angle of $d\omega$, and at an angle of θ with respect to the surface. The amount of energy entering the solid angle within a frequency range $[v, v+dv]$ in a given time of dt can be given as (Palmer & Davenhall, 2001):

$$dE_v = I_v \cos\theta dA dv d\omega dt \tag{15}$$

where I_v is the specific intensity of the EM radiation at the frequency v and in the direction of the solid angle $d\omega$, with dimensions of $Wm^{-2}Hz^{-1}sr^{-1}$.

The total intensity, I is given as (Palmer & Davenhall, 2001):

Figure 9. Diffraction pattern

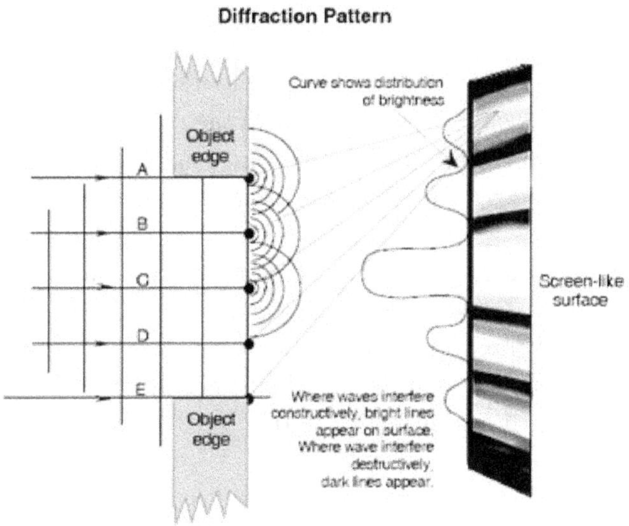

Figure 10. Diffraction pattern seen when observing a star through a telescope

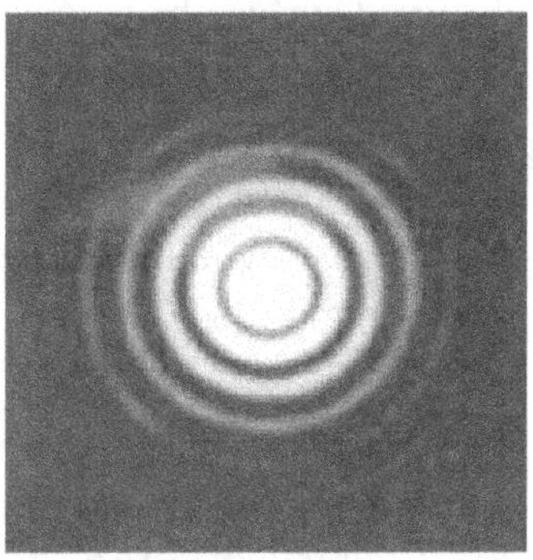

Figure 11. The relationship between EM radiation intensity (I_v) and energy passing though a surface with the area of dA, into a solid angle at an angle of θ to the surface (Palmer & Davenhall, 2001).

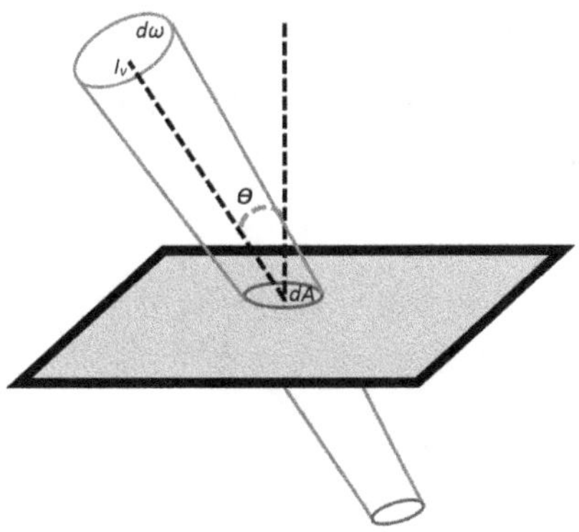

$$I = \int_0^\infty I_v dv \tag{16}$$

However, from the observational point of view, we are more interested in the energy flux (L_v, L) and the flux density (F_v, F). Flux density gives the power of the radiation per unit area and hence has the dimensions of $Wm^{-2}Hz^{-1}$. The observed flux densities are usually extremely small, especially in radio astronomy. Therefore, flux densities are often expressed in units of the Jansky (Jy), where 1 Jy = 10^{-26} $Wm^{-2}Hz^{-1}$.

Let us consider a star as the source of the EM radiation, then the flux emitted by the star into a solid angle, ω is $L = \omega r^2 F$, where F is the flux density observed at a distance r from the star. Note that it is also usual to refer to the total flux from a star as the luminosity, L. If the radiation from the star is isotropic, then radiation at a distance r will be distributed evenly on a spherical surface of area $4\pi r^2$. Hence we get the equation (Palmer & Davenhall, 2001):

$$L = 4\pi r^2 F \tag{17}$$

The situation would be more complicated for an extended luminous object such as a nebula or galaxy. The *surface brightness* is defined as the flux density per unit solid angle. The geometry of the situation results in the interesting fact that the observed surface brightness is *independent* of the distance of the observer from the extended source. This somewhat counter-intuitive phenomenon can be understood by realising that although the flux density arriving from a unit area is inversely proportional to the distance to the observer, the area on the surface of the source enclosed by a unit solid angle at the observer is directly proportional to the square of the distance. As a result, the two effects cancel each other out.

REFERENCES

Cheng, D. K. (1989). *Field and Wave Electromagnetics*. US: Addison-Wesley.

Guru, B. S., & Hiziroglu, H. R. (2004). *Electromagnetic Field Theory Fundamentals*. US: Cambridge University Press. doi:10.1017/CBO9781139165297

Palmer, J., & Davenhall, A. C. (2001). The CCD Photometric Calibration Cookbook. Retrieved from http://star-www.rl.ac.uk/star/docs/sc6.htx/sc6se5.html

Yeap, K. H., & Tham, C. Y. (2018). Optimization of an offset receiver optics for radio telescopes. *Journal of Infrared, Millimeter, and Terahertz Waves*, *39*(1), 64–76. doi:10.100710762-017-0449-z

Section 3
Radio Telescope Design Methods

Chapter 6
Physical Optics

Hirokazu Kobayashi
Osaka Institute of Technology, Japan

ABSTRACT

Physical optics (PO) is one of the fundamental and powerful high-frequency theories for electromagnetic scattering and radiation. The total field of a source (antenna) which radiates in the presence of a perfectly conducting surface may be expressed as a superposition of the incident and the scattered fields. The current fields which exist everywhere are chosen in PO to denote the electric and magnetic fields of the source, i.e., they exist as if the scatterer was "absent"; this is unlike the geometrical optics (GO) incident field, which exists in the presence of the surface of the scatterer. The scattered fields in this case can be expressed in terms of the radiation integrals over the actual currents induced on the surface of the scatterer. These currents also radiate the scattered fields in the absence of the scatterer. This chapter shows the fundamental PO formulation and calculated results, and some topics which improve the conventional PO to the extended PO such as "physical theory of diffraction (PTD)" and "PO with transition current (PTD-TC)".

PHYSICAL OPTICS

Physical optics (PO) is one of the fundamental electromagnetic high-frequency theories for scattering and radiation problem. The total field of a source (such as an antenna) which radiates in the presence of a perfectly conducting surface may be expressed as a superposition of the incident field ($\boldsymbol{E}^i$, $\boldsymbol{H}^i$) and the field ($\boldsymbol{E}^s$, $\boldsymbol{H}^s$) which is scattered by the surface. The current fields are chosen in PO to denote the electric and magnetic fields of the source which exist everywhere, i.e., they exist as if the scatterer was "absent"; this is unlike the geometrical optics (GO) incident field, which exists in the presence of the surface of the scatterer. The scattered fields in this case can be expressed in terms of the radiation integrals over the actual currents induced on the surface of the scatterer. These currents also radiate the scattered

DOI: 10.4018/978-1-7998-2381-0.ch006

fields in the absence of the scatterer, i.e., these currents are now viewed as equivalent sources, which are impressed over the mathematical boundary but with the perfecting conducting scatterer removed.

In the conventional PO method, the radiation integral for the scattered field is calculated by employing a GO approximation for the currents induced on the surface, which is assumed to be an electrically large plane; hence, the PO method is also a high-frequency method. In this chapter, the following are illustrated: the fundamental PO formulation and calculated results, and some topics which improve the conventional PO to the extended PO such as physical theory of diffraction (PTD) (Ufimtsev, 1971) and PO with transition current (PTD-TC) considering transition currents (Kobayashi, 1999). The PO method described here is the same concept as Kirchhoff's integration derived from the Huygens' principle.

Both the electromagnetic radiation and scattering which re-radiates electromagnetic waves around the scatterer from the induced currents are considered as a secondary source on the surface which is illuminated by incident waves. The PO is a typical high-frequency technology, and is a powerful calculation method with high accuracy of approximation if the size of scattering object is sufficiently large compared to the wavelength. There are two main features of PO from a practical point of view. First, there is a degree of freedom in how to express the electromagnetic current, and second, the scatterer surface which induces currents can be subdivided and calculated. As described later, the former can individually treat transition electromagnetic currents at discontinuities such as wedge, and the latter indicates that it can be divided into patches even if the target model has a complicated shape. It is well known that PO gives a finite value with a certain correctness near the focal point and focal line, and that the radiation integration includes a diffraction effect. In the general PO method, as mentioned above, the equivalent electromagnetic current of the radiation integral is obtained by the GO method. Since GO is a fundamental high-frequency theory that treats waves as ray, PO is also a high-frequency theory, so the electromagnetic current is zero in the shadow area of the scattering object. The electromagnetic current distribution in the transition region is also different from the actual one, and this is a reason that limits the application range of PO.

By analyzing from the viewpoint of the integral equation, it can be seen that the solution of PO evaluated by GO currents in the illuminated region is the result of ignoring the higher-order transition electromagnetic current term, and the contribution of the shadow region is canceled by the incident field. The general approach for determining the surface electromagnetic current using an integral equation is a matrix calculation. This method is not only time consuming but may also produce an incorrect result when the size of the scatterer is larger than the wavelength. Because of this reason, several attempts have been proposed to remove the aforementioned limitations based on PO, such as PTD and PTD-TC. As mentioned above, another feature of PO is the degree of freedom with respect to the segmentation area. A mathematical expression of the scatterer surface shape is required to perform the PO radiation integral analytically. The simplest shape is a flat plate, and the integration is evaluated assuming that the electromagnetic current flows on an infinite flat plate. Complicated and electrically large shaped objects are calculated by dividing them with a large number of flat plates, and numerical integration is performed when the shape is expressed by mathematical equations or data. In order to allow readers to have a detailed understanding on PO and PTD, the historical background of PO is first illustrated in this chapter. This is then followed by the explanation of the scattering caused by rotating curved surfaces and smooth convex flat plates.

Historical Aspect of PO

In the old days, optics meant GO (light goes straight). However, it has been recognized that wave theory is necessary to explain phenomena such as color and diffraction. There are two types of wave theory: physical optics and physiological optics. What is commonly recognized by people with normal vision is called physical optics, and those related to individual differences are called physiological optics (Mach, 1926). Schrödinger pointed out that Hamilton's equation corresponds to Huygens principle (wave theory, physical optics) and their relationship is the same as the one between classical and quantum mechanics. In Ruck, Barrick and Kirchbaum's handbook on scattering theory (Ruck, 1970), the term PO represents an approximate method for determining the scattered field from an object by assuming a specific field distribution on the surface of the object. In particular, the field on the surface is a GO field. This means that the scattering on the surface of an object illuminated geometrically behaves as if the surface is an infinite tangent plane, and the surface field is zero in the shadow region of the object. According to the book written by Bowman, Senior and Uslenghi [Bowman, 1969], on the other hand, PO is a method whose current is approximated by $\boldsymbol{J} = 2\boldsymbol{n} \times \boldsymbol{H}$ in exact radiation integral, where the current is called the GO current.

From the description above, it is natural to think that PO is not a specific method, but rather it is conceptual and has a fairly broad meaning. Since PO has such a broad meaning, it can be applied not only to perfect conductors but also to imperfect conductors. In addition, the Aperture Field Method (AFM) for such antenna analysis naturally falls within the category of the PO method. AFM is a useful technique for antenna analysis such as parabolic reflector and pyramidal horn with an aperture. In the case of analyzing scattering problems by dielectrics, the use of a polarization current is called Born's or Rytov's approximation (Born, 1964). This treatment which uses the electromagnetic field distribution on the surface of dielectrics is also in the category of the PO method.

There are relatively few books that provide detailed explanation on PO. This is thought to be one of the reasons that obscure the definition of PO. Here, an outline of PO will be given with reference to the above handbook. The term PO is often used as a synonym for Kirchhoff approximation, tangential plane approximation, and Huygens principle. These terms are still used today because of their historical development and their respective advantages. Huygens' Principle or Kirchhoff's approximation was formulated long before Maxwell showed the vector nature of the electromagnetic field. Therefore, the original form of Kirchhoff approximation is formulated based on the light wave and the scalar wave, not the vector field. Since the electromagnetic field satisfies the boundary condition of the vector, not the scalar, scalar Kirchhoff approximation is justified only when it is clearly shown to match the vector equation.

The Rayleigh-Stevenson's expansion developed for the reciprocal of wavelength, that is, the method based on the series expansion or the method of moments (MoM) for the integral equation is limited to the analysis and calculation in principle when the scattering object is small compared to the wavelength. Therefore, it is necessary for high-frequency techniques that enable efficient calculations even when the object is larger than the wavelength. There is a theory called geometrical theory of diffraction (GTD) as another typical method for high-frequency techniques, which was proposed by J. B. Keller in the 1950s. GTD extends the theory of classical geometric optics (GO) by adding diffraction ray (James, 1980 and Borovikov, 1994). Uniform asymptotic theory (UAT), uniform theory of diffraction (UTD), equivalent current method (ECM) (Pathak, 1988, Osipov, 2017), equivalent source method (ESM) (Kobayashi, 2012), etc. improved the original GTD so that a uniform field can be obtained. These ray theories are introduced to overcome the limitations at the shadow boundary (SB), transition region, and caustic region overlap, etc. On the other hand, physical theory of diffraction (PTD) is an improved extension version of PO by introducing a transition current at a discontinuous region, just as GTD is an extension of GO. In PTD, a transition current is introduced at the illuminated and shadow boundary region. The PTD was proposed by Ufimtsev, Soviet, at the same time as Keller, USA. When the PTD was first developed, it was applied only to objects with edges. Recently, the method has also been applied to smooth impedance objects (Ufimtsev, 2007). However, there are not so many application examples compared with the GTD.

In Green's theorem: $\int\left(p\partial q\,/\,\partial n - q\partial p\,/\,\partial n\right)dS$, PO and PTD can be regarded as an theory that shows what quantity to use for p or $\partial p\,/\,\partial n$. In conventional PO, p or $\partial p\,/\,\partial n$ is determined by using GO fields of incident and reflected waves in the illuminated area of a scattering object. On the other hand, PTD is considered to add non-uniform transition to p or $\partial p\,/\,\partial n$ near the edge. This turbulence, which is the difference of PTD from PO, can be obtained indirectly from exact solutions such as wedges. Since the PO method is calculated regardless of the shape and material of the object in shadow area other than the illuminated area, the same result is obtained for all shapes as shown in Figure 1(a). In

Figure 1. The problem of PO with the same scattering pattern

Figure 2. Problems of PO requiring higher-order reflected and diffracted waves

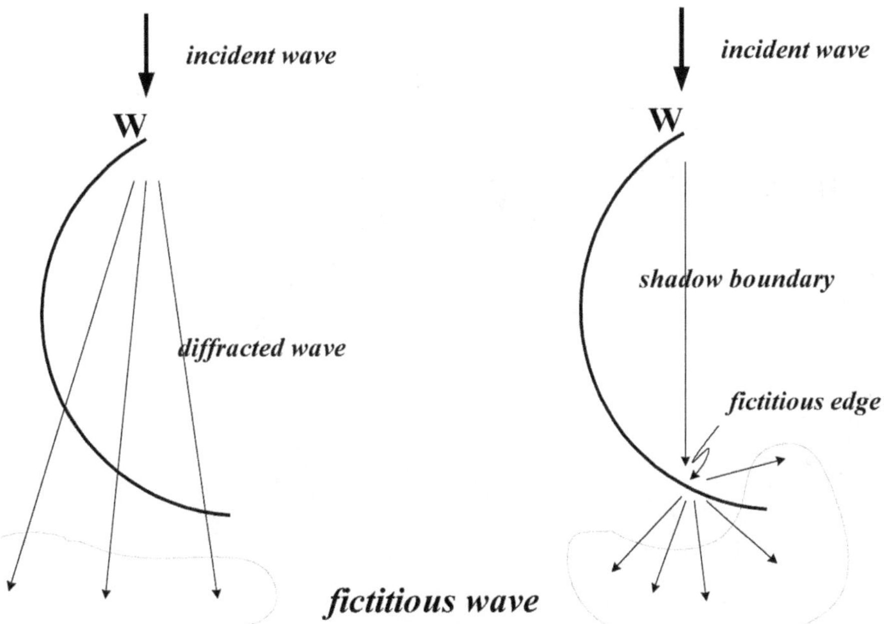

addition, in the case of Figure 1(b) where the end portion *W* of the object is blocked geometrically by other portions of the object, the diffracted wave from *W* may deteriorate the accuracy. From the above simple consideration, the following can be inferred for the PO method.

(a) PO and PTD incorporate the effect of diffracted waves in the reflected wave, and both of them give a finite solution in the vicinity of the shadow boundary (SB) which improve the accuracy of the result. Furthermore, both these two methods can be effectively applied in complex areas such as focal and caustic region where GTD is not applicable.
(b) Since integration is performed along the incident wave which illuminate the surface of the scatterer, the integration range can be divided into patches. This means that the scatterer model can be divided by plural flat plates or into geometric shapes that can be integrated analytically.
(c) There is a degree of freedom in giving electric and magnetic currents. The most common method is a GO current, which is determined by an approximated tangential plane at a reflection point. If a non-uniform transition current can be considered with respect to such a uniform current, the calculation accuracy can be improved.
(d) In the conventional PO, an induced GO current is assigned to be 0 in the shadow region of the object, so that the transition current at the end is ignored. For this reason, when the size of the scatterer is not sufficiently large compared to the wavelength, its accuracy in diffraction region may be accurate (Figure 1a).

Figure 3. Surface currents and coordinate system of scatterer and antenna

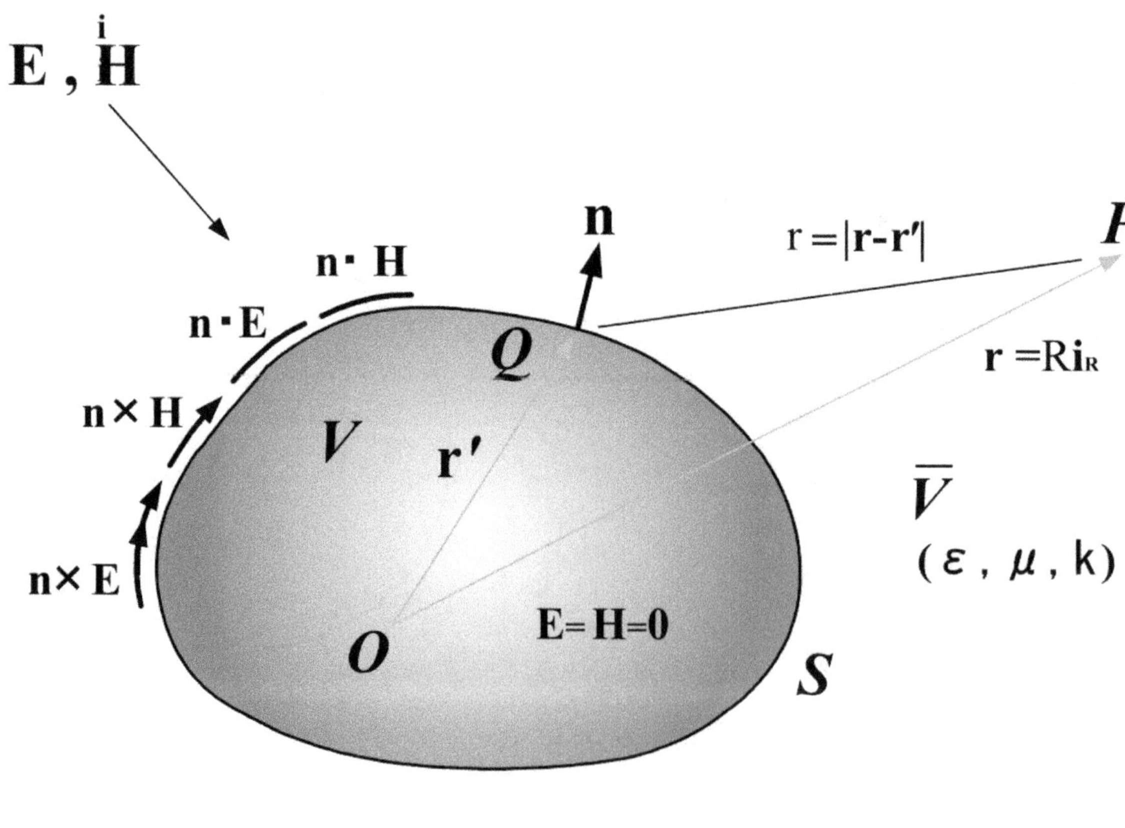

(e) When an electromagnetic wave is incident to a smooth surface at a parallel angle, the current is difficult to be evaluated (Figure 1a).

(f) When a diffracted wave or a direct wave is obstructed by other objects, it is necessary to consider a secondary or a higher-order current (multiple reflected and diffracted wave) (Figure 1b). It is required to account for these higher-order multiple reflection and diffraction during analysis.

The conditions in (d) to (f) occur when the current evaluation does not satisfy the boundary condition. For this reason, the asymptotic solution method generally employs multiple reflection or diffraction effect or higher-order current. The problems in (d) and (e) are already solved by the method which considers the current due to the canonical problem such as wedge or cylinder, This is to say that, by applying the radiation integral (PTD-TC) to solve the transition current in (c), the scattering field can therefore be sufficiently improved by considering the current in the small transition region. The conditions in (f) and (g) are particularly problematic for a concave corner-reflector like object. Usually, it is necessary to identify radiation and non-radiation regions of higher-order reflected waves in order to solve the problem.

To understand the condition in (c), consider the integral equation for perfect conductor shown in Figure 2. When the boundary condition $\boldsymbol{n} \times \boldsymbol{E} = 0$, $\boldsymbol{n} \bullet \boldsymbol{H} = 0$ on the surface S is applied to the scattered electromagnetic field expressed by surface integration and since it is necessary to consider only a tangential component on the scatterer surface, the following equation is obtained.

$$\boldsymbol{n} \times \boldsymbol{E}^{i} = \boldsymbol{n} \times \int_{S-\delta} \left\{ j\omega\mu \left(\boldsymbol{n} \times \boldsymbol{H} \right) G - \left(\boldsymbol{n} \bullet \boldsymbol{E} \right) \nabla' G \right\} dS', \tag{1a}$$

$$\boldsymbol{n} \times \boldsymbol{H} = 2\boldsymbol{n} \times \boldsymbol{H}^{i} + 2\boldsymbol{n} \times \int_{S-\delta} \left(\boldsymbol{n} \times \boldsymbol{H} \right) \times \nabla' G dS', \tag{1b}$$

where $\boldsymbol{H}(\boldsymbol{r}) \rightarrow \boldsymbol{H}(\boldsymbol{r})/2$ is employed when an observation point is on the surface S, and $\omega, G, \boldsymbol{n}$ are angular frequency, Green's function of free space and outward normal vector, respectively. The first expression contains two unknowns, and the second expression has one in the integrand. However, by using the relationship: $\boldsymbol{n} \bullet \boldsymbol{E} = j / \left(\omega\varepsilon \right) \nabla \bullet \left(\boldsymbol{n} \times \boldsymbol{H} \right)$, $\boldsymbol{n} \times \boldsymbol{H} = \boldsymbol{J}_s$, equation (1) can be expressed in termed of only one unknown variable as

$$\boldsymbol{n} \times \boldsymbol{E}^{i} = \frac{1}{j\omega\varepsilon} \boldsymbol{n} \times \int_{S-\delta} \left\{ -\omega^{2}\mu\varepsilon \boldsymbol{J}_s G + \nabla' \bullet \boldsymbol{J}_s \nabla G \right\} dS', \tag{2a}$$

$$\boldsymbol{J}_s = 2\boldsymbol{n} \times \boldsymbol{H}^{i} + 2\boldsymbol{n} \times \int_{S-\delta} \boldsymbol{J}_s \times \nabla' G dS'. \tag{2b}$$

Either the former electric or the latter magnetic field type equation may be used for the equivalent source. PTD-TC is a technique that gives a degree of freedom to electromagnetic distribution induced on the scatterer surface. Considering transition current for the conventional PO current and the discontinuity for uniform radiation region, equation (2) can be expressed as follows:

$$J_s = 2n \times H^i \text{ for illuminated uniform region} \tag{3a}$$

$$= 2n \times H^i + 2n \times \int_{\text{illuminated}} \left(2n \times H^i\right) \times \nabla' G dS' + \int_{\text{transition}} J_s \times \nabla' G dS',$$

for transition and shadow region (3b)

where the first integral of the second equation is the current obtained by the conventional PO, and the second integral is the current in the transition region or shadow region. Assuming that the first term of the second equation is a GO solution, the integral of the second term represents a higher-order solution. In particular, the accuracy can be improved by adding a transition current at a discontinuous portion such as an edge to the conventional GO term. Once the expression of the transition region is obtained, these solutions may be applied locally to an object of arbitrary shape. The integration range is depending on its incident angle and shape of the discontinuity. Hence, these two factors are to be taken into consideration assigning the integration range. Although the integration range can either be the whole region or only near the discontinuity region, integrating over the entire region is usually not necessary.

By introducing the surface impedance of a scattered object, ¶$_s$, and since the impedance boundary condition states that the magnetic current $M_s = -\zeta_s Z_0 \left(n \times J_s\right)$ [Senior, 1995], equation (3) can be expressed in terms of the surface impedance as follows:

$$J_s = 2n \times H^i + 2n \times \int_S J_s \times \nabla' G dS' + \frac{j2\zeta_s}{k} \int_S \left(k^2 + \nabla'\nabla' G\right)\left(n \times J_s\right) dS', \tag{4}$$

where k is wave-number of free space. This expression is available for a non-metal imperfect conductor.

Now, the general far-field expression of the radiation integral is obtained from Stratton-Chu formula or the vector potential theory as follows [Stratton, 1941]:

$$E^s\left(r\right) = -j\omega\mu G_0\left(R\right) \int_S \left\{n \times H - i_R \bullet \left(n \times H\right) i_R - Y_0 \left(n \times E\right) \times i_R\right\} e^{jki_R \bullet r'} dS', \tag{5a}$$

$$H^s\left(r\right) = j\omega\varepsilon G_0\left(R\right) \int_S \left\{n \times E - i_R \bullet \left(n \times E\right) i_R + Z_0 \left(n \times H\right) \times i_R\right\} e^{jki_R \bullet r'} dS', \tag{5b}$$

where $G_0(R) = \exp(-jkR)/(4\pi R)$. The relation of the co-ordinates and variables are shown in Figure 2. When the illuminated point of perfect conductor is approximated by a flat plate, the boundary condition is given as:

$$\boldsymbol{n}\times\boldsymbol{E} = \boldsymbol{n}\times\left(\boldsymbol{E}^i + \boldsymbol{E}^r\right) = 0, \qquad \boldsymbol{n}\times\boldsymbol{H} = 2\boldsymbol{n}\times\boldsymbol{H}^i. \tag{6}$$

Therefore, the radiation integral for conductor in equation (5) can be written as follows.

$$\boldsymbol{E}^s\left(\boldsymbol{r}\right) = -j2kZ_0G_0\left(R\right)\int_S \left\{\boldsymbol{n}\times\boldsymbol{H}^i - \boldsymbol{i}_R\bullet\left(\boldsymbol{n}\times\boldsymbol{H}^i\right)\boldsymbol{i}_R\right\}\mathrm{e}^{jk\boldsymbol{i}_R\bullet\boldsymbol{r}'}dS', \tag{7a}$$

$$\boldsymbol{H}^s\left(\boldsymbol{r}\right) = j2kG_0\left(R\right)\int_S \left\{\boldsymbol{n}\times\boldsymbol{H}^i\times\boldsymbol{i}_R\right\}\mathrm{e}^{jk\boldsymbol{i}_R\bullet\boldsymbol{r}'}dS', \tag{7b}$$

Scattering for Rotating Curved Surface η(*x*, *y*)

When an object is illuminated by electromagnetic waves, a current is induced on the surface of the object, and this surface current re-radiates a new radiation wave. The surface current is determined from the boundary condition that the tangential component of the electric field on the conductor is zero, however it is difficult to accurately determine this current for any shape of object. In the PO method, an object having an arbitrary shape is replaced with a flat plate at an illuminated point, and its induced current is evaluated by GO. Namely, PO can be regarded as an approximated solution of high-frequency techniques since PO is employing a GO current. Here, a generalized formula is derived based on the radiation integral of PO as much as possible.

The magnetic field of the incident plane wave is given as

$$\boldsymbol{H}^i = \boldsymbol{H}_0\exp\left(-jk\boldsymbol{k}_i\bullet\boldsymbol{r}'\right), \tag{8}$$

and the scatterer with a curved surface is expressed by

$$z' = \eta\left(x, y\right). \tag{9}$$

Then, a normal vector and an area element are given by

$$\boldsymbol{n} = \frac{1}{N}\left\{-\frac{\partial\eta}{\partial x'}\boldsymbol{i}_x - \frac{\partial\eta}{\partial y'}\boldsymbol{i}_y + \boldsymbol{i}_z\right\}, \quad dS' = Ndx'dy', \tag{10}$$

$$N = \sqrt{1 + \left(\frac{\partial \eta}{\partial x'}\right)^2 + \left(\frac{\partial \eta}{\partial y'}\right)^2}, \qquad \boldsymbol{r}' = x'\boldsymbol{i}_x + y'\boldsymbol{i}_y + \eta(x, y)\boldsymbol{i}_z,$$

where $\boldsymbol{k}_i$ and $\boldsymbol{i}_x$ are a wave-number vector of an incident wave and a unit direction vector toward to x direction. Using the above descriptions, the equation (7b) is modified to

$$\boldsymbol{H}^s(\boldsymbol{r}) = j2kG_0(R)\boldsymbol{i}_R \times \boldsymbol{H}_0 \times \int_S \frac{1}{N}\left\{-\frac{\partial \eta}{\partial x'}\boldsymbol{i}_x - \frac{\partial \eta}{\partial y'}\boldsymbol{i}_y + \boldsymbol{i}_z\right\} e^{jk(\boldsymbol{i}_R - \boldsymbol{k}_i)\bullet \boldsymbol{r}'} dx'dy'. \tag{11}$$

When a plane wave incidents to an infinite flat plate, the reflected wave is a plane wave. The reflection coefficient depends on the polarization, so that it is decomposed into parallel $E_{\parallel}$ and perpendicular $E_{\perp}$ components. The reflected wave for each component is given by

$$E_{\perp}^r = R_{\perp}E_{\perp}^i, \qquad \boldsymbol{k}_r \times \boldsymbol{E}_{\parallel}^r = R_{\parallel}\boldsymbol{k}_i \times \boldsymbol{E}_{\parallel}^i, \tag{12}$$

where $\boldsymbol{k}_i$, $\boldsymbol{k}_r$ are unit direction vectors of an incident and a reflected wave, respectively, and also correspond to unit wave-number vectors. The parallel $R_{\parallel}$ and perpendicular $R_{\perp}$ reflective coefficients are expressed as follows:

$$R_{\parallel} = \frac{\varepsilon_r \cos\theta_i - \sqrt{\mu_r \varepsilon_r - \sin^2\theta_i}}{\varepsilon_r \cos\theta_i + \sqrt{\mu_r \varepsilon_r - \sin^2\theta_i}}, \qquad R_{\perp} = \frac{\mu_r \cos\theta_i - \sqrt{\mu_r \varepsilon_r - \sin^2\theta_i}}{\mu_r \cos\theta_i + \sqrt{\mu_r \varepsilon_r - \sin^2\theta_i}}. \tag{13}$$

Here, θi is an incident angle. The total fields are given by $E_{\parallel} = E_{\parallel}^i + E_{\parallel}^i, E_{\perp} = E_{\perp}^i + E_{\perp}^i$.

In the case of backscattering, the integral for the scattered wave in equation (11) is significantly simplified, where the incident direction and the scattering direction are reversed, $\boldsymbol{i}_R = -\boldsymbol{k}_i$. Taking the direction of $\mathbf{i}_R$ to z-axis and assuming that the scatterer is a perfect conductor, the radiation integral in equation (11) is derived as follows:

$$\boldsymbol{H}^s(\boldsymbol{r}) = j2kG_0(R)\boldsymbol{H}_0 \int_S (\boldsymbol{n}\bullet \boldsymbol{i}_z)\exp(j2k\eta)\, dS'$$

$$= j2kG_0(R)\boldsymbol{H}_0 \int_S \exp(j2k\eta)\, dx'dy'. \tag{14}$$

As shown in Figure 3, $(\boldsymbol{n}\bullet \boldsymbol{i}_z)dS'$ is the projection of $\boldsymbol{n}dS'$ to the plane of $(x' - y')$. S_z is the projected area in range $z > \eta$ given as

$$dS_z = \frac{\partial S_x}{\partial \eta} d\eta \tag{15}$$

Therefore, the following expression is obtained:

$$\boldsymbol{H}^s\left(\boldsymbol{r}\right) = j2kG_0\left(R\right)\boldsymbol{H}_0 \int_L^0 \exp\left(j2k\eta\right)\frac{\partial S_x}{\partial \eta} d\eta. \tag{16}$$

Here, L is the maximum value of the surface satisfying $z > \eta$, and when η is in the $z' = 0$ plane and increases, L approaches 0. For example, considering the case of a spherical surface of the radius a: $S_z = \pi\left(a^2 - \eta^2\right)$, so the following equation can be calculated easily.

$$\boldsymbol{H}^s\left(\boldsymbol{r}\right) = \frac{jk\exp\left[-jk\left(R-a\right)\right]}{R}\left\{\frac{a}{j2k} + \frac{1}{\left(2k\right)^2}\left[1 - \exp\left(-j2ka\right)\right]\right\}. \tag{17}$$

Figure 4. Application of PO method to rotating object

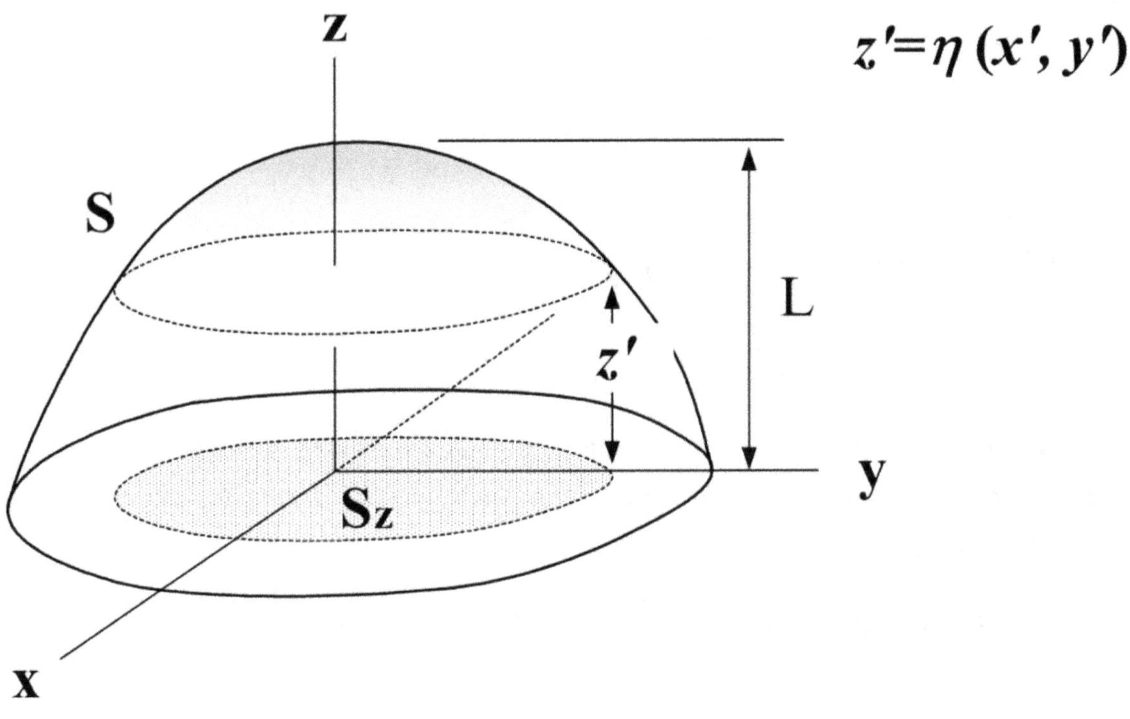

Since only the first term is sufficient for high-frequencies, the radar cross-section for conducting sphere σ is derived as $\sigma = \lim_{R\to\infty} 4\pi R^2 \left(\boldsymbol{H}^s / \boldsymbol{H}^i\right)^2 = \pi a^2$. This means that the electrical reflection size of a conducting sphere equals to its physical size.

Now, for the incomplete conductor, the backscattered wave for a linearly polarized plane wave incident along the z-axis can be arranged as follows:

$$\boldsymbol{E}_X^s\left(\boldsymbol{r}\right) = -j2kG_0\left(R\right)\left[a_{11}\boldsymbol{i}_x + a_{21}\boldsymbol{i}_y\right], \qquad \boldsymbol{E}_Y^s\left(\boldsymbol{r}\right) = -j2kG_0\left(R\right)\left[a_{12}\boldsymbol{i}_x + a_{22}\boldsymbol{i}_y\right], \tag{18a}$$

where

$$a_{11} = \int_{S_p} \frac{\left(\frac{\partial \eta}{\partial x'}\right)^2 R_{\parallel} - \left(\frac{\partial \eta}{\partial y'}\right)^2 R_{\perp}}{\left(\frac{\partial \eta}{\partial x'}\right)^2 + \left(\frac{\partial \eta}{\partial y'}\right)^2} \exp\left(j2k\eta\right) dx'dy', \tag{18b}$$

$$a_{12} = a_{21} = \int_{S_p} \left(R_{\parallel} + R_{\perp}\right) \frac{\left(\frac{\partial \eta}{\partial x'}\right)^2 \left(\frac{\partial \eta}{\partial y'}\right)^2}{\left(\frac{\partial \eta}{\partial x'}\right)^2 + \left(\frac{\partial \eta}{\partial y'}\right)^2} \exp\left(j2k\eta\right) dx'dy', \tag{18c}$$

$$a_{22} = \int_{S_p} \frac{\left(\frac{\partial \eta}{\partial y'}\right)^2 R_{\parallel} - \left(\frac{\partial \eta}{\partial x'}\right)^2 R_{\perp}}{\left(\frac{\partial \eta}{\partial x'}\right)^2 + \left(\frac{\partial \eta}{\partial y'}\right)^2} \exp\left(j2k\eta\right) dx'dy'. \tag{18d}$$

Here, $\boldsymbol{E}_X^s\left(\boldsymbol{r}\right)$ and $\boldsymbol{E}_Y^s\left(\boldsymbol{r}\right)$ are scattering fields by x-polarized incident wave $\boldsymbol{E}^i = \boldsymbol{i}_x E_{0x} \exp\left(jkz\right)$ and y-polarized incident wave $\boldsymbol{E}^i = \boldsymbol{i}_y E_{0y} \exp\left(jkz\right)$, respectively. For circular polarization, as the same manner, the following expressions are obtained.

$$\boldsymbol{E}_R^s\left(\boldsymbol{r}\right) = -j2kG_0\left(R\right)\left[a_{RR}\boldsymbol{i}_R + a_{LR}\boldsymbol{i}_L\right], \quad \boldsymbol{E}_L^s\left(\boldsymbol{r}\right) = -j2kG_0\left(R\right)\left[a_{RL}\boldsymbol{i}_R + a_{LL}\boldsymbol{i}_L\right], \tag{19a}$$

$$\boldsymbol{i}_R = \frac{1}{\sqrt{2}}\left(\boldsymbol{i}_x \mp j\boldsymbol{i}_y\right), \qquad \boldsymbol{i}_L = \frac{1}{\sqrt{2}}\left(\boldsymbol{i}_x \pm j\boldsymbol{i}_y\right) \qquad \text{for wave to} \begin{Bmatrix} +z \text{ direction} \\ -z \text{ direction} \end{Bmatrix}, \tag{19b}$$

$$a_{RR} = \int_{S_p} \left(R_{\parallel} + R_{\perp}\right) \frac{\left(\frac{\partial \eta}{\partial x'}\right)^2 - j2\frac{\partial \eta}{\partial x'}\frac{\partial \eta}{\partial y'} - \left(\frac{\partial \eta}{\partial y'}\right)^2}{\left(\frac{\partial \eta}{\partial x'}\right)^2 + \left(\frac{\partial \eta}{\partial y'}\right)^2} \exp\left(j2k\eta\right) dx'dy', \tag{19c}$$

$$a_{RL} = a_{LR} = \frac{1}{2}\int_{S_p} \left(R_{\perp} - R_{\parallel}\right) \exp\left(j2k\eta\right) dx'dy', \tag{19d}$$

$$a_{LL} = \int_{S_p} \left(R_{\parallel} + R_{\perp}\right) \frac{\left(\frac{\partial \eta}{\partial x'}\right)^2 + j2\frac{\partial \eta}{\partial x'}\frac{\partial \eta}{\partial y'} - \left(\frac{\partial \eta}{\partial y'}\right)^2}{\left(\frac{\partial \eta}{\partial x'}\right)^2 + \left(\frac{\partial \eta}{\partial y'}\right)^2} \exp\left(j2k\eta\right) dx'dy'. \tag{19e}$$

Here, $\boldsymbol{E}_R^s\left(\boldsymbol{r}\right)$ and $\boldsymbol{E}_L^s\left(\boldsymbol{r}\right)$ are the scattering fields of the right-handed circular-polarized incident wave $\boldsymbol{E}^i = \boldsymbol{i}_R E_{0R} \exp\left(jkz\right)$ and left-handed wave $\boldsymbol{E}^i = \boldsymbol{i}_L E_{0L} \exp\left(jkz\right)$, respectively. Since these integrations are performed in the cross-section perpendicular to the incident wave direction (in $x' - y'$ plane), the reflection coefficients $R_{\parallel}$ and $R_{\perp}$ are required to be expressed by x' and y'. In this case, it can be obtained by using the following equation:

$$\cos\theta_i = \left[1 + \left(\frac{\partial \eta}{\partial x'}\right)^2 + \left(\frac{\partial \eta}{\partial y'}\right)^2\right]^{-\frac{1}{2}}, \quad \sin\theta_i = \left(1 - \cos^2\theta_i\right)^{-\frac{1}{2}}. \tag{20}$$

In the case of normal incidence $R_{\parallel} = -R_{\perp}$, the term which consists of $R_{\parallel} + R_{\perp}$ is 0. In other words, if the incident wave is circularly polarized in the clockwise direction, the reflected wave is then in left-handed circular polarization. This is because the direction of polarization is defined when viewed from the back of the propagation wave.

Next, the PO integration for backscattering is evaluated by the stationary phase method. The direction of the backscattering wave is taken the z-axis and the normal vector $\boldsymbol{n}$ on the surface is also oriented towards the z-direction. The x'- and y'-axes are optional except that they are perpendicular to the z-axis. The object is assumed to be a perfect conductor and the reflection point is $[0, 0, \eta(0, 0)]$. In order to evaluate the stationary phase point, Maclaurin's expansion is performed for $\eta(x', y')$ as follows:

$$\eta\left(x', y'\right) = \eta\left(0, 0\right) + x'\frac{\partial \eta}{\partial x'} + y'\frac{\partial \eta}{\partial y'} + \frac{x'^2}{2}\frac{\partial^2 \eta}{\partial x'^2} + x'y'\frac{\partial^2 \eta}{\partial x' \partial y'} + \frac{y'^2}{2}\frac{\partial^2 \eta}{\partial y'^2} + \cdots, \tag{21}$$

where the partial differential coefficient of η with respect to x', y' is calculated at $x' = y' = 0$. Since normal vector at $x' = y' = 0$,

$$\boldsymbol{n} = \nabla\eta = \frac{\partial\eta}{\partial x'}\boldsymbol{i}_x + \frac{\partial\eta}{\partial y'}\boldsymbol{i}_y + \frac{\partial\eta}{\partial z'}\boldsymbol{i}_z, \tag{22}$$

where the direction of the wave is assumed to be oriented towards $\boldsymbol{i}_z$ and $\partial\eta / \partial x' = 0, \partial\eta / \partial y' = 0$. For convenience, the variables in equation (23) below are substituted into equation (21):

$$p = \frac{\partial^2\eta}{\partial x'^2}_{|x'=y'=0}, \quad q = \frac{\partial^2\eta}{\partial y'^2}_{|x'=y'=0}, \quad r = \frac{\partial^2\eta}{\partial x'\partial y'}_{|x'=y'=0}. \tag{23}$$

Then, equation (21) modified as shown below

$$\eta\left(x', y'\right) = \frac{p}{2}x' + rx'y' + \frac{q}{2}y'. \tag{24}$$

Now, by rotating (x', y') coordinates by α and transforming them to a new set of coordinates (x_1, y_1) and also letting the bilinear term of (x_1, y_1) to be 0, the following equation is obtained

$$\eta\left(x', y'\right) = K_1x_1^2 + K_2y_1^2, \quad K_{1,2} = \frac{p+q}{4} \pm \frac{\left(p-q\right)r}{2}\left[\left(\frac{p-q}{2}\right)^2 + r^2\right]^{-1/2}. \tag{25}$$

This method is called principal-axis transformation, and the coefficients $K_{1,2}$ have simple geometric optical meaning, that is

$$\rho_{1,2} = \frac{1}{\left|K_{1,2}\right|} \tag{26}$$

are principal curvature radiuses of surface η, which give maximum and minimum curvature radiuses. From the above discussion, the equation (14) can be modified as follows:

$$\boldsymbol{H}^s\left(\boldsymbol{r}\right) = j2k\exp\left[j2k\eta\left(0,0\right)\right]G_0\left(R\right)\boldsymbol{H}_0\int_{x_1}\int_{x_2}\exp\left[j2k\left(K_1x_1^2 + K_2x_2^2\right)\right]dx_1dx_2. \tag{27}$$

When the limit of wavelength approaches 0, i.e. $\lambda \to 0$, K_1 and K_2 approach a very large value. The integration range of x_1, x_2 are not specifically determined except when η is too small to be expanded by Maclaurin's expansion. For example, if $kK_1x_1^2$ approaches around 10λ, the contribution of the ignored terms in Maclaurin's expansion is negligible. Also, if $kK_1x_1^2$ has the above-mentioned size, the integration value is hardly changed even if the range of x_1 is extended to infinity. From this consideration, equation (27) can be expressed as follows:

$$\boldsymbol{H}^s(\boldsymbol{r}) = j2k\exp\left[j2k\eta(0,0)\right]G_0(R)\boldsymbol{H}_0\int_{-\infty}^{\infty}\exp\left(j2kK_1x_1^2\right)dx_1\int_{-\infty}^{\infty}\exp\left(j2kK_2x_2^2\right)dx_2$$

$$= \frac{\exp\left[-jk\left(R-2\eta(0,0)\right)\right]}{2R\sqrt{K_1K_2}}\boldsymbol{H}_0 = \frac{\exp\left[-jk\left(R-2\eta(0,0)\right)+j\delta\right]}{2R}\sqrt{\rho_1\rho_2}\boldsymbol{H}_0, \tag{28}$$

where δ is 0 when K_1 and K_2 are both positive, $\pi/4$ when either K_1 and K_2 is negative, and $\pi/2$ when both K_1 and K_2 are negative. In this relation, the radar cross-section defined by $\sigma = 4\pi \lim_{R\to\infty} R^2\boldsymbol{H}^{s2} / \boldsymbol{H}^{i2}$ can be expressed in terms of ρ_1 and ρ_2 as $\sigma = \pi\rho_1\rho_2$. If $\rho_{1\,\text{or}\,2} \to \infty$, it can be reduced to a flat plate and equation (27) can be applied to solve the magnetic field. If $\rho_{1\text{ or }2}$ is an object given by a function of x, one can use the equation (27) or (28) depending on the conditions. Additionally, if the object is a sphere, $\rho_1 = \rho_2 = \text{a}$ and $\eta(0, 0) = a$. Then, equation (28) is calculated as follows:

$$\boldsymbol{H}^s(\boldsymbol{r}) = \frac{a}{2R}\exp\left[-jk\left(R-2a\right)\right]\boldsymbol{H}_0. \tag{29}$$

This coincides with the first term of equation (17). Evaluating the radiation integral at the stationary phase point is nothing more than having the first term corresponding to the result of GO.

Scattering by Flat Plate (I): PO for General Conductors

In this section and the following section, we apply PO or PTD to plane wave diffraction to formulate the scattering field expressions on a conducting plate with surface impedance. The perimeter of the flat plates analyzed here has smooth and convex boundary – two of such examples are the triangular and square flat plates.

First, the scattering fields of plane waves incident on a smooth convex flat plate are derived. For simplicity, the case of a flat plate with perfect conductor is first discussed; the analysis is then expanded to the case of a flat plate with surface impedance in the next section. When the integrand of the radiation integral has only a first-order phase term, the area integral can be converted into a contour integral along the perimeter of the plate. If the plate size is sufficiently large compared to the wavelength, the asymptotic solution of the integral for the convex plate can be obtained using the stationary phase method. For discs and ellipses, which are special cases of convex plates, the exact asymptotic solution using PO is completely consistent. If this asymptotic solution is compared with the exact PO solution of a circular and elliptical plate, a caustic correction expression for a convex plate of an arbitrary shape is obtained. Furthermore, the equivalent electromagnetic current method (ECM) is applied to the edge current of Ufimstev to derive the PTD solution for the convex plate.

In order to formulate the radiation integral of a perfect conducting flat plate, a simple shape such as an ellipse is to be solved analytically. When a plane wave incidents on a flat plate, the area integral can be converted into a line integral along the perimeter using Stokes' theorem. Then, the problem is generalized to the case of a convex flat plate, and the radiation integral is calculated by introducing local

Figure 5. Diffraction of electromagnetic plane wave by a smooth convex plate

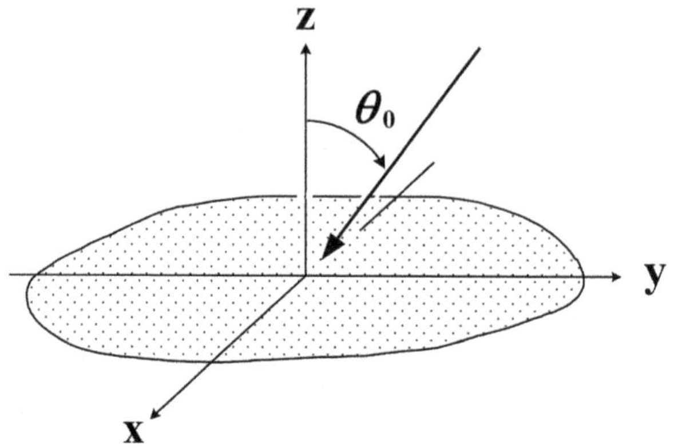

coordinates. From the obtained asymptotic expression, the physical interpretation becomes clear that the diffraction field is composed of contributions from two diffraction points on the perimeter of a convex plate, its magnitude depends on the radius of curvature at the diffraction point, and so on. This solution becomes inaccurate at a caustic point. However, by examining the relationship between the Airy's pattern directly solved by PO for a disk etc. and its asymptotic solution, an effective modified formula can be derived even at a caustic point.

A smooth convex plate is a flat plate surrounded by a closed curve where the gradient or curvature does not change suddenly. This coordinate system is shown in Figure 4. The incident plane is $y-z$ plane ($\phi = \pi/2$), and incident wave is given by the following equations:

E-polarization: $$E_x^i = E_{0x}\exp\left\{jk\left(y\sin\theta_0 + z\cos\theta_0\right)\right\}, \tag{30a}$$

$$H_y^i = -Y_0E_{0x}\cos\theta_0\exp\left\{jk\left(y\sin\theta_0 + z\cos\theta_0\right)\right\}, \tag{30b}$$

H-polarization: $$H_x^i = H_{0x}\exp\left\{jk\left(y\sin\theta_0 + z\cos\theta_0\right)\right\}, \tag{31a}$$

$$E_y^i = Z_0H_{0x}\cos\theta_0\exp\left\{jk\left(y\sin\theta_0 + z\cos\theta_0\right)\right\}, \tag{31b}$$

where $Z_0 = Y_0^{-1} = \sqrt{\mu / \varepsilon}$ is the free space impedance, and θ_0 is the incident angle measured from the positive z-axis. The PO current induced on the plate is given as follows:

$$\boldsymbol{J} = 2\boldsymbol{n}\times\boldsymbol{H}^i = 2\boldsymbol{i}_x\times\boldsymbol{H}^i = 2\left(Y_0E_{0x}\cos\theta_0\boldsymbol{i}_x + H_{0x}\cos\theta_0\boldsymbol{i}_y\right)\exp\left(jk\sin\theta_0\right). \tag{32}$$

On the other hand, the magnetic vector potential for this current is derived as:

$$A_x = 2Y_0\mu E_{0x}G_0\left(R\right)\cos\theta_0\int_S \exp\left\{jk\left(ux' + vy'\right)\right\}dx'dy', \tag{33a}$$

$$A_y = 2\mu H_{0x}G_0\left(R\right)\int_S \exp\left\{jk\left(ux' + vy'\right)\right\}dx'dy', \tag{33b}$$

$$G_0\left(R\right) = \frac{\exp\left(-jkR\right)}{4\pi R}, \quad u = \sin\theta\cos\theta, \quad v = \sin\theta_0 + \sin\theta\sin\theta, \tag{33c}$$

where $\boldsymbol{i}_x$ and $\boldsymbol{i}_y$ are respectively the unit vectors toward the x and y-axis, u and v are angular parameters according to the diffraction direction, and $R = \sqrt{x^2 + y^2 + z^2}$ is the distance from the origin to the observation point. Using far-field approximation, the vector potentials are expressed by electrical field $\boldsymbol{E} \cong -j\boldsymbol{A}$, so that the fields by PO are calculated as follows:

$$E_\theta^{PO} = -j\omega\left\{A_x\cos\theta + A_y\sin\theta\right\}\cos\theta$$

$$= -j2kG_0\left(R\right)\cos\theta\left\{E_{0x}\cos\theta_0\cos\theta + Z_0H_{0x}\sin\theta\right\}P\left(\theta,\phi\right) \tag{34a}$$

$$E_\phi^{PO} = -j2kG_0\left(R\right)\left\{-E_{0x}\cos\theta_0\sin\theta + Z_0H_{0x}\cos\theta\right\}P\left(\theta,\phi\right) \tag{34b}$$

$$P\left(\theta,\phi\right) = P\left(u,v\right) = \int_S \exp\left\{jk\left(ux' + vy'\right)\right\}dS' \tag{34c}$$

In this way, the electromagnetic field of the plane wave diffracted by the conducting flat plate is reduced to the radiation integral $P(u, v)$, so that it is only necessary to pay attention to this integral thereafter. This integral can be converted to a line integral using Stokes' theorem:

$$\oint_C \boldsymbol{V}\bullet d\boldsymbol{l} = \int_S \nabla \times \boldsymbol{V}\bullet \boldsymbol{n}dS. \tag{35}$$

Here, vector $\boldsymbol{V}$ is assigned by

$$\boldsymbol{V} = \left\{-u\boldsymbol{i}_x + v\boldsymbol{i}_y\right\}\exp\left\{jk\left(ux' + vy'\right)\right\}, \tag{36}$$

then, the pattern function $P(u, v)$ is given by the line integral

$$P\left(u,v\right) = \frac{1}{j2kuv}\oint_C \left\{-u\boldsymbol{i}_x + v\boldsymbol{i}_y\right\}\bullet\boldsymbol{i}_l\exp\left\{jk\left(ux' + vy'\right)\right\}dl. \tag{37}$$

The integration path C is the perimeter of the plate and i_y is a tangential unit vector along C. For the special case of u or $v = 0$, the pattern function is calculated as follows:

$$P(u,0) = \frac{1}{j2ku}\oint_C \{\boldsymbol{i}_y \bullet \boldsymbol{i}_l\} \exp\{jkux'\}\, dl, \tag{38a}$$

$$P(0,v) = -\frac{1}{j2kv}\oint_C \{\boldsymbol{i}_x \bullet \boldsymbol{i}_l\} \exp\{jkuy'\}\, dl. \tag{38b}$$

and (0, 0) = S is the area of the plate. This integral can be easily performed if C is explicit. In order to show this, calculations for a rectangle, an isosceles triangle, a circle, and an ellipse are performed below.

(i) Rectangular Plate (2A × 2B)

Considering a rectangular plate of size $2A \times 2B$, the integral along each side is calculated as follows:

$$P(\theta,\varphi) = 4AB\mathrm{sinc}(kAu)\,\mathrm{sinc}(kBv), \quad \mathrm{sinc}(x) = \frac{\sin x}{x}. \tag{39}$$

If a complex shaped object can be expressed by many flat patches, the scattering field from this object can also be easily calculated by using this simple sampling function, which is a standard method for the commercialized software system [Adana, 2011]. However, an object with a large curvature or a large electrical length needs to be divided into a large number of flat patches. For example, a large aircraft needs to be divided into approximately the number of 100,000 to 1 million patches at X-band, depending on the calculation accuracy. Basically, when the size of the flat patch is approximately a half wavelength, the result is almost the same as that obtained by integrating the curved surface.

(ii) Isosceles Triangle (Base 2A, Base Angle α)

Let us consider an isosceles triangle with a base length of $2A$ along the x-axis and two angles of α. Since its vertex is at (0, $A\tan\alpha$), the integration along each side yields as follows:

$$P(\theta,\phi) = P_1(\theta,\phi) + P_2(\theta,\phi) + P_3(\theta,\phi) \tag{40a}$$

$$P_1(\theta,\phi) = -\frac{A}{jkv}\mathrm{sinc}(kAu) \tag{40b}$$

$$P_2(\theta,\phi) = -\frac{A}{j2kuv}(v\tan\pm - u)\ \exp\left\{j\frac{kA}{2}(v\tan\pm - u)\right\}\mathrm{sinc}\left\{\frac{kA}{2}(v\tan\pm + u)\right\} \tag{40c}$$

$$P_3(\theta,\phi)=\frac{A}{j2kuv}(v\tan\pm+u)\quad \exp\left\{j\frac{kA}{2}(v\tan\pm+u)\right\}\mathrm{sinc}\left\{\frac{kA}{2}(u-v\tan\pm)\right\} \tag{40d}$$

(iii) Circular Disk With Radius a

Considering $u=\sin\theta\cos\theta$, $v=\sin\theta_0+\sin\theta\sin\theta$, and $dl=\left(-\mathbf{i}_x\sin\varphi'+\mathbf{i}_y\cos\phi'\right)ad\phi'$, then by performing variable conversion $x'=a\cos\phi'$, $u=w\cos'$, $v=w\sin'$, the pattern function $P(\theta,\phi)$ is obtained as follows:

$$P(\theta,\phi)=\frac{a}{j2kw\cos'\sin'}\int_0^{2\pi}\exp\left\{jkaw\cos\left('-\phi'\right)\right\}\sin\left('+\phi'\right)d\phi'. \tag{41}$$

Assuming $\phi'=\delta+\psi$ and employing Bessel's integral formula and the relation $d\left[xJ_1(x)\right]/dx=xJ_0(x)$, the following expression is obtained.

$$P(\theta,\phi)=\pi a^2\frac{J_1(kaw)}{kaw},\quad w=\sqrt{\sin^2\theta\cos^2\phi+\left(\sin\theta_0+\sin\theta\sin\phi\right)^2}. \tag{42}$$

This expression shows the well-known Airy's pattern, and it is finite at even caustic point $w=0$. Using the asymptotic form of Bessel's function for $kaw\gg1$

$$J_1(z=kaw)\cong\sqrt{\frac{2}{\pi z}}\left\{\cos\left(z-\frac{3\pi}{4}\right)-\frac{3}{8z}\sin\left(z-\frac{3\pi}{4}\right)\right\}\approx\sqrt{\frac{2}{\pi z}}\cos\left(z-\frac{3\pi}{4}\right), \tag{43}$$

$P(\theta,\varphi)$ reduces to

$$P(\theta,\varphi)=\sqrt{2\pi}\frac{a^2}{(kaw)^2}\left\{\exp\left(jkaw-j\frac{3\pi}{4}\right)+\exp\left(-jkaw+j\frac{3\pi}{4}\right)\right\}. \tag{44}$$

From this expression, the main contribution of a far-field is found from two points on the plate perimeter.

(iv) Elliptical Plate With Axial Length 2A, 2B

The elliptical plate is calculated in the same way as the circular plate. Using relational and transformation expressions below:

$$x' = A\cos\phi',\ y' = B\sin\phi',\ dl = dx'\mathbf{i}_x + dy'\mathbf{i}_y = \left(-A\sin\phi'\mathbf{i}_x + B\cos\phi'\mathbf{i}_y\right)d\phi', \tag{45a}$$

$$u = w\cos\delta, \qquad v = w\sin\delta,\ \left[-ui_x + vi_y\right]\Delta dl = wC\sin\left(\varphi' + \Phi\right)d\phi', \tag{45b}$$

$$C = \sqrt{A^2\cos^2\delta + B^2\sin^2\delta},\quad \tan\Phi = \frac{B\sin\delta}{A\cos\delta},\quad \Phi = y\sin\theta_0 + z\cos\theta_0, \tag{45c}$$

the following expression is obtained.

$$P\left(\theta,\varphi\right) = \frac{C}{j2kw\cos\delta\sin\delta}\int_0^{2\pi}\exp\left\{jkCw\cos\left(\varphi' - \Phi\right)\right\}\sin\left(\varphi' + \Phi\right)d\phi'$$

$$= \pi AB\frac{2J_1\left(kCw\right)}{kCw}. \tag{46}$$

Substituting $A = B = C$, it is found that equation (46) reduces to equation (44), i.e. from an ellipse to a circular disk.

(v) Monostatic Radar Cross-Section of Area S Conducting Plate

The radar scattering cross section (RCS) of 3D object is defined by

$$\sigma = 4\pi\lim_{R\to\infty} R^2\frac{\boldsymbol{E}^{s}\bullet\boldsymbol{E}^{s*}}{\boldsymbol{E}^{i}\bullet\boldsymbol{E}^{i*}}. \tag{47}$$

Since $\theta = \theta_0 = 0^{\circ}$ for a plane wave which is incident from the normal direction to an arbitrary shaped conductor plate with area S, equation (48) below can be derived for both polarizations

$$4\pi R^2E^s\bullet E^{s*} = \frac{4\pi}{\lambda^2}\left(\frac{E_{0x}^2}{Z_0^2H_{0x}^2}\right)P^2\left(\theta = 0,\phi\right) \tag{48}$$

Therefore, σ is generally given as follows:

$$\sigma = \frac{4\pi}{\lambda^2}P^2\left(w = 0\right) = 4\pi\left(\frac{S}{\lambda}\right)^2 \tag{49}$$

This is the formula for the backscattering cross-section at normal incidence when $k \to \infty^2$, that is, in high-frequency.

Scattering by Flat Plate (II): PO for Conducting Smooth Convex

Now, let us asymptotically find the PO integral by using the stationary phase method for a convex conductor plate with a smooth edge (Kobayashi, 1996). The stationary phase method is an approximate technique that evaluates the integration in a region where the integration variable does not change a lot, and is an analytical method that is frequently used in the field of electromagnetic wave engineering (Borovikov, 1994). As a matter of course, it is assumed that the integral value in the stationary region is the main integral evaluation. Physically, a reflection and a diffraction point of electromagnetic waves correspond to the stationary point. As mentioned earlier, since the main contribution of the diffraction field is given in the vicinity of the edge of perimeter, it is convenient to introduce a local curved coordinate system having the edge as an origin for representing a convex shape.

A point on the perimeter is expressed by using the parameter t as follows:

$$x' = f(t), \qquad y' = g(t). \tag{50}$$

Here, the PO integration in equation (37) is modified to:

$$P(\theta, \phi) = \frac{1}{j2kuv} \oint_C \left\{-f'(t)u + g'(t)v\right\} \exp\left\{jk\left[f(t)u + g(t)v\right]\right\} dt, \tag{51}$$

where $u = \sin\theta\cos\phi = w\cos'$, $v = \sin\theta_0 + \sin\theta\sin\phi = w\sin'$. By introducing function $H(t)$ and γ, the differential coefficients of $f(t)$ and $g(t)$ are expressed by

$$f'(t) = H(t)\cos\gamma, \quad g'(t) = H(t)\sin\gamma, \quad H(t) = \sqrt{f'^2(t) + g'^2(t)}. \tag{52}$$

Then, by substituting equation (52) intoquation (51), $P(\theta, \phi)$ is modified to:

$$P(\theta, \phi) = -\frac{w}{j2kuv} \oint_C H(t)\cos(\gamma + \delta) \exp\left\{jkw\left[f(t)\cos(\delta) + g(t)\sin(\delta)\right]\right\} dt. \tag{53}$$

The stationary phase point is evaluated as:

$$f'(t)\cos(\delta) + g'(t)\sin(\delta) = H(t)\cos(\gamma - \delta) = 0 \quad \text{or } \gamma - \delta = \pm\frac{\pi}{2}. \tag{54}$$

In order to examine the physical meaning of these equations, the following unit vectors are introduced

$$\boldsymbol{i}_\delta = \cos\delta\boldsymbol{i}_x + \sin\delta\boldsymbol{i}_y, \ \boldsymbol{i}_t = \cos\gamma\boldsymbol{i}_x + \sin\gamma\boldsymbol{i}_y. \tag{55}$$

It is found out that the stationary phase point is a point where vector $\boldsymbol{i}_\delta$ is perpendicular to tangential vector $\boldsymbol{i}_t$. This is shown in Figure 5.

Next, in order to find an asymptotic solution, the exponential part $p(t)$ of equation (53) is expanded around the stationary phase point given in equation (54) into Taylor's series, and the expression up to the second term is given as follows:

$$p(t) = \{f(t_s)\cos(\delta) + g(t_s)\sin(\delta)\} + \frac{1}{2}\{f''(t_s)\cos(\delta) + g''(t_s)\sin(\delta)\}(t - t_s)^2, \quad (56)$$

where t_s is the value of t at the stationary phase point. The perimeter of the convex plate is a closed curve, so that there are two phase points, $t_s = t_1, t_2$ as aforementioned, given by

$$\gamma - \delta = +\frac{\pi}{2} \quad \mathit{for} \quad t_s = t_1,\ \gamma - \delta = -\frac{\pi}{2} \text{ for } \quad t_s = t_2. \quad (57)$$

Therefore, the second term of equation (56) is calculated as

$$\frac{1}{2}\{f''(t_1)\sin\gamma - g''(t_1)\cos\gamma\} = -\frac{1}{2}\frac{H^2(t_1)}{\kappa(t_1)}, \quad (58a)$$

$$\frac{1}{2}\{-f''(t_2)\sin\gamma + g''(t_2)\cos\gamma\} = \frac{1}{2}\frac{H^2(t_2)}{\kappa(t_2)}, \quad (58b)$$

where the relations in equation (59) below

$$H(t_s)\sin\delta = -f'(t_1) = f'(t_2),\ H(t_s)\cos\delta = g'(t_1) = -g'(t_2) \quad (59)$$

are employed. The variable $\kappa(t)$ is a radius of curvature along the perimeter at point t defined by

$$\kappa(t) = \frac{H^3(t)}{f'(t)g''(t) - g'(t)f''(t)}. \quad (60)$$

The first term in equation (56) is a projection in the i_δ direction with magnitude $\sqrt{f^2(t_s) + g^2(t_s)}$, so this is given by

$$f(t_1)\cos(\delta) + g(t_1)\sin(\delta) = d_1,\quad f(t_2)\cos(\delta) + g(t_2)\sin(\delta) = -d_2. \quad (61)$$

Figure 6. Relationship between stationary phase point and vector $\mathbf{i}_t$ and $\mathbf{i}_\delta$

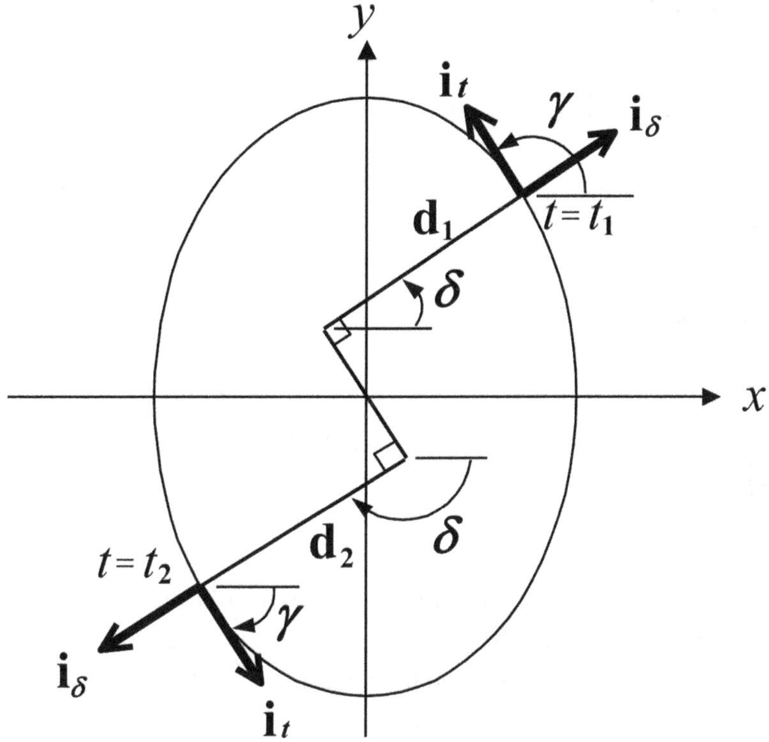

The exponential part of equation (53) can be calculated by equations (58), (60), and (61).

Substituting the above expressions into equation (53) and calculating the asymptotic solution by the standard stationary phase method, the pattern function is obtained as follows:

$$P(\theta,\phi)=\frac{\sqrt{2\pi}}{(kw)^{3/2}}\left\{\sqrt{\kappa(t_1)}\exp\left(jkd_1w-j\frac{3}{4}\pi\right)+\sqrt{\kappa(t_2)}\exp\left(-jkd_2w+j\frac{3}{4}\pi\right)\right\}. \tag{62}$$

The same result can be obtained by variable transformation for equation (53) in radial and tangential directions, using partial integration and the stationary phase method. For circular disc that are special cases of a convex plate, by letting $\kappa(t_1) = \kappa(t_2) = $ a, $d_1 = d_2 = a$, it can be directly calculated as

$$P(\theta,\phi)=\frac{\sqrt{2\pi a}}{(kw)^{3/2}}\left\{\exp\left(jkaw-j\frac{3}{4}\pi\right)+\exp\left(-jkaw+j\frac{3}{4}\pi\right)\right\}. \tag{63}$$

This is consistent with the asymptotic solution in equation (44) for the circular disc.

The above derived asymptotic solution diverges in a caustic area where adjacent rays overlap. Therefore, when considering the relationship between asymptotic solutions and exact solutions in the case of a circle or an ellipse, a corrected Airy's pattern can be presumed. By applying the following asymptotic first term of Bessel's function,

$$J_1(z) \cong \sqrt{\frac{2}{\pi z}} \cos\left(z - \frac{3}{4}\pi\right),\ J_2(z) \cong \sqrt{\frac{2}{\pi z}} \sin\left(z - \frac{3}{4}\pi\right), \tag{64}$$

the expression that is also effective in caustic is obtained as follows:

$$P(\theta,\phi) = \pi\left(d_{av}\right)^{3/2} \exp\left(jkw\frac{d_1 - d_2}{2}\right)$$

$$\cdot\left\{\left(\sqrt{\kappa(t_1)} + \sqrt{\kappa(t_2)}\right)\frac{J_1\left(kd_{av}w\right)}{kd_{av}w} + j\left(\sqrt{\kappa(t_1)} - \sqrt{\kappa(t_2)}\right)\frac{J_2\left(kd_{av}w\right)}{kd_{av}w}\right\}, \tag{65}$$

where $d_{av} = (d_1 + d_2)/2$. Naturally, the above asymptotic solution becomes equation (63). Thus, equation (65) can be regarded as a generalization of a circular disk and an ellipsoid plate. It is conceivable that a partial pattern correction is performed by the term of $J_2\left(kd_{av}w\right)$ resulted from the difference in radius of a curvature. For the case of an ellipse,

$$d_1 = d_1 = \sqrt{A^2\cos^2\delta + B^2\sin^2\delta} = C,\ \kappa(t_1) = \kappa(t_2) = \kappa = \frac{A^2B^2}{C^3}, \tag{66}$$

are obtained. Therefore, equation (65) is reduced to coincide exactly with equation (46).

Now, let us examine the validity of equation (65) numerically. The expression of the convex conductor plate is given as

$$f(t) = a\left(\cos t + 0.1\cos 2t - 0.01\cos 4t\right), g(t) = a\left(\sin t + 0.1\sin 2t - 0.01\sin 4t\right). \tag{67}$$

The calculation process is as follows: (i) Determine the value of the stationary phase point t for observation point (θ,ϕ) from equation (54) and (ii) Calculate equation (65) using the two stationary points. Then, (iii) Equation (53) is directly numerically integrated for comparison. Figure 6 shows the case where the parameters are $ka = 10$, θ-polarization, and $\phi_0 = 0^\circ$. The shape expressed by equation (67) is close to an ellipse, and considering that the caustic correction expression derived here is close to the exact PO solution (numerical integration) of the ellipsoid, its consistency can be confirmed sufficiently. Next, as a general example of a convex flat plate, two types of shape are shown in Figure 7. The perimeter coordinates are given numerically. The corresponding scattering field given by equation (65) (solid line) are shown in Figure 7(c) and (d) where the diamond symbol is the result of the numerical integration.

Scattering by Flat Plate (III): PTD for Conducting Smooth Convex

The PO solution U^{PO} derived in the previous section is more accurate in the specular direction, however GTD solution U^{GTD} is known to be more accurate in other areas. By using the difference between these two solutions $\Delta U = U^{GTD} - U^{PO}_{asy}$, the expression $U^{PTD} = U^{PO} + \Delta U$ is called PTD solution. Here, U^{PO}_{asy} is the asymptotic expression of PO solution U^{PO}. In other words, the solution for an object with an edge such as a wedge is given by $U = U^{PO} + \left(U^{GTD} - U^{PO}_{asy}\right)$. Therefore, it approaches asymptotically like $U^{PTD} \to U^{PO}$ in the vicinity of the shadow boundary (SB) such as a mirror surface direction and $U^{PTD} \to U^{GTD}$ in other diffraction regions. In this way, the PTD can be regarded as a method that incorporates the advantage of both the PO and GTD methods, i.e. the PO solution can be applied in the main lobe direction and the GTD solution in other side lobe regions. In this section, the PTD solutions for (i) a conducting strip, (ii) a rectangular conducting plate, and (iii) a convex conducting plate are discussed.

Figure 7. Scattering pattern by modified caustic expression

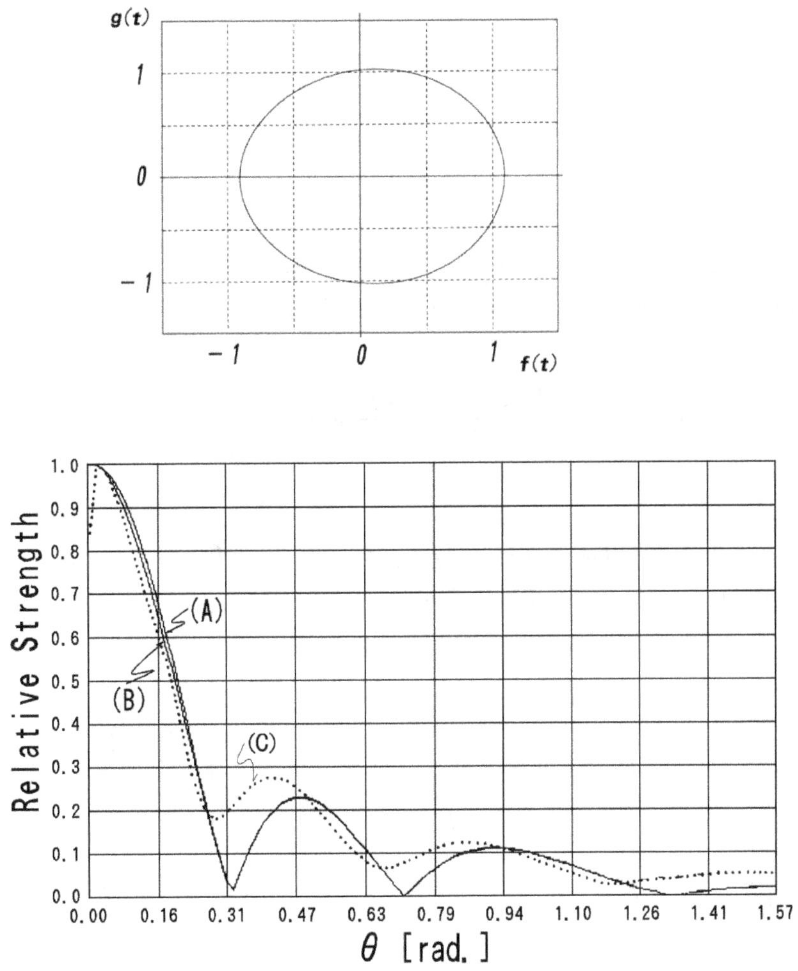

(a). Perimeter shape of convex plate ($a = 1$) according to the equation (67)
(b). Solid line A: numerical integration, solid line B: caustic correction, dashed line C: asymptotic solution

(i) Conducting Strip

Diffraction of planar electromagnetic waves by a two-dimensional conductor strip is one of the PTD canonical problems. An incident wave is given as

$$E_x^i = E_0 \exp\left\{jk\left(x\cos\phi_0 + y\sin\phi_0\right)\right\},\quad H_z^i = H_0 \exp\left\{jk\left(x\cos\phi_0 + x\sin\phi_0\right)\right\}. \tag{68}$$

Here, the incident plane lies in x-y plane, and ϕ_0 and ϕ are an incident and an observation angle, respectively. The width of the strip in the x-z plane is $2a$. The induced current on this strip is calculated as follows:

$$J_z = -2H_x = 2Y_0 E_0 \sin\phi_0 \exp(jkx\cos\phi_0),\quad J_x = 2H_z = 2H_0 \exp(jkx\cos\phi_0). \tag{69}$$

(a). Convex plate model 1
(b). Convex plate model 2
(c). Scattering pattern by model 1
(d). Scattering pattern by model 2

Employing a standard stationary phase method to a radiation integral calculated from vector potential with these currents, the asymptotic fields by PO are obtained as:

$$E_z^{PO} = -j\omega A_z = -jE_0 C\left(kR\right)\frac{\sin\left\{ka\left(\cos\phi + \cos\phi_0\right)\right\}}{\cos\phi + \cos\phi_0}\sin\phi_0, \tag{70a}$$

Figure 8. Scattering pattern by caustic corrected expression: θ-polarization, $\phi_0 = 0^\circ$, solid line: caustic correction, diamond ◊: numerical integration

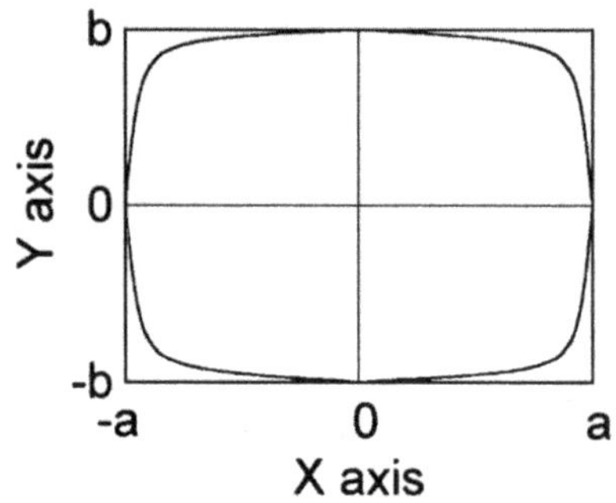

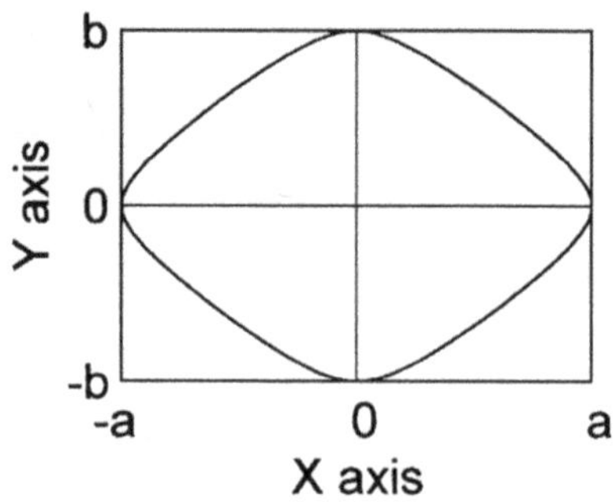

$$H_z^{PO} = \frac{jk}{\mu}\sin\phi A_x = jH_0 C(kR)\frac{\sin\{ka(\cos\phi + \cos\phi_0)\}}{\cos\phi + \cos\phi_0}\sin\phi, \tag{70b}$$

where

$$C(kR) = \sqrt{\frac{2}{\pi kR}} \qquad \exp\left\{-j\left(kR + \frac{\pi}{4}\right)\right\} \tag{70c}$$

is a function expressing the two-dimensional wave.

On the other hand, the GTD solution for the same problem is given as follows:

$$E_z^{GTD} = E_0 C(kR)\left\{-\frac{\cos\frac{\phi_0}{2}\cos\frac{\phi}{2}}{\cos\phi + \cos\phi_0}e^{jka(\cos\phi+\cos\phi_0)} + \frac{\sin\frac{\phi_0}{2}\sin\frac{\phi}{2}}{\cos\phi + \cos\phi_0}e^{-jka(\cos\phi+\cos\phi_0)}\right\}, \tag{71a}$$

$$H_z^{GTD} = H_0 C(kR)\left\{\frac{\sin\frac{\phi_0}{2}\sin\frac{\phi}{2}}{\cos\phi + \cos\phi_0}e^{jka(\cos\phi+\cos\phi_0)} + \frac{\cos\frac{\phi_0}{2}\cos\frac{\phi}{2}}{\cos\phi + \cos\phi_0}e^{-jka(\cos\phi+\cos\phi_0)}\right\}. \tag{71b}$$

Scattering fields by PTD are constructed by GTD–PO_{asy}, where PO_{asy} is given by (70), then the PTD solution for a conducting strip is arranged as follows:

$$E_z^{PTD} = -jkE_0 C(kR)\frac{\sin\{ka(\cos\phi + \cos\phi_0)\}}{\cos\phi + \cos\phi_0}\sin\phi_0 + E_0 C(kR)$$

$$\bullet\left\{D^s(\pi - \phi, \pi - \phi_0)e^{jka(\cos\phi+\cos\phi_0)} + D^s(\phi, \phi_0)e^{-jka(\cos\phi+\cos\phi_0)}\right\}, \tag{72a}$$

$$H_z^{PTD} = jkH_0 C(kR)\frac{\sin\{ka(\cos\phi + \cos\phi_0)\}}{\cos\phi + \cos\phi_0}\sin\phi + H_0 C(kR)$$

$$\bullet\left\{D^h(\pi - \phi, \pi - \phi_0)e^{jka(\cos\phi+\cos\phi_0)} + D^h(\phi, \phi_0)e^{-jka(\cos\phi+\cos\phi_0)}\right\}, \tag{72b}$$

where GTD diffraction coefficients $D^{s,h}$ for a half-plane are defined by

$$D^s\left(\psi,\psi_0\right)=-\frac{1}{2}\frac{\sin\dfrac{\psi_0}{2}}{\sin\dfrac{\psi}{2}+\cos\dfrac{\psi_0}{2}},\quad D^h\left(\psi,\psi_0\right)=-\frac{1}{2}\frac{\cos\dfrac{\psi_0}{2}}{\sin\dfrac{\psi}{2}+\cos\dfrac{\psi_0}{2}}. \tag{72c}$$

(ii) Rectangular Plate

Next, let us find a PTD solution for a rectangular conducting plate. The rectangular plate only needs to apply the result of the above-described strip to two pairs of facing sides, and the solution to this problem is given by the formulation below. At first, $\Delta U = U^{GTD} - U_{asy}^{PO}$ is calculated as follows:

$$E_\theta^{\Delta U} = -jk\frac{Z_0}{4\pi}\bullet \tag{73a}$$

$$\left\{\begin{array}{l} 2B\mathrm{sinc}\left(kBv\right)\left[\left(I_{y1}\cos\theta\sin\phi+Y_0M_{y1}\cos\phi\right)e^{-jkAu}+\left(I_{y3}\cos\theta\sin\phi+Y_0M_{y3}\cos\phi\right)e^{jkAu}\right] \\ +2A\mathrm{sinc}\left(kAu\right)\left[\left(I_{x2}\cos\theta\cos\phi-Y_0M_{x2}\sin\phi\right)e^{-jkBv}+\left(I_{x4}\cos\theta\cos\phi-Y_0M_{x4}\sin\phi\right)e^{jkBv}\right]\end{array}\right\},$$

$$E_\phi^{\Delta U} = jk\frac{Z_0}{4\pi}\bullet \tag{73b}$$

$$\left\{\begin{array}{l} 2B\mathrm{sinc}\left(kBv\right)\left[\left(I_{y1}\cos\phi-Y_0M_{y1}\cos\theta\sin\phi\right)e^{-jkAu}+\left(I_{y3}\cos\phi-Y_0M_{y3}\cos\theta\sin\phi\right)e^{jkAu}\right] \\ -2A\mathrm{sinc}\left(kAu\right)\left[\left(I_{x2}\sin\phi+Y_0M_{x2}\cos\theta\cos\phi\right)e^{-jkBv}+\left(I_{x4}\sin\phi+Y_0M_{x4}\cos\theta\cos\phi\right)e^{jkBv}\right]\end{array}\right\}.$$

Here, the equivalent currents are given as

$$\begin{pmatrix} I_{y1} \\ I_{y3}\end{pmatrix} = j\frac{4}{k}\frac{2H_{0x}\cos\theta_0}{\sqrt{\sin\beta_{01}\sin\beta_1}}\left\{\begin{array}{c} D^s\left(\psi_{01},\psi_1\right) \\ D^s\left(\pi-\psi_{01},\pi-\psi_1\right)\end{array}\right\}, \tag{74a}$$

$$\begin{pmatrix} I_{y2} \\ I_{y3}\end{pmatrix} = j\frac{4}{k}\frac{2E_{0x}}{\sqrt{\sin\beta_2}}\left\{\begin{array}{c} D^s\left(\psi_{02},\psi_2\right) \\ D^s\left(\pi-\psi_{02},\pi-\psi_2\right)\end{array}\right\}, \tag{74b}$$

$$\begin{pmatrix} M_{y1} \\ M_{y3} \end{pmatrix} = j\frac{4}{k}\frac{2E_{0x}\cos\theta_0}{\sqrt{\sin\beta_{01}\sin\beta_1}} \begin{Bmatrix} D^h\left(\psi_{01}, \psi_1\right) \\ D^h\left(\pi - \psi_{01}, \pi - \psi_1\right) \end{Bmatrix}, \tag{74c}$$

$$\begin{pmatrix} M_{y2} \\ M_{y3} \end{pmatrix} = j\frac{4}{k}\frac{2H_{0x}}{\sqrt{\sin\beta_2}} \begin{Bmatrix} D^h\left(\psi_{02}, \psi_2\right) \\ D^h\left(\pi - \psi_{02}, \pi - \psi_2\right) \end{Bmatrix}, \tag{74d}$$

where β_{01} and β_1 are diffraction angles incident to a diffraction point on side-1 and an observation point measured from the tangential direction of the edge, respectively. When the observation point is in the plane of $\phi = \pi / 2$, the PTD solution is as follows:

$$E_\theta^{PO} + E_\theta^{\Delta U} = -\frac{jkZ_0H_{0x}}{2\pi}G_0\left(R\right)\frac{4A\cos\theta}{k\left(\sin\theta + \sin\theta_0\right)}\sin\left[kB(\sin\theta + \sin\theta_0)\right] \tag{75a}$$

$$+\frac{jkZ_0}{2\pi}\left\{-Y_0B\operatorname{sinc}\left(kBv\right)\cos\theta\left(I_{y1} + I_{y3}\right) + Y_0A\left(M_{x2}e^{-jkBv} + M_{y4}e^{jkBv}\right)\right\},$$

$$E_\phi^{PO} + E_\phi^{\Delta U} = \frac{jkE_{0x}}{2\pi}G_0\left(R\right)\frac{4A\cos\theta_0}{k\left(\sin\theta + \sin\theta_0\right)}\sin\left[kB(\sin\theta + \sin\theta_0)\right] \tag{75b}$$

$$-\frac{jkZ_0}{2\pi}\left\{Y_0B\operatorname{sinc}\left(kBv\right)\cos\theta\left(M_{y1} + M_{y3}\right) - A\left(I_{x2}e^{-jkBv} + I_{y4}e^{jkBv}\right)\right\}.$$

(iii) Convex Conducting Plates

Finally in this section, let us derive the PTD solution for a convex conductor plate described by PO. The corrected term of PTD described by electromagnetic current is given as follows [Pathak, 1988]:

$$\boldsymbol{E}_d = \frac{jkZ_0}{4\pi}\oint_C\left[\boldsymbol{i}_R \times \boldsymbol{i}_R \times \boldsymbol{i}_tI_t\left(t'\right) + Y_0\boldsymbol{i}_R \times \boldsymbol{i}_tM_t\left(t'\right)\right]\frac{e^{-jkr}}{r}H\left(t'\right)dt', \tag{76}$$

where function $H(t')$ is defined by equation (52). $I_t\left(t'\right)$ and $M_t\left(t'\right)$ are an electric and a magnetic edge current of Ufimtsev, respectively, given by

In this expression, $\tilde{D}^{s}_{nu}$ and $\tilde{D}^{h}_{nu}$ are diffraction coefficients considering the non-uniform edge current, and obtained as follows by subtracting a PO asymptotic solution from GTD.

$$\tilde{D}^{s,h}_{nu} = \frac{\sin\beta_0}{\sqrt{\sin\beta\sin\beta_0}}\left\{D^{s,h}_{GTD}\left(\psi,\psi_0;\beta_0\right) - D^{s,h}_{PO}\left(\psi,\psi_0;\beta_0\right)\right\}$$

$$= \frac{2\exp\left(-j\pi/4\right)}{\sqrt{2\pi k}\sin\beta_0}\begin{Bmatrix}D^{s}_{GTD}\left(\psi,\psi_0\right)\\ D^{h}_{GTD}\left(\psi,\psi_0\right)\end{Bmatrix}, \tag{78}$$

where $D^{s,h}_{GTD}$ and $D^{s,h}_{PO}$ are Keller's diffraction coefficients of GTD and PO, respectively, and are given as follows:

$$D^{s,h}_{GTD}\left(\phi,\phi_0;\beta_0\right) = \frac{\exp\left(-j\pi/4\right)}{2\sqrt{2\pi k}\sin\beta_0}\left\{-\frac{1}{\cos\dfrac{\phi-\phi_0}{2}} \pm \frac{1}{\cos\dfrac{\phi+\phi_0}{2}}\right\}$$

$$= \frac{2\exp\left(-j\pi/4\right)}{\sqrt{2\pi k}\sin\beta_0}\frac{1}{\cos\phi+\cos\phi_0}\begin{Bmatrix}\sin\dfrac{\phi}{2}\sin\dfrac{\phi_0}{2}\\ -\cos\dfrac{\phi}{2}\cos\dfrac{\phi_0}{2}\end{Bmatrix}, \tag{79a}$$

$$D^{s,h}_{PO}\left(\phi,\phi_0;\beta_0\right) = -\frac{\exp\left(-j\pi/4\right)}{2\sqrt{2\pi k}\sin\beta_0}\left\{\tan\frac{\phi-\phi_0}{2} \mp \tan\frac{\phi+\phi_0}{2}\right\}$$

$$= \frac{2\exp\left(-j\pi/4\right)}{\sqrt{2\pi k}\sin\beta_0}\frac{1}{2}\frac{1}{\cos\phi+\cos\phi_0}\begin{Bmatrix}-\sin\phi_0\\ \sin\phi\end{Bmatrix}, \tag{79b}$$

Now, $\boldsymbol{i}_t\bullet\boldsymbol{E}^i\left(Q\right)$ and $\boldsymbol{i}_t\bullet\boldsymbol{H}^i\left(Q\right)$ in equation (77) are values of the incident wave at the diffracted point Q. In this case, it is calculated as follows:

$$\boldsymbol{i}_t\bullet\boldsymbol{E}^i\left(Q\right) = \frac{1}{H\left(t'\right)}\left\{f'\left(t'\right)E_{0x} + g'\left(t'\right)Z_0H_{0x}\cos\theta_0\right\}\exp\left[jkg\left(t'\right)\sin\theta_0\right], \tag{80a}$$

$$\boldsymbol{i}_t \bullet \boldsymbol{H}^i(Q) = \frac{1}{H(t')}\left\{f'(t')H_{0x} - g'(t')Y_0 E_{0x}\cos\theta_0\right\}\exp\left[jkg(t')\sin\theta_0\right]. \tag{80b}$$

Since the wave-number vectors of an incident and a diffracted wave are given by $\boldsymbol{k}_i = -\boldsymbol{i}_y\sin\theta_0 - \boldsymbol{i}_z\cos\theta_0$ and $\boldsymbol{k}_d = \boldsymbol{i}_x\sin\theta\cos\phi - \boldsymbol{i}_y\sin\theta\sin\phi + \mathbf{i}_z\cos\theta$, so by calculating the inner-product between the tangential vector $\boldsymbol{i}_t$ and $\boldsymbol{k}_i$, and between $\boldsymbol{i}_t$ and $\boldsymbol{k}_d$, the relationships are obtained as

$$\cos\beta_0 = -\frac{g'(t')}{H(t')}\sin\theta_0, \quad \cos\beta = \frac{f'(t')\cos\phi + g'(t')\sin\phi}{H(t')}\sin\theta, \tag{81}$$

then, $\sin\beta_0$ and $\sin\beta$ can be calculated. When an equivalent current method (ECM) is employed, it is necessary for ψ and ψ_0 in equation (78) to measure in the perpendicular plane to obtain tangential vector $\boldsymbol{i}_t$. The projections of vector $\boldsymbol{k}_i$ and $\boldsymbol{k}_d$ into this plane are given as follows:

$$\boldsymbol{i}_t \times \boldsymbol{k}_i \times \boldsymbol{i}_t = \frac{g'(t')}{H(t')}\sin\theta_0\boldsymbol{i}_b - \cos\theta_0\boldsymbol{i}_z = \cos\gamma\sin\theta_0\boldsymbol{i}_b - \cos\theta_0\boldsymbol{i}_z, \tag{82a}$$

$$\boldsymbol{i}_t \times \boldsymbol{k}_d \times \boldsymbol{i}_t = \frac{f'(t')\cos\phi + g'(t')\sin\phi}{H(t')}\sin\theta_0\boldsymbol{i}_b + \cos\theta_0\boldsymbol{i}_z = \sin(\gamma-\phi)\sin\theta\boldsymbol{i}_b - \cos\theta\boldsymbol{i}_z, \tag{82b}$$

where vector $\boldsymbol{i}_b$ is defined by

$$\boldsymbol{i}_b = \frac{g'(t')}{H(t')}\boldsymbol{i}_x - \frac{f'(t')}{H(t')}\boldsymbol{i}_y = \sin\gamma\boldsymbol{i}_x - \cos\gamma\boldsymbol{i}_y. \tag{82c}$$

Let us consider a special case where an observation point lies in the $\phi = \pi/2$ plane. In this case, $\gamma = \pi$ for $t = t_1$ and $\gamma = 0$ for $t = t_2$, from which the contribution of the non-uniform current to the diffracted field is obtained. First, the diffracted field in equation (76) is separated into θ and ϕ components as follows:

$$E_\theta^d = -\frac{jkZ_0}{4\pi}\oint_C\left[(\boldsymbol{i}_\theta\bullet\boldsymbol{i}_t)I_t(t') + (\boldsymbol{i}_\phi\bullet\boldsymbol{i}_t)Y_0M_t(t')\right]H(t')\frac{e^{-jkr}}{r}dt'$$

$$\approx -\frac{jkZ_0}{4\pi}G_0(R)\oint_C U_\theta(t')H(t')e^{jk[f(t')u+g(t')v]}dt', \tag{83a}$$

$$E_{\phi}^{d} = -\frac{jkZ_0}{4\pi}\oint_C \left[(\boldsymbol{i}_{\phi}\bullet\boldsymbol{i}_t)I_t(t') + (\boldsymbol{i}_{\theta}\bullet\boldsymbol{i}_t)Y_0M_t(t')\right]H(t')\frac{e^{-jkr}}{r}dt'$$

$$\approx -\frac{jkZ_0}{4\pi}G_0(R)\oint_C U_{\phi}(t')H(t')e^{jk[f(t')u+g(t')v]}dt'. \tag{83b}$$

Here, the second equation in each component is approximated to the far-field. $U_{\theta}(t')$ and $U_{\phi}(t')$ are obtained by substituting $I_t(t')$ and $M_t(t')$ in equation (77) into the first expression in equation (83), as follows:

$$U_{\theta}(t') = \frac{j4}{k\sqrt{\sin\beta\sin\beta_0}}\left\{M_{t0}\cos(\gamma-\phi)\cos\theta D^s(\psi,\psi_0) + Z_0J_{t0}\sin(\gamma-\phi)D^h(\psi,\psi_0)\right\}, \tag{84a}$$

$$U_{\phi}(t') = \frac{j4}{k\sqrt{\sin\beta\sin\beta_0}}\left\{M_{t0}\sin(\gamma-\phi)\cos\theta D^s(\psi,\psi_0) - Z_0J_{t0}\cos(\gamma-\phi)D^h(\psi,\psi_0)\right\}, \tag{84b}$$

$$M_{t0} = E_{0x}\cos\gamma + Z_0H_{0x}\cos\theta\sin\gamma,\ J_{t0} = H_{0x}\cos\gamma - Y_0E_{0x}\cos\theta_0\sin\gamma, \tag{84c}$$

$$\psi_0 = \frac{\pi}{2} + \tan^{-1}(\tan\theta_0\cos\gamma),\ \psi = \frac{\pi}{2} - \tan^{-1}\left[\tan\theta\sin(\gamma-\phi)\right], \tag{84d}$$

$$\sin\beta_0 = \sqrt{1-\sin^2\theta_0\sin^2\gamma},\ \sin\beta = \sqrt{1-\sin^2\theta\cos^2(\gamma-\phi)}, \tag{84e}$$

From the above discussion, the following asymptotic solution is obtained by applying the stationary phase method to the integral in the second equation of (83) as follows:

$$E_{\theta}^{ud} = \sqrt{\frac{k}{8\pi w}}G_0(R)U_{\theta}(t_1)\sqrt{\kappa(t_1)}e^{jkd_1w-\frac{j3\pi}{4}} - U_{\theta}(t_2)\sqrt{\kappa(t_2)}e^{-jkd_2w+\frac{j3\pi}{4}}, \tag{85a}$$

$$E_{\phi}^{ud} = \sqrt{\frac{k}{8\pi w}}G_0(R)U_{\phi}(t_1)\sqrt{\kappa(t_1)}e^{jkd_1w-\frac{j3\pi}{4}} - U_{\phi}(t_2)\sqrt{\kappa(t_2)}e^{-jkd_2w+\frac{j3\pi}{4}}. \tag{85b}$$

Therefore, the final PTD solution is given as

$$E_\theta^{PO} + E_\theta^{ud} = -\frac{jk}{2\pi} Z_0 H_{0x} G_0\left(R\right)\cos\theta \bullet P\left(\theta, \phi = \frac{\grave{A}}{2}\right) + \frac{j2}{\sqrt{2\pi k w}} Z_0 H_{0x} G_0\left(R\right) \tag{86a}$$

$$\bullet\left\{\sqrt{\kappa\left(t_1\right)} D^h\left(\frac{\pi}{2}-\theta, \frac{\pi}{2}-\theta_0\right) e^{jkd_1 w - j\frac{3\pi}{4}} - \sqrt{\kappa\left(t_2\right)} D^h\left(\frac{\pi}{2}+\theta, \frac{\pi}{2}+\theta_0\right) e^{-jkd_2 w + j\frac{3\pi}{4}}\right\},$$

$$E_\phi^{PO} + E_\phi^{ud} = \frac{jk}{2\pi} E_{0x} G_0\left(R\right)\cos\theta_0 \bullet P\left(\theta, \phi = \frac{\grave{A}}{2}\right) + \frac{j2}{\sqrt{2\pi k w}} E_{0x} G_0\left(R\right) \tag{86b}$$

$$\bullet\left\{\sqrt{\kappa\left(t_1\right)} D^s\left(\frac{\pi}{2}-\theta, \frac{\pi}{2}-\theta_0\right) e^{jkd_1 w - j\frac{3\pi}{4}} - \sqrt{\kappa\left(t_2\right)} D^s\left(\frac{\pi}{2}+\theta, \frac{\pi}{2}+\theta_0\right) e^{-jkd_2 w + j\frac{3\pi}{4}}\right\}.$$

The PO solution for plane wave diffracted by a convex conducting plate is given by equations (34) and (65), and the corresponding PTD solution is $E_\theta^{PO} + E_\theta^{ud}$ and $E_\phi^{PO} + E_\phi^{ud}$ as described above. Figure 8(a) shows the result for a two-dimensional strip of infinite length in the z-axis direction with a width of $x = (-a, a)$. The horizontal axis is the azimuth angle ϕ taken from the positive x-axis. The solid line is the conventional PO solution, the diamond ◇ mark is that of the PTD, and the broken line is the GTD result. For the thin strip with $ka = 5.0$, it can be seen that PTD and GTD are completely consistent. Similar results are obtained at $ka = 3.0$. Comparing with MoM, it is found that PTD improves PO in both main and side lobe regions. Figure 8(b) shows the result of planar cutting in a plane $\phi = 0°$ for a square plate of $2A \times 2B$. The length of side is the same size as in the strip and the same result is obtained. Figure 9(a) is the far-field for a circular disk with radius $ka = 5.0$, and indicates the results of PO (solid line) and PTD (diamond ◇). Figure 9(b) shows the case of an elliptical plate with $kA = 10.0$, and $kB = 20.0$. Noting that, if the size is larger than the circular disk, it can be found that the PO results are close to those of PTD even in the side lobe region.

(a). Conducting strip ($\phi_0 = 90^\circ$)
(b). Conducting rectangular plate ($\theta_0 = 0^\circ$)
(a). Conducting circular disk ($ka = 5$)
(b). Conducting ellipse plate ($kA = 10$, $kB = 20$)

Scattering by Flat Plate (III): PO for Surface Impedance Convex

Let us consider the scattering of a plane wave by an impedance plate such as a dielectric loaded on a conducting plate. If the electromagnetic field distribution in the vicinity of a flat plate can be incorporated into the radiation integral in consideration of a reflection coefficient, the scattered field in a far-region can be induced. Furthermore, because the relationship between the electromagnetic currents induced on the impedance surface can be directly obtained using the surface impedance, this approach is also effective. In this section, two methods: reflection coefficients and surface impedance are examined.

For the method using a reflection coefficient, the novel expression for the far-field of the total electromagnetic field contributed to the reflected and incident waves is given as (Kobayashi, 2011)

$$E_{\theta,\phi} = -jkG_0\left(R\right)\int_S \{Z_0\left(\boldsymbol{n}\times\boldsymbol{H}\right)\bullet\boldsymbol{i}_{\theta,\phi} \pm \left(\boldsymbol{E}\times\boldsymbol{n}\right)\bullet\boldsymbol{i}_{\phi,\theta}\}\exp\left(jk\boldsymbol{r}'\bullet\boldsymbol{i}_R\right)dS', \tag{87}$$

where S is an illuminated area of the curved surface including the plate, $E_{\theta,\phi} = A \pm B$ means $E_\theta = A + B$ and $E_\phi = A - B$, and $G_0\left(R\right) = \exp\left(-jkR\right)/\left(4\pi R\right)$. Introducing coefficients P, Q and vector $\boldsymbol{B}$, the electromagnetic fields near the plate or its equivalent electromagnetic currents can be described by

$$\boldsymbol{n}\times\boldsymbol{H} = \boldsymbol{n}\times\left(\boldsymbol{H}^{\mathrm{i}} + \boldsymbol{H}^{\mathrm{r}}\right) = P^{\mathrm{i}}\left(1 + R_{\|}\right)\boldsymbol{n}\times\boldsymbol{B} + Q^{\mathrm{i}}\left(1 - R_{\perp}\right)\cos\theta_0\boldsymbol{B}, \tag{88a}$$

$$\boldsymbol{E}\times\boldsymbol{n} = \left(\boldsymbol{E}^{\mathrm{i}} + \boldsymbol{E}^{\mathrm{r}}\right)\times\boldsymbol{n} = P^{\mathrm{i}}\left(1 - R_{\|}\right)\cos\theta_0\boldsymbol{B} - Q^{\mathrm{i}}\left(1 - R_{\perp}\right)\boldsymbol{n}\times\boldsymbol{B}. \tag{88b}$$

Then, equation (86) is modified as follows:

$$E_{\theta,\phi} = -jkG_0\left(R\right)\int_S D_{\theta,\phi}\left(R_{\|}, R_{\perp}, \theta, \phi\right)\exp\left(jk\boldsymbol{r}'\bullet\boldsymbol{i}_R\right)dS', \tag{89a}$$

$$D_\theta\left(R_{\|}, R_{\perp}, \theta, \phi\right) = \left\{P^i\left(1 + R_{\|}\right)\boldsymbol{n}\times\boldsymbol{B} + Q^i\left(1 - R_{\perp}\right)\cos\theta_0\boldsymbol{B}\right\}\bullet\boldsymbol{i}_\theta$$

$$+\left\{P^i\left(1 - R_{\|}\right)\cos\theta_0\boldsymbol{B} - Q^i\left(1 - R_{\perp}\right)\boldsymbol{n}\times\boldsymbol{B}\right\}\bullet\boldsymbol{i}_\phi, \tag{89b}$$

$$D_\phi\left(R_{\|}, R_{\perp}, \theta, \phi\right) = \left\{P^i\left(1 + R_{\|}\right)\boldsymbol{n}\times\boldsymbol{B} + Q^i\left(1 - R_{\perp}\right)\cos\theta_0\boldsymbol{B}\right\}\bullet\boldsymbol{i}_\phi$$

$$+\left\{P^i\left(1 - R_{\|}\right)\cos\theta_0\boldsymbol{B} - Q^i\left(1 - R_{\perp}\right)\boldsymbol{n}\times\boldsymbol{B}\right\}\bullet\boldsymbol{i}_\theta, \tag{89c}$$

Figure 9. Scattering pattern by a conducting strip and rectangular plate, θ - polarization, solid line: PO, diamond ◊ : PTD, dotted line: GTD

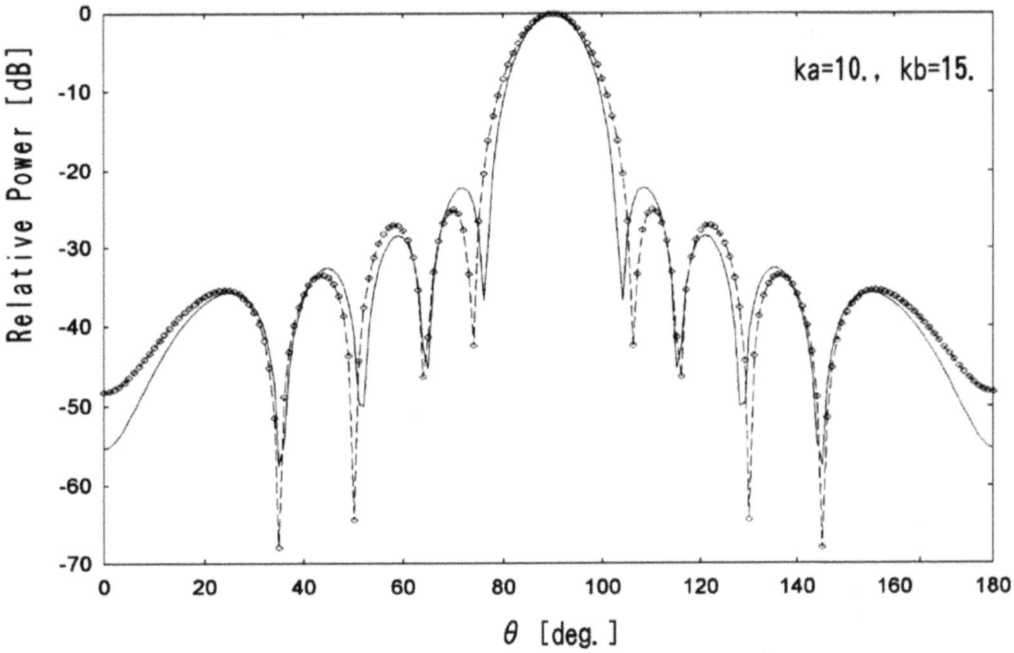

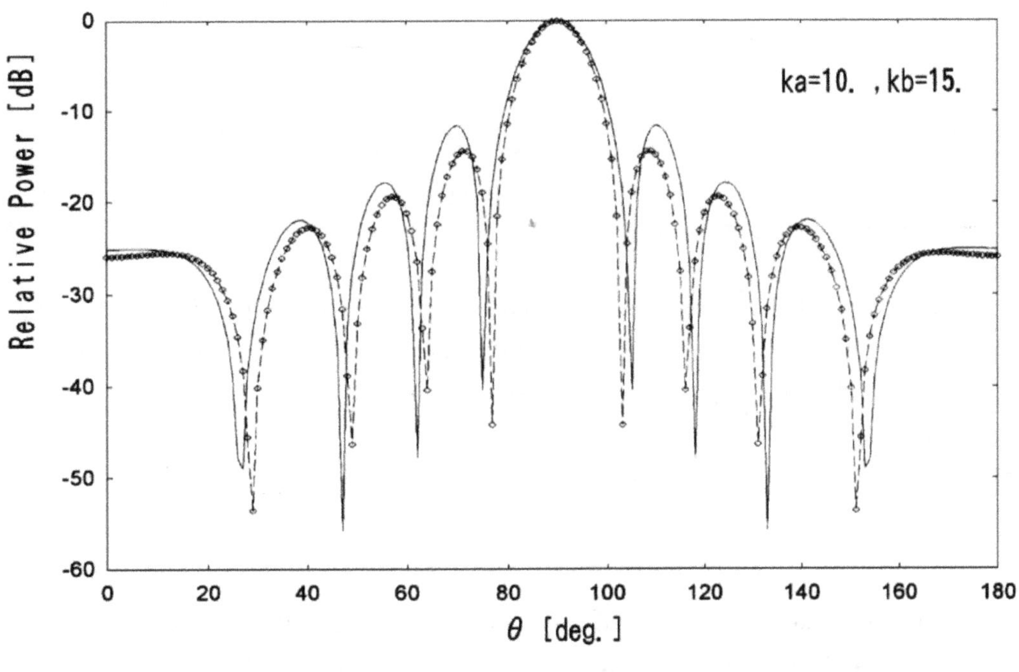

Figure 10. Scattering pattern by a conducting circular and rectangular plate, θ - polarization, $\theta_0 = 0°$ *solid line: PO, diamond ◊ : PTD*

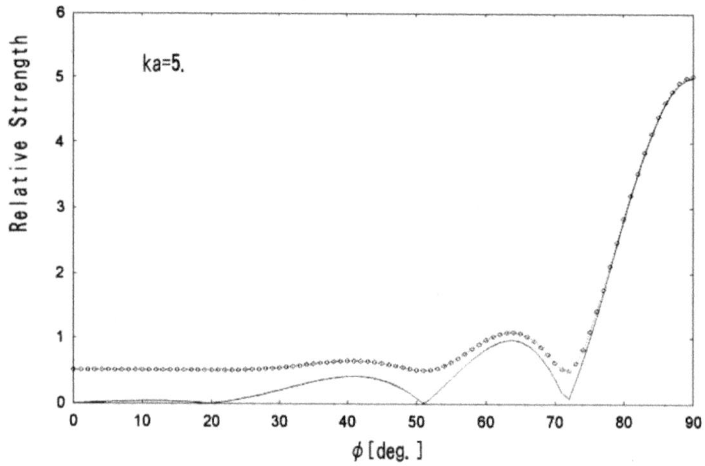

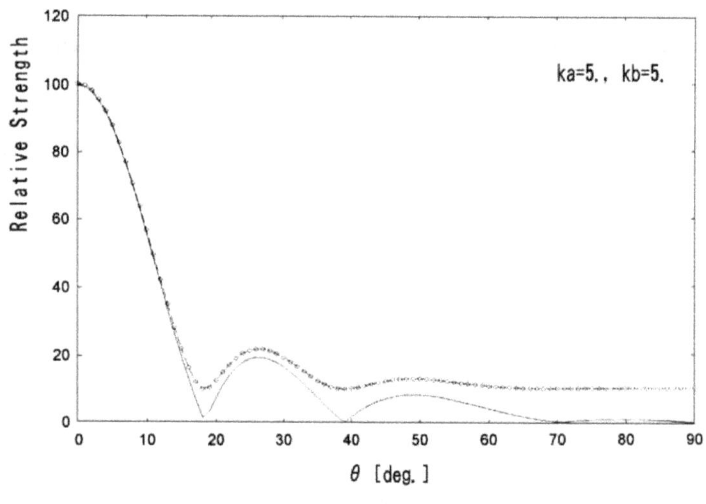

where vector $\boldsymbol{B}$ is perpendicular to the incident plane given by $\boldsymbol{B} = \boldsymbol{k}_i \times \boldsymbol{n} / \sin\theta_0 = \boldsymbol{k}_r \times \boldsymbol{n} / \sin\theta_0$ using wave-number vector $\boldsymbol{k}_{i,r}$ and θ_0 is produced by an angle between $\boldsymbol{k}_i$ and normal vector $\boldsymbol{n}$ on the surface (Figure 10). P^{i} and Q^{i} are the amplitude and a phase of an incident plane wave expressed as:

$$\boldsymbol{E}^{i,r} = P^{i,r}\boldsymbol{T}^{i,r} + Q^{i,r}\boldsymbol{B}, \quad Z_0\boldsymbol{H}^{i,r} = -Q^{i,r}\boldsymbol{T}^{i,r} + P^{i,r}\boldsymbol{B}, \tag{90}$$

where P^{r} and Q^{r} are related to a reflected wave. Given incident wave, P^{r} and Q^{r} are derived from $P^{\mathrm{r}} = R_{\parallel}P^{\mathrm{i}}$ and $Q^{\mathrm{r}} = R_{\perp}Q^{\mathrm{r}}$ using the reflection coefficient and they are dependent on the polarization.

Figure 11. Scattering by three-dimensional curved surface with surface impedance

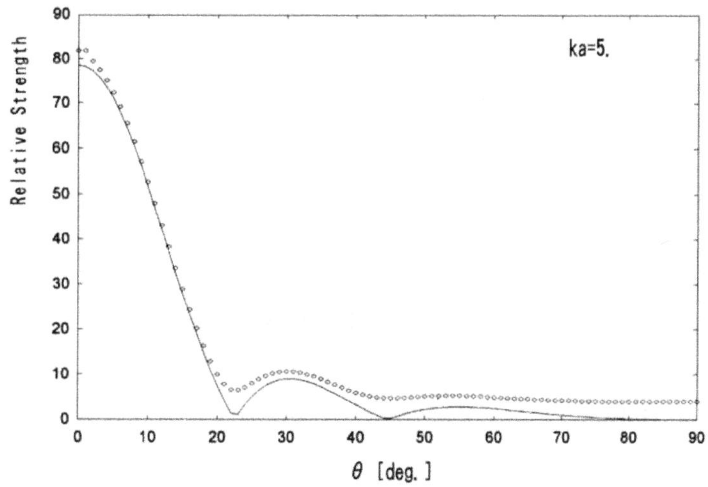

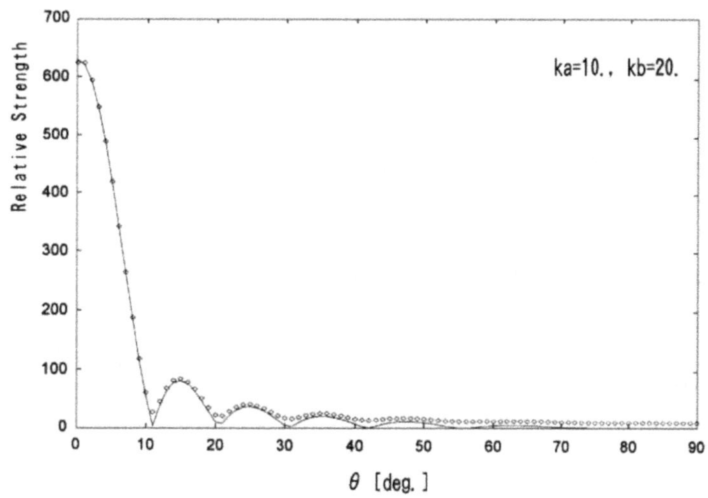

Of course, the reflected wave can be obtained directly from each problem using Maxwell's equation without using such an idea.

For this scattering problem, an incident plane lies in the x-z plane and the plate in the x-y plane. Then, since the unit normal vector of the plate is $\boldsymbol{n} = \boldsymbol{i}_z$, $\boldsymbol{B} = -\boldsymbol{i}_x$ is obtained. Therefore, the incident wave in equations (30) and (31) are modified as follows:

$$\boldsymbol{E}^i = -(E_{0x}\boldsymbol{B} + Z_0 H_{0x}\boldsymbol{T}^i \exp\left[jk\left(y\sin\theta_0 + z\cos\theta_0\right)\right]. \tag{91}$$

Alternatively, the incident wave can also be found as equation (92) below when $\boldsymbol{T}^i = -\cos\theta_0 \boldsymbol{i}_y P^i + \sin\theta_0 \boldsymbol{i}_z$, is substituted into equation (91),

$$\boldsymbol{E}^i = \left\{E_{0x}\boldsymbol{i}_x + Z_0 H_{0x}\left(\cos\theta_0 \boldsymbol{i}_y P^i - \sin\theta_0 \boldsymbol{i}_z\right)\right\}\exp\left[jk\left(y\sin\theta_0 + z\cos\theta_0\right)\right]. \tag{92}$$

The above incident wave is represented by overlapping two polarized waves in equations (30) and (31). Now, comparing this incident wave with that in equation (90), P^{i} and Q^{i} can be obtained as:

$$P^{\mathrm{i}}\,|_{z=0} = -Z_0 H_{0x}\exp\left(jky\sin\theta_0\right), \quad Q^{\mathrm{i}}\,|_{z=0} = -E_{0x}\exp\left(jky\sin\theta_0\right). \tag{93}$$

The unit direction vectors in spherical coordinates are given as

$$\boldsymbol{i}_\theta = \cos\theta\cos\phi\boldsymbol{i}_x + \cos\theta\sin\phi\boldsymbol{i}_y - \sin\theta\boldsymbol{i}_z, \quad \boldsymbol{i}_\phi = -\sin\phi\boldsymbol{i}_x + \cos\phi\boldsymbol{i}_y. \tag{94}$$

Furthermore, using the relationship

$$\boldsymbol{n}\times\boldsymbol{B}\bullet\boldsymbol{i}_{\theta,\phi} = \begin{pmatrix}\cos\theta\sin\phi \\ -\cos\phi\end{pmatrix}, \ \boldsymbol{B}\bullet\boldsymbol{i}_{\theta,\phi} = \begin{pmatrix}-\cos\theta\cos\phi \\ \sin\phi\end{pmatrix}, \tag{95}$$

$D_{\theta,\phi}\left(R_\parallel, R_\perp, \theta, \phi\right)$ in equation (89) are obtained as follows:

$$D_\theta\left(R_\parallel, R_\perp, \theta, \phi\right) = Z_0 H_{0x}\left\{\left(1+R_\parallel\right)\cos\theta - \left(1-R_\parallel\right)\cos\theta_0\right\}\sin\phi$$

$$+E_{0x}\left\{\left(1-R_\perp\right)\cos\theta_0\cos\theta - \left(1+R_\perp\right)\right\}\cos\phi, \tag{96a}$$

$$D_\phi\left(R_\parallel, R_\perp, \theta, \phi\right) = Z_0 H_{0x}\left\{\left(1+R_\parallel\right) - \left(1-R_\parallel\right)\cos\theta_0\cos\theta\right\}\cos\phi$$

$$+E_{0x}\left\{\left(1+R_\perp\right)\cos\theta - \left(1-R_\perp\right)\cos\theta_0\right\}\sin\phi. \tag{96b}$$

Since TE-wave (ϕ-polarization) and TM-wave (θ-polarization) correspond to $H_{0x} = 0$ and $E_{0x} = 0$, respectively, and $R_\parallel = +1$, $R_\perp = -1$ for perfect electrical conductor (pec), equation (96) reduces to

$$D_\theta^{pec}\left(R_\parallel, R_\perp, \theta, \phi\right) = 2\left\{Z_0 H_{0x}\cos\theta\sin\phi + E_{0x}\cos\theta_0\cos\theta\cos\phi\right\}, \tag{97a}$$

$$D_{\phi}^{pec}\left(R_{\|}, R_{\perp}, \theta, \phi\right) = 2\left\{Z_0 H_{0x}\cos\phi - E_{0x}\cos\theta_0\sin\phi\right\}, \tag{97b}$$

which completely coincides with equation (34). Thus, the reflection coefficient, which includes the dielectric loading and surface impedance, can be taken out from the integration, and it is independent on the perimeter of the plate. Therefore, equation (34) is changed to the expression including the case of imperfect conductor as follows:

$$E_{\theta,\phi} = -j4\pi kG_0\left(R\right)D_{\theta,\phi}\int_S \exp\left(jkux' + jkvy'\right)dS' = -j4\pi kG_0\left(R\right)D_{\theta,\phi}P\left(u,v\right), \tag{98a}$$

$$P\left(u,v\right) = \int_S \exp\left\{jk\left[x'\sin\theta\cos\phi + y'(\sin\theta_0 + \sin\theta\sin\phi)\right]\right\}dS'. \tag{98b}$$

All contributions due to the convex shape are included in the integral $P(u, v)$, so that the effects of surface impedance or dielectric loading will depend on the angle function $D_{\theta,\phi}$. In other words, the discussion of the conducting plate that includes the convex shape so far can be applied as it only considers coefficients $D_{\theta,\phi}$.

In the second method which is based on the concept of surface impedance, the radiation integral is derived from the approximated relationship of electromagnetic currents induced on an impedance surface. The electromagnetic current flowing on the plate due to the presence of surface impedance is approximated as:

$$\boldsymbol{M}_s = -Z_s\boldsymbol{n}\times\boldsymbol{J}_s. \tag{99}$$

Also, the relationship between the reflection coefficients and surface impedance ζ_s normalized by characteristic impedance in free-space Z_0 are given as:

$$R_{\|} \cong \frac{\cos\theta_0 - \zeta_s^{\|}}{\cos\theta_0 + \zeta_s^{\|}}, \quad \text{or} \quad \zeta_s^{\|} \cong \cos\theta_0\frac{1-R_{\|}}{1+R_{\|}}, \tag{100a}$$

$$R_{\perp} \cong \frac{\zeta_s^{\perp}\cos\theta_0 - 1}{\zeta_s^{\perp}\cos\theta_0 + 1}, \quad \text{or} \quad \zeta_s^{\perp} \cong \cos\theta_0\frac{1+R_{\perp}}{1-R_{\perp}}, \quad \zeta_s = \frac{Z_s}{Z_0}. \tag{100b}$$

Applying equation (99), equation (87) is modified to

$$E_{\theta,\phi} = -jkG_0\left(R\right)\int_S \{Z_0\boldsymbol{J}_s\bullet\boldsymbol{i}_{\theta,\phi} \mp Z_s\left(\boldsymbol{n}\times\boldsymbol{J}_s\right)\bullet\boldsymbol{i}_{\phi,\theta}\}\exp\left(jk\boldsymbol{r}'\bullet\boldsymbol{i}_R\right)dS', \tag{101}$$

where Z_s is the surface impedance of the plate. Since the scatterer here is a plate, the electric current in equation (101) can be calculated using equation (88) as:

$$Z_0 \boldsymbol{J}_s = -Q^i \left(1 - R_\perp\right) \cos\theta_0 \boldsymbol{i}_x - P^i \left(1 + R_\parallel\right) \boldsymbol{i}_y, \tag{102a}$$

$$Z_0 \boldsymbol{n} \times \boldsymbol{J}_s = P^i \left(1 + R_\parallel\right) \boldsymbol{i}_x - Q^i \left(1 - R_\perp\right) \cos\theta_0 \boldsymbol{i}_y. \tag{102b}$$

Then, the integrand of equation (101) is calculated as:

$$D_\theta^{IBC} = Z_0 \boldsymbol{J}_s \bullet \boldsymbol{i}_\theta - Z_s \left(\boldsymbol{n} \times \boldsymbol{J}_s\right) \bullet \boldsymbol{i}_\phi$$

$$= \left(\zeta_s - \cos\theta\right) \left\{P^i \left(1 + R_\parallel\right) \sin\phi + Q^i \left(1 - R_\perp\right) \cos\theta_0 \cos\phi\right\}, \tag{103a}$$

$$D_\phi^{IBC} = Z_0 \boldsymbol{J}_s \bullet \boldsymbol{i}_\phi + Z_s \left(\boldsymbol{n} \times \boldsymbol{J}_s\right) \bullet \boldsymbol{i}_\theta$$

$$= \left(\zeta_s \cos\theta - 1\right) \left\{P^i \left(1 + R_\parallel\right) \cos\phi - Q^i \left(1 - R_\perp\right) \cos\theta_0 \sin\phi\right\}. \tag{103b}$$

When the plate is perfectly conducting ($\zeta_s = 0$), $R_\parallel = +1$ and $R_\perp = -1$. Equation (103) is found to coincide exactly with equation (97) as expected. The derived impedance coefficient $D_{\theta,\phi}^{IBC}$ employ the reflection coefficient in addition to Z_s. However, it can be expressed only by Z_s if $R_{\parallel,\perp}$ is given by equation (100). It has been pointed out that the surface impedance is equal to the intrinsic impedance of a medium at normal incidence and its approximating accuracy is high. Since the normal incidence $\theta_0 = 0$, equation (103) reduces to

$$D_\theta^{IBC} = \left(\zeta_s - \cos\theta\right) \left\{P^i \left(1 + R_\parallel\right) \sin\phi + Q^i \left(1 - R_\perp\right) \cos\phi\right\}, \tag{104a}$$

$$D_\phi^{IBC} = \left(\zeta_s \cos\theta - 1\right) \left\{P^i \left(1 + R_\parallel\right) \cos\phi - Q^i \left(1 - R_\perp\right) \sin\phi\right\}. \tag{104b}$$

For normal incidence, the surface impedance given in equation (100) can be written as:

$$\zeta_s^\parallel = \frac{1 - R_\parallel}{1 + R_\parallel}, \qquad \zeta_s^\perp = \frac{1 + R_\perp}{1 - R_\perp}. \tag{105}$$

At this time, by solving $P^i = 0$ (TE-wave) and $Q^i = 0$ (TM-wave) separately, the coefficient $D_{\theta,\phi}^{IBC}$ is found to coincide with $D_{\theta,\phi}$. What described in this section is the formulation for a flat plate, however it can be naturally generalized for a three-dimensional curved surface object as expected in equation (87).

REFERENCES

Adana, F. S. D., Gutiérrez, O., Conzález, I., Cátedra, M. F., & Lozano, L. (2011). *Practical applications of asymptotic techniques in electromagnetics* (pp. 73–118). Artech House.

Born, M., & Wolf, E. (1964). *Principles of Optics. 2nd ed.* New York: Pergamon Press.

Borovikov, V. A. (1994). Uniform Stationary Phase Method. IEE Electromagnetic Waves Series 40, UK.

Borovikov, V. A., & Kimber, B. Y. (1994). Geometrical theory of diffraction. IEE Electromagnetic Waves Series 37, UK. doi:10.1049/PBEW037E

Bowman, J. J., Senior, T. B. A., & Uslenghi, P. L. E. (1969). *Electromagnetic and Acoustic Scattering by Simple Shapes*. Amsterdam, The Netherlands: North-Holland Publishing Company.

James, G. L. (1980). *Geometrical Theory of Diffraction for Electromagnetic Waves* (2nd ed.). London, UK: Peter Peregrinus.

Kobayashi, H. (2011). *Electromagnetic Wave in Space-Domain*. Tokyo, Japan: Sakura-tech. (in Japanese)

Kobayashi, H. (2012). Geometrical Theory of Diffraction and Related Topics. Press of Niigata University. (in Japanese)

Kobayashi, H., & Hongo, K. (1996). Scattering of Electromagnetic Plane Wave by Conducting Convex Plate. In *Proceedings Symposium of Antenna Technology and Applied Electromagnetics (ANTEM96)*, pp. 663-667.

Kobayashi, H., Tokumitsu, Y., & Hongo, K. (1999). RCS Calculation Software System using PTD-TC. In *Proceedings International Conference on Computational Electromagnetics and Its Applications (ICCEA'99)*, pp. 569-572.

Mach, E. (1926). *The Principle of Physical Optics*. Dover Pub.

Osipov, A. V., & Tretyakov, A. A. (2017). *Modern Electromagnetic Scattering Theory with Applications*. John Wiley & Sons. doi:10.1002/9781119004639

Pathak, P. H. (1988). *Techniques for High-Frequency Problems. Antenna Handbook* (Y. T. Lo, & S. W. Lee, Eds.). Van Nostrand Reinhold.

Ruck, G. T., Barrick, D. E., Stuwart, W. D., & Kirchbaum, C. K. (1970). *Radar Cross Section Handbook* (Vol. I). Plenum Press. doi:10.1007/978-1-4899-5324-7

Senior, T. B. A., & Volakis, J. L. (1995). Approximate Boundary Conditions in Electromagnetics. IEE Electromagnetic Wave Series 41, UK. doi:10.1049/PBEW041E

Stratton, J. A. (1941). *Electromagnetic Theory*. McGraw-Hill Book Co.

Ufimtsev, P. Y. (1971). *Method of Edge Waves in the Physical Theory of Diffraction, Translated from Russian*. Prepared by the U.S. Air Force Foreign Tech. Div.

Ufimtsev, P. Y. (2007). *Fundamentals of the Physical Theory of Diffraction*. IEEE Press. doi:10.1002/0470109017

Chapter 7
Gaussian Optics

Mey Chern Loh
Universiti Tunku Abdul Rahman, Malaysia

ABSTRACT

Analysis of Terahertz waves comes in three main forms, physical optics, geometrical optics, and Gaussian optics. Physical optics has the highest accuracy but it is time consuming when it is applied in the design of large radio telescopes. Also, it is only capable of computing radiation characteristics. Geometrical optics, on the other hand, reduces computational time significantly. But it does not give accurate results when designing telescopes which are to operate at Terahertz frequencies. Gaussian optics is a good trade-off between these two methods and it is a popular approach used in the design of large radio telescopes — particularly those which operate near/in the Terahertz band. Since it accounts for the effects of diffraction, this method produces reasonably accurate results. This chapter describes Gaussian optics, with emphasis given on its application in the design of radio telescopes.

GAUSSIAN OPTICS

Waves in the THz band are rich with spectral and spatial information in the field of astrophysics and is one of the few regions of the electromagnetic spectrum yet to be fully available to astronomy. THz systems are typically analyzed using physical optics, geometrical optics or Gaussian beam techniques.

Physical optics is very time consuming while offering high accuracy but the trade-off becomes apparent in the analysis of multiple reflector antennas. Another limitation of physical optics is that it is only capable of computing radiation characteristics, such as beam patterns and field contours, of the reflector antenna based on a set of predetermined antenna design parameters and it could not be applied to calculate optimum antenna design parameters which would give the highest aperture efficiency. To overcome the limitations of physical optics, the antenna design parameters are usually calculated first before applying the physical optics approach to determine radiation characteristics.

Geometrical optics approach on the other hand, ignores diffraction and polarization effects entirely to reduce computational time which in turn, reduces its accuracy. As cross polarization, distortion, and diffraction cannot be ignored in the THz region, the analysis of THz band is unable to utilize both tech-

DOI: 10.4018/978-1-7998-2381-0.ch007

niques for analysis in the optical and microwave bands directly. Therefore, an intermediate approach is required to accelerate the analysis process and to provide sufficient accuracy in THz optical system designs.

A popular approach is the Gaussian beam modes. The Gaussian beam modes include the effects of diffraction within reasonable limit and are reasonably accurate for most applications. As performing accurate calculations for real systems can be very time consuming (Chou & Pathak, 2004), approximations are necessary to speed up the analysis process especially when speed is particularly important when a large number of optical elements are involved or when optimization techniques are required.Derived based on paraxial wave equations (Goldsmith, 1998), the simplicity of the Gaussian beam modes can also provide a starting point for more diffraction calculations when necessary and can also be used to calculate the parameters of the reflector (Tham, Yassin & Carter, 2007; Yeap &T ham, 2018). On average, the Gaussian beam technique requires 4.84s to find the radiation pattern at 626 field points as opposed to 164.3s required by conventional numerical physical optics approach using the same computer (Tham, Yassin & Carter, 2007).This means that the Gaussian beam analysis is approximately 40 times faster than the conventional numerical physical optics approach. A study showed that Gaussian beam analysis is able to shorten solution time significantly compared to that performed using physical optics (Chou & Pathak, 2004). However, inaccuracies can be observed due to inabilities to account for distortion and polarization (Yeap & Tham, 2018) and hence, could only provide approximation for the design parameters.

As the Gaussian beam propagation is based on the paraxial wave equation, it has limited transverse variation compared to a plane wave. It originates from a region of finite extent which is different from a beam originating from a source in geometrical optics which is from an infinitesimal point source.

FUNDAMENTAL QUASIOPTICAL GAUSSIAN BEAM PROPAGATION

Quasioptical propogation is important for millimeter and submillimeter wavelength systems. To obtain the paraxial wave equation that is of use for millimeter and submillimeter wavelength systems, the Helmholtz wave equation is needed (Goldsmith, 1998). A single component, ψ, of an electromagnetic wave propagating in a uniform medium satisfies the Helmholtz wave equatin

$$\left(\nabla^2 + k^2\right)\psi = 0\,. \tag{1}$$

where,

$$\nabla^2 = \text{Laplacian} = \frac{1}{h_x h_y h_z}\left[\frac{\partial^2}{\partial x^2}\left(\frac{h_y h_z}{h_x}\right) + \frac{\partial^2}{\partial y^2}\left(\frac{h_x h_z}{h_y}\right) + \frac{\partial^2}{\partial z^2}\left(\frac{h_x h_y}{h_z}\right)\right]$$

k = wave number

ψ = amplitude ofany component of E (electric field) or H (magnetic field).

Time variation at angular frequency is assumed to be exp(jωt). The wave number, k, is equal to $2\pi/\lambda$ so that

$k = \omega(\epsilon_r\mu_r)^{0.5}/c$

where,

ϵ_r = relative permittivity of the medium

μ_r = relative permeability of the medium

For a plane wave, the amplitudes of the electric and magnetic fields are constant and their directions are mutually perpendicular. The directions of the electric and magnetic fields are also perpendicular to the propagation vector. When a beam of radiation is similar to a plane wave but has some variation perpendicular to the axis of propagation, it is assumed that the electric and magnetic fields are mutually perpendicular to each other and both are perpendicular to the direction of propagation. Letting the direction of propagation be in the positive z direction, the distribution for any component of the electric field (suppressing the time dependence) can be written s

$$E(x, y, z) = u(x,y,z) \exp(-jkz). \tag{2}$$

where

u = complex scalar function that defines the non-plane wave part of the beam in terms of x, y and z.

In rectangular coordinates, the Helmholtz equation is

$$\frac{\partial^2 E}{\partial x^2} + \frac{\partial^2 E}{\partial y^2} + \frac{\partial^2 E}{\partial z^2} + k^2 E = 0. \tag{3}$$

If Eq. 2 is substituted into Eq. 3, the reduced wave equation) is obtained

$$\frac{\partial^2 u}{\partial x^2} + \frac{\partial^2 u}{\partial y^2} + \frac{\partial^2 u}{\partial z^2} - 2jk\frac{\partial u}{\partial z} = 0. \tag{4}$$

The paraxial approximation consists of assuming that the variation along the direction of propagation if the amplitude u, due to diffraction, will be small over a distance comparable to a wavelength, and that the axial variation will be small compared to the variation perpendicular to this direction. The first statement ($\P^2u/\P x^2$) implies that in magnitude, $[\Delta(\P u/\P z)/\Delta z]\lambda$ is very much smaller than $\P u/\P z$, which means the third term ($\P^2u/\P z^2$) in Eq. 4 is small compared to the fourth term $[2jk(\P u/\P z)]$. As the direction of propagation is assumed to be in the positive z direction, the third term is small compared to the first two thus, the third term can be drop, obtaining finally the paraxial wave equation in rectangular coordinates

$$\frac{\partial^2 u}{\partial x^2} + \frac{\partial^2 u}{\partial y^2} - 2jk\frac{\partial u}{\partial z} = 0. \tag{5}$$

Solutions to the paraxial wave equation are the Gaussian beam modes that form the basis of quasi-optical system design. There is no rigorous "cutoff" for the application of the paraxial approximation, but it is generally reasonably good as long as the angular divergence of the beam is confined or largely confined to within 0.5 radians or about 30 degrees of the z axis.

FUNDAMENTAL GAUSSIAN BEAM MODE SOLUTION IN CYLINDRICAL COORDINATES

In cylindrical coordinates, r represents the perpendicular distance from the axis of propagation, taken again to be the z axis, and the angular coordinate is represented by φ. In cylindrical coordinate systems, the paraxial wave equation is

$$\frac{\partial^2 u}{\partial r^2}+\frac{1}{r}\frac{\partial u}{\partial r}+\frac{1}{r}\frac{\partial^2 u}{\partial \varphi^2}-2jk\frac{\partial u}{\partial z}=0 \tag{6}$$

where u ° u(r, φ, z). For the moment, axial symmetry, u, is assumed to be independent of φ which makes the third term, $\left(\frac{1}{r}\frac{\partial^2 u}{\partial \varphi^2}\right)$, equal to zero,obtaining the axially symmetric paraxial wave equation

$$\frac{\partial^2 u}{\partial r^2}+\frac{1}{r}\frac{\partial u}{\partial r}-2jk\frac{\partial u}{\partial z}=0\,. \tag{7}$$

The simplest solution of the axially symmetric paraxial wave equation can be written in the form

$$u(r, z) = A(z)\exp\left[\frac{-jkr^2}{2q(z)}\right] \tag{8}$$

where A(z) and q(z) are two complex functions of z only, which remain to be determined. To obtain the unknown terms in Eq. 8, this expression is substituted for u into the axially symmetric paraxial wave Eq. 7 and obtain

$$-2jk\left(\frac{A}{q}+\frac{\partial A}{\partial z}\right)+\frac{k^2r^2A}{q^2}\left(\frac{\partial q}{\partial z}-1\right)=0\,. \tag{9}$$

Since this equation must be satisfied for all r as well as all z, and given that the first part depends only on z while the second part depends on r and z, the two parts must individually be equal to zero. This produces two relationships that must be simultaneously satisfied:

$$\frac{\partial q}{\partial z}=1 \tag{10a}$$

and

$$\frac{\partial A}{\partial z}=-\frac{A}{q} \tag{10b}$$

Eq. 10a has the solution

$$q(z) = q(z_0) + (z\text{-}z_0). \tag{11a}$$

Without loss of generality, the reference position along the z axis is defined to be $z_0 = 0$, which yields

$$q(z) = q(0) + z. \tag{11b}$$

The function q is called the complex beam parameter (since it is complex), but it is often referred to simply as the beam parameter or Gaussian beam parameter. Since it appears in Eq. 8 as 1/q, it is reasonable to write

$$\frac{1}{q} = \left(\frac{1}{q}\right)_r - j\left(\frac{1}{q}\right)_i . \tag{12}$$

where the subscripted terms are the real and imaginary parts of the quantity 1/q, respectively. Substituting into Eq. 8, the exponential term becomes

$$exp\left(\frac{-jkr^2}{2q}\right) = exp\left[\left(\frac{-jkr^2}{2q}\right)\left(\frac{1}{q}\right)_r - \left(\frac{kr^2}{2}\right)\left(\frac{1}{q}\right)_i\right]. \tag{13}$$

The imaginary term has the form of the phase variation produced by a spherical wave front in the paraxial limit. This starts with an equiphase surface having radius of curvature, R, and defining $\phi(r)$ to be the phase variation relative to a plane for a fixed value of z as a function of r. In the limit $r \ll R$, the phase delay incurred is approximately equal to

$$\phi(r) \cong \frac{\pi r^2}{\lambda R} = \frac{kr^2}{2R} . \tag{14}$$

Thus the important identification of the real part of 1/q with the radius of curvature of the beam

$$\left(\frac{1}{q}\right)_r = \frac{1}{R} . \tag{15}$$

Since q is a function of z, it is evident that the radius of curvature of the beam will depend on the position along the axis of propagation. It is important not to confuse the phase shift ϕ (depends on z) with azimuthal coordinate φ.

The second part of the exponential in Eq. 13 is real and has a Gaussian variation as a function of the distance from the axis of propagation. Taking the standard form for a Gaussian distribution to be

$$f(r) = f(0)exp\left[-\left(\frac{r}{r_0}\right)^2\right]. \tag{16}$$

The quantity r_0 represents the distance to the 1/e point relative to the on-axis value. To make the second part of Eq. 13 have this form,

$$\left(\frac{1}{q}\right)_i = \frac{2}{k\omega^2(z)} = \frac{\lambda}{\pi\omega^2} \tag{17}$$

and thus define the beam radius, ω, which is the value of the radius at which the field falls to 1/e relative to its on-axis value. Since q is a function of z, the beam radius as well as the radius of curvature will depend on the position along the axis of propagation.

With these definitions, the function q is given by

$$\frac{1}{q} = \frac{1}{R} - \frac{j\lambda}{\pi\omega^2} \tag{18}$$

where both R and ω are functions of z.

At z = 0, from Eq. 8, u(r, 0) = A(0)exp[-jkr^2/2q(0)], and if ω_0 such that $\omega_0 = [\lambda q(0)/j\pi]^{0.5}$, the relative field distribution at z=0 will be

$$u(r,0) = u(0,0)\exp\left(\frac{-r^2}{\omega_0^2}\right) \tag{19}$$

where ω_0 denotes the beam radius at z = 0, which is called the beam waist radius. With this definition, from Eq. 11b a second important expression for q is obtained:

$$q = \frac{j\pi\omega_0^2}{\lambda} + z. \tag{20}$$

Eqs. 18 and 20 together obtains the radius of curvature, R, and the beam radius, ω, as a function of position along the axis of propagation:

$$R = z + \frac{1}{z}\left(\frac{\pi\omega_0^2}{\lambda}\right)^2 \tag{21a}$$

$$\omega = \omega_0\left[1 + \left(\frac{\lambda z}{\pi\omega_0^2}\right)^2\right]^{0.5}. \tag{21b}$$

It can be seen that the beam waist radius is the minimum value of the beam radius and that it occurs at the beam waist, where the radius of curvature is infinite, characteristic of a plane wave front. The relationships given in Eqs. 21a and 21b are fundamental for Gaussian beam propagation. The confocal distance, $\pi\omega_0^2/\lambda$, plays a prominent role.

To complete the analysis of the basic Gaussian beam equation, the second of the pair of equations obtained from substituting our trial solution in the paraxial wave equation must be used. Rewriting Eq. 10b, dA/A = -dz/q is obtained, and from Eq. 10a, dz = dq so that dA/A = -dq/q. Hence, A(z)/A(0) = q(0)/ q(z), and substituting q from Eq. 20,

$$\frac{A(z)}{A(0)} = \frac{1 + j\lambda z / \pi\omega_0^2}{1 + \left(\lambda z / \pi\omega_0^2\right)^2}. \tag{22}$$

It is convenient to express this in terms of a phasor, and defining

$$\tan\phi_0 = \frac{\lambda z}{\pi\omega_0^2}. \tag{23}$$

It can be seen that

$$\frac{A(z)}{A(0)} = \frac{\omega_0}{\omega} exp(j\phi_0). \tag{24}$$

The Gaussian beam phase shift, ϕ_0, also is discussed in more detail below. If the amplitude on-axis at the beam waist is unity, the complete expression for the fundamental Gaussian beam mode becomes

$$u(r,z) = \frac{\omega_0}{\omega} exp\left(\frac{-r^2}{\omega^2} - \frac{j\pi r^2}{\lambda R} + j\phi_0\right). \tag{25a}$$

The expression for the electric field can be obtained immediately using Eq. 2, and differs only due to the plane wave phase factor, to get

$$E(r,z) = \left(\frac{\omega_0}{\omega}\right) exp\left(\frac{-r^2}{\omega^2} - jkz - \frac{j\pi r^2}{\lambda R} + j\phi_0\right) \tag{25b}$$

with the variation in ω, R, and ϕ_0 as a function of z being given by Eqs. 21a, 21b and 23.

To relate the expression for the electric field given above to the total power in a propagating Gaussian beam, the electric and magnetic field components in the paraxial limit are assumed to be related to each other like those in a plane wave. Thus, the total power is proportional to the square of the electric field integrated over the area of the beam. A convenient normalization is to set the integral (extending from radius 0 to ¥) to unity, namely, òlEl2 • 2πr dr = 1. Using the electric field distribution from Eq. 25b, this integral, evaluated at z = 0, gives $\pi\omega_0^2/2$. Consequently, the normalized electric field distribution at any distance along the axis of propagation is given by Eq. 26.

GAUSSIAN BEAM MODE SOLUTIONS IN CYLINDRICAL COORDINATES

Many encounters in practice are axially symmetry. This makes cylindrical coordinates the popular choice to solve the paraxial wave equation. Based on the solution of the paraxial wave equation with a scalar field, fundamental Gaussian beam quasioptics is given by

$$E(r,z)=\sqrt{\frac{2}{\pi\omega^2}}exp\left(\frac{-r^2}{\omega^2}-jkz-\frac{j\pi r^2}{\lambda R}+j\phi_0\right) \tag{26}$$

where R is the radius of curvature Equation (21a) and ω is the beam radius Equation (21b), and ϕ_0 can be derive from Eqatuion (23) to get

$$\phi_0=\tan^{-1}\left(\frac{\lambda z}{\pi\omega_0^2}\right). \tag{27}$$

The radius of curvature,R, beam radius,ω, and phase,ϕ_0, completely characterize the beam along the z-axis which is the propagation path. This is the first order solution of the paraxial wave equation. The general solution includes higher order modes and may be expanded in terms of Gauss-Laguerre polynomials as

$$E_{pm}(r,\varphi,z)=\sqrt{\frac{2p!}{\pi(p+m)!}}\frac{1}{\omega(z)}\left[\frac{\sqrt{2}r}{\omega(z)}\right]^m L_{pm}\left(\frac{2r^2}{\omega^2(z)}\right)\exp\left[\frac{-r^2}{\omega^2(z)}-jkz-\frac{j\pi r^2}{\lambda R(z)}-j(2p+m+1)\phi_0(z)\right]\exp(jm\varphi) \tag{28}$$

where L_{pm} is the generalized Laguerre polynomials, p is the radial index, m the angular index, and φ is the polar angle. The beam radius ω, radius of curvature,R, and the phase shift ϕ_0 are exactly the same as for the fundamental mode by an amount that depends on the mode parameters.

The Laguerre polynomials can be obtained from the expression (Goubau, 1969)

$$L_{pm}(u)=\sum_{l=0}^{l=p}\frac{(p+m)!(-u)^l}{(m+l)!(p-l)!l!}. \tag{29}$$

In the case of corrugated feedhorns, the systems are azimuthally symmetric but are not exactly described by the fundamental Gaussian beam mode. Modes that are axially symmetric, which means they are independent of φ, need to be considered. These modes can be written as

$$E_{p0}(r,z)=\sqrt{\frac{2r^2}{\pi\omega^2}}L_{p0}\left(\frac{2r^2}{\omega^2}\right)\exp\left[-\frac{r^2}{\omega^2}-jkz-\frac{j\pi r^2}{\lambda R}+j(2p+1)\phi_0\right] \tag{30}$$

where,

r = the perpendicular distance from the axis of propagation

ω = beam radius

λ = wavelength.

Explicit dependence of the various quantities on distance along the axis of propagation (z-axis) have been omitted and L_{p0} functions are the ordinary Laguerre polynomials given by

$$L_p(u)=\frac{e^u}{p!}\frac{d^p}{du^p}\left(e^{-u}u^p\right). \tag{31}$$

GAUSSIAN BEAM MODE SOLUTIONS IN RECTANGULAR COORDINATES

At times, solving Gaussian beam modes in rectangular coordinates is required. To solve one dimensional Gaussian beam mode in rectangular coordinate, a beam is considered to have variation in one coordinate perpendicular to the axis of propagation (z axis). The general solution for variation along the x axis only, is

$$E(x,z)=\sqrt[4]{\frac{2}{\pi\omega_x^2}}\exp\left(-\frac{x^2}{\omega_x^2}-jkz-\frac{j\pi x^2}{\lambda R_x}+\frac{j\phi_{0x}}{2}\right) \tag{32}$$

where$\phi_{0x}=\phi_0$,$R_x=R$ and $\omega_x=\omega$ as shown in Eq. 21a,Eq. 21b and Eq. 27.

In order to solve two dimensional Gaussian beam mode in rectangular coordinate, a similar approach to solve the paraxial wave is employed. It is desirable to keep the solution independent in the two orthogonal coordinates as long as the separate solutions are valid independently for all x and y axes. This produces the general solution

$$E(x,y,z)=\sqrt{\frac{2}{\pi\omega_x\omega_y}}\exp\left(-\frac{x^2}{\omega_x^2}-\frac{y^2}{\omega_y^2}-\frac{j\pi x^2}{\lambda R_x}-\frac{j\pi y^2}{\lambda R_y}+\frac{j\phi_{0x}}{2}+\frac{j\phi_{0y}}{2}\right) \tag{33}$$

where,

$$\omega_x=\omega_{0x}\sqrt{1+\left(\frac{\lambda z}{\pi\omega_{0x}^2}\right)^2}, \tag{34}$$

$$\omega_y = \omega_{0y}\sqrt{1+\left(\frac{\lambda z}{\pi\omega_{0y}^2}\right)^2}, \tag{35}$$

$$R_x = z + \frac{1}{z}\left(\frac{\pi\omega_{0x}^2}{\lambda}\right)^2, \tag{36}$$

$$R_y = z + \frac{1}{z}\left(\frac{\pi\omega_{0y}^2}{\lambda}\right)^2, \tag{37}$$

$$\phi_{0x} = \tan^{-1}\left(\frac{\lambda z}{\pi\omega_{0x}^2}\right), \tag{38}$$

$$\phi_{0y} = \tan^{-1}\left(\frac{\lambda z}{\pi\omega_{0y}^2}\right). \tag{39}$$

By keeping the solution independent in the two orthogonal coordinates (x and y axes), the solution is independent of the beam waist radii along the orthogonal coordinates and reference positions can be chosen along the z-axis for complex beam parameters to be different. As the critical parameters describing variation of the Gaussian beam in the orthogonal coordinates perpendicular to its axis of propagation are independent, asymmetric Gaussian beams can be dealt with and focusing of a Gaussian beam along a single axis independent of its variation in the orthogonal direction can be considered.

Eq. 33 becomes identical to Eq. 26 when the beam waist radii ω_{0x} and ω_{0y} are equal and are located at the same value of z. When considering higher order modes in rectangular coordinates, the expression for the one-dimensional Gaussian beam mode of order m with variation along the x-axis is

$$E_m(x,z) = \sqrt[4]{\frac{2}{\pi}}\sqrt{\frac{1}{\omega_x 2^m m!}}H_m\left(\frac{\sqrt{2}x}{\omega_x}\right)\exp\left[-\frac{x^2}{\omega_x^2} - jkz - \frac{j\pi x^2}{\lambda R_x} + \frac{j(2m+1)\phi_{0x}}{2}\right] \tag{40}$$

where H_m is the Hermite polynomial of the order m and can be found from the expression (Margenau and Murphy, 1956)

$$H_m(u) = (-1)^m e^{u^2}\frac{d^m}{du^m}\left(e^{-u^2}\right). \tag{41}$$

When dealing with two-dimensional Gaussian beam mode with order m along the x-axis and order n along the y-axis, the expression for the mn Gauss-Hermite beam mode is

$$E_{mn}(x,y,z)=\sqrt{\frac{1}{\pi\omega_x\omega_y 2^{m+n-1}m!n!}}H_m\left(\frac{\sqrt{2}x}{\omega_x}\right)H_n\left(\frac{\sqrt{2}y}{\omega_y}\right)\exp\left[-\frac{x^2}{\omega_x^2}-\frac{y^2}{\omega_y^2}-jkz-\frac{j\pi x^2}{\lambda R_x}-\frac{j\pi y^2}{\lambda R_y}+\frac{j(2m+1)\phi_{0x}}{2}+\frac{j(2n+1)\phi_{0y}}{2}\right] \tag{42}$$

where H_n utilizes Eq.41 to find the Hermite polynomial for the order n. As the orthogonal coordinates are again independent of each other, it is possible to deal with higher order modes having unequal beam waist radii and different beam waist locations. When beams in x and y with equal beam waist radii are located at the same value of z, taking $\omega_x = \omega_y \circ \omega$, $R_x = R_y \circ R$, and $\phi_{0x} = \phi_{0y} \circ \phi_0$, the equation below is obtain

$$E_{mn}(x,y,z)=\sqrt{\frac{1}{\pi\omega^2 2^{m+n+1}m!n!}}H_m\left(\frac{\sqrt{2}x}{\omega}\right)H_n\left(\frac{\sqrt{2}y}{\omega}\right)\exp\left[-\frac{(x^2+y^2)}{\omega^2}-jkz-\frac{j\pi(x^2+y^2)}{\lambda R}+j(m+n+1)\phi_0\right]. \tag{43}$$

This expression can be useful when the beam of interest is not simply the fundamental Gaussian beam mode but have two equal waist radii in the two coordinates ($\omega_x = \omega_y$). When $m = n = 0$, the fundamental Gaussian beam can be obtained with purely Gaussian distribution.

The simplicity of Gaussian beam analysis can provide a starting point for more rigorous diffraction calculations and can be used to calculate the parameters of the reflector (Tham, Yassin&Carter, 2007). The computation time of Gaussian beam analysis is shorter by a factor of 72 compared to physical optics approach in a study by (Chou &Pathak, 2004). Examples of Gaussian optics application for Cassegrain antennas can be found in (Lamb, 1986; Padman, Murphy & Hills, 1987; Withington, Murphy, Egan & Hills, 1992; Kempen, Corder, Lucas &Mauersberger, 2012; Gonzalez, Uzawa, Fujii&Kmeko, 2011; Candotti, Baryshev& Trappe, 2009; Tham&Withington, 2003; Tham, Yassin& Carter, 2007). These examples include application of Gaussian optics for the ALMA and James Clerk Maxwell Telescopes. To go in deeper into the subject of Gaussian optics, readers are recommended to read (Goldsmith, 1998).

REFERENCES

Candotti, M., Baryshev, A. M., & Trappe, N. (2009). Quasi-optical assessment of the ALMA band 9 frond-end. *Infrared Physics & Technology*, *52*(5), 174–179. doi:10.1016/j.infrared.2009.07.007

Chou, H. T., & Pathak, P. H. (2004). Fast Gaussian beam-based synthesis of shaped reflector antennas for contoured beam applications. *IEEE Proceedings Microwaves, Antennas, and Propagation,* 151, 13 – 20. 10.1049/ip-map:20040250

Goldsmith, P. F. (1998). *Quasioptical systems; Gaussian beam, quasioptical propagation and applications.* New York: IEEE Press.

Gonzalez, A., Uzawa, Y., Fujii, Y., & Kaneko, K. (2011). ALMA band 10 tertiary optics. *Infrared Physics & Technology*, *54*(6), 488–496. doi:10.1016/j.infrared.2011.07.007

Goubau, G. (1969). Optical or quasi-optical transmission schemes. In F. A. Benson (Ed.), *Millimetre and Submillimetre Waves*. London, UK: Illife Press.

Kempen, T. V., Corder, S., Lucas, R., & Mauersberger, R. (2012). How ALMA is calibrated: I. Antenna-based pointing, focus and amplitude calibration. *ALMA Newsletter*, *9*, 8–16.

Lamb, J. W. (1986). Quasioptical coupling of Gaussian beam systems to large Cassegrain antennas. *International Journal of Infrared and Millimeter Waves*, *7*(10), 1511–1536. doi:10.1007/BF01010755

Margenau, H., & Murphy, G. (1956). *The mathematics of physics and chemistry* (pp. 120–126). New York: Van Nostrand.

Padman, R., Murphy, J., & Hills, R. (1987). Gaussian mode analysis of Cassegrain antenna efficiency. *IEEE Transactions on Antennas and Propagation*, *35*(10), 1093–1103. doi:10.1109/TAP.1987.1143983

Tham, C. Y., & Withington, S. (2003). *Receiver optics design electromagnetic analysis, second report (Bands 3, 6, 7, and 9). Technical Report*. Cambridge University.

Tham, C. Y., Yassin, G., & Carter, M. (2007). Analysis techniques for the optics in millimetre/submillimetre wave radio telescopes receivers. *JurnalFizik Malaysia*, *28*, 49–53.

Withington, S., Murphy, J., Egan, A., & Hills, R. (1992). On the design of broadband quasioptical systems for submillimter-wave radio-astronomy receivers. *International Journal of Infrared and Millimeter Waves*, *13*(10), 1515–1537. doi:10.1007/BF01009233

Yeap, K. H., & Tham, C. Y. (2018). Optimization of an offset receiver optics for radio telescopes. *Journal of Infrared, Millimeter, and Terahertz Waves*, *39*(1), 64–76. doi:10.100710762-017-0449-z

ADDITIONAL READING

Ediss, G. A., Carter, M., Cheng, J., Effland, J. E., Grammer, W., Horner, M., . . . Sullivan, M. (2004). ALMA band 6 prototype cartridge: Design and performance. 15th International Symposium on Space Terahertz Technology, Northampton, MA.

Hsi-Tseng, C., Pathak, P. H., & Burkholder, R. J. (2001). Novel Gaussian beam method for the rapid analysis of large reflector antennas. *IEEE Transactions on Antennas and Propagation*, *49*(6), 880–893. doi:10.1109/8.931145

Lamb, J. W. (2003). Low-noise, high-efficiency optics design for ALMA receivers. *IEEE Transactions on Antennas and Propagation*, *51*(8), 2035–2047. doi:10.1109/TAP.2003.814743

Lamb, J. W., Baryshev, A., Carter, M. C., D'Addario, L. R., Ellison, B. N., Grammer, W., Lazareff, B., Sekimoto, Y &Tham, C. Y. (2001). ALMA receiver optics design. ALMA Memo, 362.

Murphy, J. A. (1988). Distortion of a simple Gaussian beam on reflection from off-axis ellipsoid mirrors. *International Journal of Infrared and Millimeter Waves*, *8*(9), 1165–1187. doi:10.1007/BF01010819

Rieckmann, C., Rayner, M. R., Parini, C. G., Martin, D. H., & Donnan, R. S. (2002). Novel modular approach based on Gaussian beam diffraction for analysing quasi-optical multi-reflector antennas. *IEE Proceedings. Microwaves, Antennas and Propagation*, *149*(3), 160–167. doi:10.1049/ip-map:20020390

Yeap, K. H., Tham, C. Y., Yeong, K. C., Chong, K. H., Rizman, Z. I., & Yang, C. C. (2011). Analysis of normal and superconducting coplanar waveguides in radio astronomy. *International Journal of Electronics, Computer, and Communications Technologies*, *2*, 9–12.

Section 4
Antennas

Chapter 8
Horn Antenna

Hirokazu Kobayashi
Osaka Institute of Technology, Japan

ABSTRACT

One of the simple and most widely used microwave antennas is the horn as a feed element for large radio telescopes, satellite tracking, and communication reflector, which are found installed throughout the world. In addition to its utility as a feed for reflectors and lenses, it is a common element of phased arrays and serves as a universal standard for calibration and gain measurement of other high gain antennas. Its widespread applicability stems from its simplicity in construction, ease of excitation, large gain, wide-band characteristics, and preferred overall performance. An electromagnetic horn can take many different forms, such as basic pyramidal, conical, corrugated, double-ridged, and dual polarized horns, as well as horns with lens and so on. The horn is nothing more than a hollow pipe of different cross-sections, which has been tapered to a larger opening aperture. This chapter explains the fundamentals of the pyramidal horn antenna in detail using aperture field method. Numerical and measured examples, are also shown.

HORN ANTENNA

Various types of primary feed antennas for microwave and millimeter wave reflector etc. are employed according to their application and purpose (Rudge, 1986; Chang, 1989). For example, the broadband conical spiral or the logarithmic periodic antenna, which is unthinkable a while ago, has been commercialized as a feeding antenna. However, for normal purposes, the horn antennas, probably the pyramidal horns, are most popular. There is a legitimate reason for this. At first the design is easy, the performance is almost equal to the theoretical calculation, and its manufacturability is excellent. Even if it is a horn antenna, they have many variations because it is such a basic antenna. Typical examples include the corrugated horn, the dielectric lens horn, the diagonal horn, and the like. Each of these also has a purpose and characteristics.

DOI: 10.4018/978-1-7998-2381-0.ch008

Figure 1. Basic types of pyramidal horn antenna

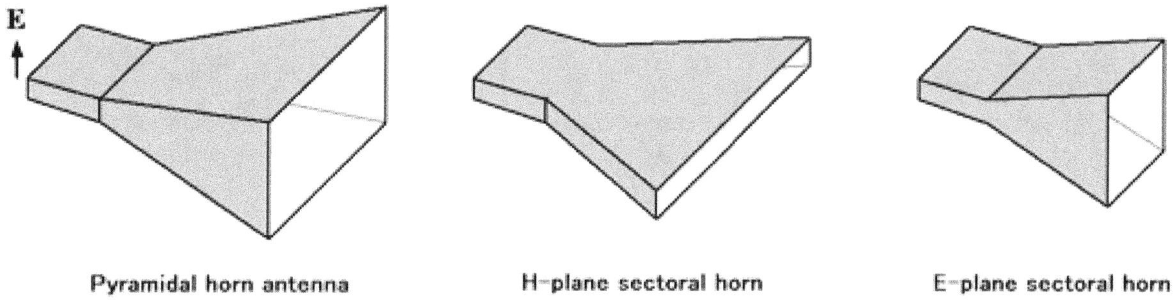

In this chapter, let us investigate a mathematical wave theory approach to show how to analyze antenna radiation characteristics as well as its applications. Preferentially the horn with a rectangular aperture, so-called pyramidal horn, should be paid attention as a typical boundary value problem in electromagnetics. Since this antenna is a three-dimensional object, it is also necessary to analyze the vector quantity in three-dimensions. However, this direct calculation is quite difficult, so that it is usually analyzed in two-dimensions. At this time, the three-dimensional pyramidal horn is reduced to a sectoral horn. The term "sectoral" here means a fan shape. Even though it is two-dimensions, the antenna is fed by a waveguide, so its height remains (see Figure 1). The electromagnetic field variation in this direction is assumed to be uniform, so that it can be a two-dimensional problem. The theoretical expressions of H- and E-planes obtained in this way have orthogonal relationship from Fourier transform theory, so the result of a pyramidal horn will be obtained by their superimposing. Horn antenna is a kind of the aperture antenna. In this case, if an electromagnetic field at a horn aperture is known, it is transformed by Fourier theory to obtain a spatial spectrum in a far-region, namely antenna pattern. The spatial spectrum of the far-field is the result of the response due to the aperture distribution as a secondary source. This method to predict the far-field from an aperture distribution is called the aperture field method (AFM) (Balanis, 1982; Rudge, 1986; Lo, 1988; Kobayashi, 2011).

At first, let us show how to calculate electromagnetic field distributions at horn aperture. Intuitively, it can be somehow calculated from the field in a feeding waveguide. In this chapter, therefore, starting with the wave-guide theory, the aperture distribution of the sectoral waveguide is searched. Then, an approximate distribution is calculated by cutting into a finite length. On the other hand, the radiation field should be formulated from the aperture distribution or its equivalent electromagnetic current distribution. Using the obtained sectoral aperture distribution, the far-field characteristics and directivity function of the sectoral horn are calculated. Finally, the radiation field of a pyramidal horn can be obtained by superimposing these two sectoral fields which have the orthogonal relation each other. In addition, since the obtained results of the pyramidal horn can be easily converted to those of a diagonal horn, let us examine this type of horn. At the same time, fabricated examples of a small horn with the double-ridge and the corrugation will be discussed. An aperture size of the double-ridged horn is less than a half of a wavelength. This special horn is applicable for active phased array antennas as an antenna element. The corrugated conical horn is fabricated by using a 3D-printer.

Figure 2. Geometrical parameters of the sectoral horn

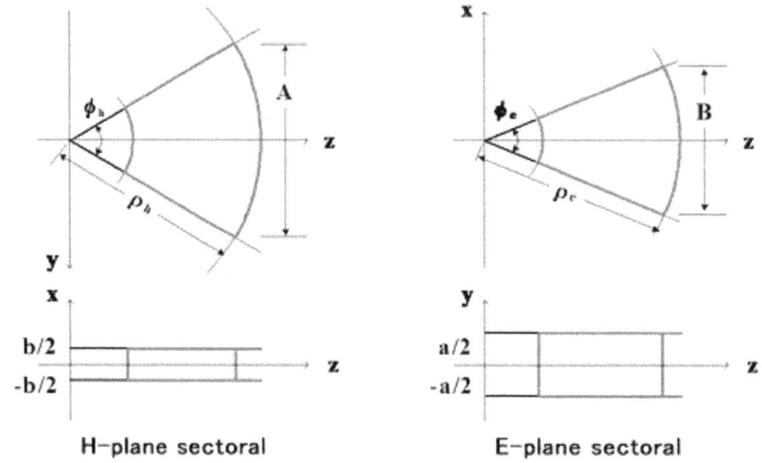

Propagation Mode of Sectoral Waveguide

Figure 2 shows geometrical parameters of the sectoral horn. As can be seen from the figure, any of opposing surfaces of the feeding waveguide has a fan shape. Desired performance is obtained by a combination of the flare length and angle, which are indicated as $\rho_{h,e}$ and $\phi_{h,e}$, in the figure. A waveguide surface in the electric field direction is called electric-wall, and a magnetic field orthogonal to an electric-field is called a magnetic-wall. Correspondingly, H- and E-plane sectoral horns are independently constructed, then their geometric parameters are arranged symmetrically on x- and y-axes, respectively. In addition, cylindrical coordinates is suitable for this boundary value problem from the viewpoint of the geometric shape, so (ρ, ϕ, x) and (ρ, ϕ, y) are employed, respectively. The independent variables here are $\rho_{h,e}$ and $\phi_{h,e}$, and horn shapes are distinguished by the subscripts.

The following analysis is based on the assumption that fundamental mode of the feeding rectangular waveguide propagates directly to the sectoral-shaped waveguide. If sectoral horn extends to infinite length, its propagation mode can be determined from the boundary conditions. Then, an electromagnetic field where an aperture of $\rho = \rho_{h,e}$ reaches a finite length is calculated. Therefore, since the electromagnetic field at the aperture, which is considered to have been cut, is obtained by using an electromagnetic field that has been continuously extended, naturally the boundary condition at this portion is not exactly satisfied. In other words, the influence of higher-order modes or edge currents due to discontinuities in the aperture is ignored. This results in a cause of deterioration of calculation accuracy of side-lobes.

Now, let us expand Maxwell's equation in the cylindrical coordinate system. Here, since the electromagnetic source currents need not be considered, Maxwell's equations for the time dependency $\exp(j\omega t)$ is expressed as $\nabla \times \boldsymbol{E} = -j\omega\mu \boldsymbol{H}$, $\nabla \times \boldsymbol{H} = j\omega\varepsilon \boldsymbol{H}$. Using the formula, curl in the cylindrical coordinate of vector $\boldsymbol{V}$

$$\nabla \times \boldsymbol{V} = \boldsymbol{i}_\rho \left\{ \frac{1}{\rho}\frac{\partial V_x}{\partial \phi} - \frac{\partial V_\phi}{\partial x} \right\} + \boldsymbol{i}_\phi \left\{ \frac{\partial V_\rho}{\partial x} - \frac{\partial V_x}{\partial \varrho} \right\} + \boldsymbol{i}_x \frac{1}{\rho} \left\{ \frac{\partial}{\partial \rho}\left(\rho V_\phi\right) - \frac{\partial V_x}{\partial \phi} \right\}, \tag{1}$$

and Gauss's law

$$\nabla \bullet \boldsymbol{E} = 0, \nabla \bullet \boldsymbol{H} = 0, \nabla \bullet \boldsymbol{V} = \frac{1}{\rho}\frac{\partial}{\partial \rho}\left(\rho V_{\rho}\right) + \frac{1}{\rho}\frac{\partial V_{\phi}}{\partial \phi} + \frac{\partial V_{x}}{\partial x}, \tag{2}$$

then Maxwell's equations in the cylindrical coordinate for an E-plane horn is expanded as follows:

$$j\omega\varepsilon E_{\rho} = \frac{1}{\rho}\frac{\partial H_{x}}{\partial \phi} - \frac{\partial H_{\phi}}{\partial x}, j\omega\varepsilon E_{\phi} = \frac{\partial H_{\rho}}{\partial x} - \frac{\partial H_{x}}{\partial \rho}, j\omega\varepsilon E_{x} = \frac{1}{\rho}\frac{\partial}{\partial \rho}\left(\rho H_{\phi}\right) - \frac{1}{\rho}\frac{\partial H_{\rho}}{\partial \phi}, \tag{3}$$

$$-j\omega\varepsilon H_{\rho} = \frac{1}{\rho}\frac{\partial E_{x}}{\partial \phi} - \frac{\partial E_{\phi}}{\partial x}, -j\omega\varepsilon H_{\phi} = \frac{\partial E_{\rho}}{\partial x} - \frac{\partial E_{x}}{\partial \rho}, -j\omega\varepsilon H_{x} = \frac{1}{\rho}\frac{\partial}{\partial \rho}\left(\rho E_{\phi}\right) - \frac{1}{\rho}\frac{\partial E_{\rho}}{\partial \phi} \tag{4}$$

$$\frac{1}{\rho}\frac{\partial}{\partial \rho}\left(\rho H_{\rho}\right) + \frac{1}{\rho}\frac{\partial H_{\phi}}{\partial \phi} + \frac{\partial H_{x}}{\partial x} = 0, \frac{1}{\rho}\frac{\partial}{\partial \rho}\left(\rho E_{\rho}\right) + \frac{1}{\rho}\frac{\partial E_{\phi}}{\partial \phi} + \frac{\partial E_{x}}{\partial x} = 0. \tag{5}$$

These equations are solved by boundary conditions of waveguide propagation modes. Let propagation mode of rectangular waveguide be basic TE_{10} mode. This mode is transmitted to a sectoral-shaped waveguide, and the lowest-order sectoral mode becomes the (pseudo) main mode after mode-converted. This mode propagates through an infinitely long sectoral horn. All other higher-order modes occurred near discontinuity at the junction of the feeding waveguide and horn, should not propagate so far. Table 1 shows boundary conditions and possible electromagnetic field components of each sectoral horn. The propagation mode of this horn is basically the same as that of the rectangular waveguide. In the E-plane horn, electric field lines arc perpendicular to the electric field wall, and the azimuth component E_{ϕ} oscillates sinusoidally depending on its mode, and zero at the parallel magnetic wall. This is a consequence of the electromagnetic field boundary conditions. On the other hand, in the H-plane, considering cyclic symmetry with the E-plane sectoral, just replacing $x \leftrightarrow y$ with the E-plane expression, the basic H-plane field in cylindrical coordinates is obtained.

Table 1. Boundary conditions for sectoral waveguide

	Boundary Condition	**Component**
E-plane	$E_y = E_\rho = 0$ for $\phi = \pm\phi_e/2$ $E_\phi = E_\rho = 0$ for $y = \pm a/2$	E_ϕ, H_y, H_ρ
H-plane	$E_x = E_\rho = 0$ for $\phi = \pm\phi_h/2$ $E_\phi = E_\rho = 0$ for $y = \pm b/2$	E_x, H_ϕ, H_ρ

Now, considering the component $E_\rho = E_\phi = H_\phi = 0$ in the H-plane sectoral from Table 1, the following equation will be obtained easily:

H-plane sectoral waveguide $\frac{\partial E_\rho}{\partial x} = \frac{\partial E_\phi}{\partial x} = \frac{\partial E_x}{\partial x} = 0,$

$$j\omega\varepsilon E_x = \frac{1}{\rho}\frac{\partial}{\partial \rho}\left(\rho H_\phi\right) - \frac{1}{\rho}\frac{\partial H_\rho}{\partial \phi}, j\omega\mu H_\rho = -\frac{1}{\rho}\frac{\partial E_x}{\partial \phi}, \tag{6}$$

$$j\omega\mu H_\phi = \frac{\partial E_x}{\partial \rho}, \frac{\partial}{\partial \rho}\left(\rho H_\rho\right) + \frac{1}{\rho}\frac{\partial H_\phi}{\partial \phi} = 0.$$

Similarly, when $E_y = E_\rho = H_\phi = 0$ in the coordinate system (ρ, ϕ, y), the relationship for the E-plane sectoral waveguide is calculated as follows:

E-plane sectoral waveguide $\frac{\partial E_\phi}{\partial \phi} = \frac{\partial H_\rho}{\partial \phi} = \frac{\partial H_y}{\partial \phi} = 0,$

$$j\omega\varepsilon E_\phi = \frac{\partial H_\rho}{\partial y} - \frac{\partial H_y}{\partial \rho}, j\omega\mu H_\rho = \frac{\partial E_\phi}{\partial y}, \tag{7}$$

$$j\omega\mu H_y = -\frac{1}{\rho}\frac{\partial}{\partial \rho}\left(\rho E_\phi\right), \frac{1}{\rho}\frac{\partial}{\partial \rho}\left(\rho H_\rho\right) + \frac{\partial H_y}{\partial y} = 0.$$

Considering the electric field pattern here from Table 1, the E_x component on the H-plane and E_ϕ on the E-plane should be focused on. Other components can be easily determined from E_x, E_ϕ. In this way, after somewhat complicated calculation, the basic equations for the H- and the E-plane sectoral waveguide combined into one variable can be obtained using equations (6) and (7) as:

H-plane

$$\frac{\partial^2 E_x}{\partial \rho^2} - \frac{1}{\rho}\frac{\partial E_x}{\partial \rho} + \frac{1}{\rho^2}\frac{\partial^2 E_x}{\partial \phi^2} + \omega^2\mu\varepsilon E_x = 0, \tag{8a}$$

E-plane

$$\frac{\partial^2 E_\phi}{\partial \rho^2} - \frac{1}{\rho}\frac{\partial E_\phi}{\partial \rho} + \frac{\partial^2 E_\phi}{\partial y^2} + \left(\omega^2\mu\varepsilon - \frac{1}{\rho^2}\right)E_\phi = 0. \tag{8b}$$

Thereafter, it is only necessary to solve these differential equations considering the boundary conditions.

Let us start with the H-plane sectoral waveguide in equation (8a). From Table 1, the boundary condition is found as

$$E_x = 0 \quad \text{at} \quad \phi = \pm\frac{\phi_h}{2}, \tag{9}$$

which is a necessary-condition that the tangential component of the electric field is 0 on the metal wall at flare angle $\phi = \pm\phi_h / 2$. When the TE_{10} mode is fed to the sectoral horn, an infinite number of higher-order modes with inherent propagation constants are generally produced. However, ignoring these higher-modes, the dominant mode is considered only TE_{10} mode here. That is, E_x expression can be assumed to be as

$$E_x = f_h(\rho)\cos\frac{\pi\phi}{\phi_h}, \tag{10}$$

which satisfies the boundary condition. All that remains is dependency on the radial direction ρ. To determine this, the means of a variable separation method is employed. This is a conventional method, solving differential equation by creating a separate variable or function like equation (10), assigning it to equation (8), and enclosing the function $f_h(\rho)$, which leads a relational expression of only ρ as follows:

$$\frac{\partial^2 f_h(\rho)}{\partial(k\rho)^2} - \frac{1}{k\rho}\frac{\partial f_h(\rho)}{\partial(k\rho)} + \left\{1 - \left(\frac{\frac{\pi}{\phi_h}}{k\rho}\right)^2\right\} f_h(\rho) = 0, k = \omega^2\mu\varepsilon = \frac{2\pi}{\lambda}. \tag{11}$$

This differential equation has a solution of a cylindrical function with order (π/ϕ_h). In other words, the above equation is Bessel's differential equation, and the solution is already given properly as Bessel's function. There is a lot of discussion about this Bessel's function, so the details is omitted here. The Bessel's differential equation has two independent solutions. The first kind of Hankel's function $H_n^{(1)}, n = \pi / \phi_h$ which shows the backward wave and the second kind of Hankel's function $H_n^{(2)}$ showing the forward wave. Each may be referred to as a convergent and divergent wave, or a reflected and incident wave. In thinking a real problem, the backward wave refers to the wave traveling in the backward direction (the negative direction of *z*). Combining these two waves and further considering equation (10), the general solution of the differential equation (8a) is calculated as follows:

$$E_x = C_1\cos(n\phi)\left\{H_n^{(2)}(k\rho) + \alpha H_n^{(1)}(k\rho)\right\}, \qquad n = \frac{\pi}{\phi_h}. \tag{12a}$$

Magnetic components are obtained from equation (6) as follows:

$$H_\rho = C_1 \frac{n}{j\omega\mu} \frac{\sin(n\phi)}{\rho} \left\{ H_n^{(2)}(k\rho) + \alpha H_n^{(1)}(k\rho) \right\}, \tag{12b}$$

$$H_\phi = C_1 \frac{\cos(n\phi)}{j\omega\mu} \left\{ H_n^{(2)}(k\rho) + \alpha H_n^{(1)}(k\rho) \right\}. \tag{12c}$$

The constants C_1 and α are usually complex numbers. The strength of α is less than 1 because it gives the ratio of incident to reflected wave. The above expressions are propagation modes in a sectoral waveguide. If a flare angle of a sectoral is not so large, the effect of higher-order modes can be ignored. However a problem is the effect of discontinuity between a horn cut to a finite length and free space.

Next, for the E-plane sectoral waveguide, a E_ϕ component oscillates sinusoidally along the *y*-axis and behaves zero on the wall. Therefore, this field can be expressed as:

$$E_\phi = f_e(\rho) \cos\frac{\pi y}{a}. \tag{13}$$

Substituting equation (13) to (8b), then

$$\frac{\partial^2 f_e(\rho)}{\partial(\beta\rho)^2} - \frac{1}{\beta\rho}\frac{\partial f_e(\rho)}{\partial(\beta\rho)} + \left\{ 1 - \left(\frac{1}{\beta\rho}\right)^2 \right\} f_e(\rho) = 0, \qquad \beta = k^2 - \left(\frac{\pi}{a}\right)^2 \tag{14}$$

is obtained. This ordinary differential equation is also a Bessel's differential equation with variable βρ, and the components can be calculated using the Hankel's function in the same manner to the H-plane sectoral waveguide as follows:

$$E_\phi = C_2 \cos\left(\frac{\pi y}{a}\right) \left\{ H_1^{(2)}(\beta\rho) + \alpha H_1^{(1)}(\beta\rho) \right\}, \tag{15a}$$

$$H_\rho = C_2 \frac{j\pi}{\omega\mu a} \sin\frac{\pi y}{a} \left\{ H_1^{(2)}(\beta\rho) + \alpha H_1^{(1)}(\beta\rho) \right\}, \tag{15b}$$

$$H_\rho = C_2 \frac{j\beta}{\omega\mu} \cos\frac{\pi y}{a} \left\{ H_1^{(2)}(\beta\rho) + \alpha H_1^{(1)}(\beta\rho) \right\}. \tag{15c}$$

Figure 3. Tow type sectoral horn in cylindrical coordinates

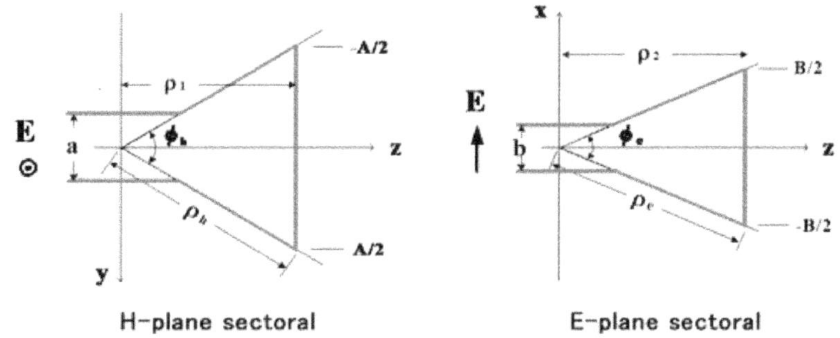

Here, the recursion formula $d\left\{t^n Z_n(t)\right\}/dt = t^n Z_{n-1}(t)$ is used, which is the common feature to all Bessel's functions. The propagating wavelength $\lambda = 2\pi/\beta$ is the same as the rectangular feeding waveguide. Therefore, the cutoff wavelength is also $\lambda_c = 2a$, and it is predicted that the attenuation of higher-order mode field occurs at this junction. Comparing the results of the H- and the E-planes, the order of the H-plane depends on the flare angle ϕ_h and smaller angle, higher-order $n = \pi/\phi_h$. When argument $\beta\rho$ is $\beta\rho \gg 1$, an equi-phase front of the H-plane is determined by the free space wavelength as opposed to the E-plane.

Aperture Field of Horn and Far-Field by Equivalent Current

Following the electromagnetic field along an infinitely sectoral waveguide, next task is to cut it to a finite length and determine aperture field distributions there. Considering actual manufacturability, it is better to make the aperture cutting part perpendicular to the *z*-axis as shown in Figure 3. However, since it cannot be denied that modes of the aperture will be destroyed by the cutting, the phase front of the aperture is taken into account the second order term. The equi-phase front propagating in a sectoral waveguide cylindrically diverges as if a wave source is at an origin where the sectoral flare surface extends. Therefore, it does not coincide with the aperture plane perpendicular to the *z*-axis. Thus, the phase variation in the flat aperture is not constant, but varies through the aperture. At first, let us consider this correction.

Now, the asymptotic expression of Hankel's function introduced in the previous section is given as follows:

$$H_n^{(2)}(k\rho) \sim \sqrt{\frac{2}{\pi k\rho}} \exp\left[-j\left(k\rho - \frac{n}{2} + \frac{\pi}{4}\right)\right]. \tag{16}$$

It is found that the phase of propagation mode varies with exp(–*jk*ρ). In order to hold this asymptotic assumption, the condition $k\rho \gg n/2$ is necessary. Consider that $\phi_{h,e}$ is almost constant and manipulate the aperture distribution as follows. In the case of an H-plane horn, the aperture plane varies along the *y*-axis, so the phase plane deviation can be corrected (see Figure 4). Therefore, $k\rho \gg n/2$ is replaced to $y \ll n/2$. The compensated phase amount Δ*y* is given as [Silver, 1949]:

Figure 4. Compensation of aperture phase in H-plane

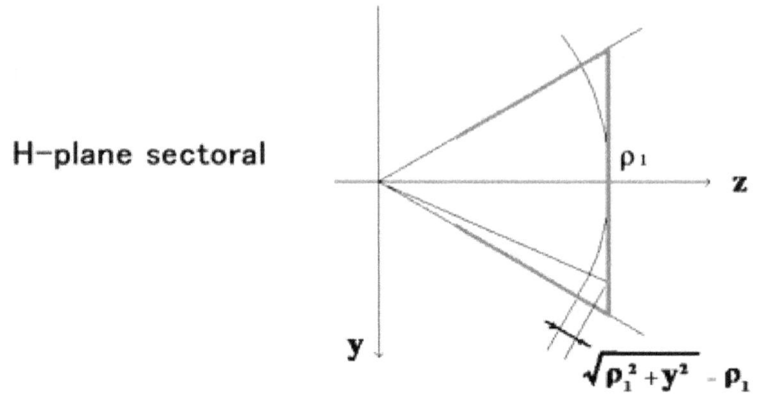

$$\Delta y = \{\text{Relative phase at } y \text{ with respect to } y = 0\}$$

$$= -k\left\{\sqrt{\rho_1^2 + y^2} - \rho_1\right\} \cong -k\frac{y^2}{2\rho_1}. \tag{17}$$

Since this phase distribution of the error compensation considers the second-order term of y, so it is called a secondary phase distribution. On the other hand, the amplitude error can be regarded to be affected not so much and a reflected wave is also ignored as $\alpha = 0$. Thus, equation (12) is modified as follows:

$$E_x(x, y) = C_1 \cos\left(\frac{\pi y}{A}\right) \exp\left(-jk\Phi_y\right),\ \Phi_y = \frac{y^2}{2\rho_1}. \tag{18}$$

Similarly, for the E-plane sectoral horn,

$$E_\phi(x, y) = C_2 \cos\left(\frac{\pi y}{a}\right) \exp\left(-jk\Phi_x\right),\ \Phi_x = \frac{x^2}{2\rho_2}. \tag{19}$$

The above discussion is the result considering aperture distribution to the secondary phase from a transmitting mode of a sectoral-shaped waveguide. By taking up to the second order, it should be closer to an actual distribution than by the first order. Then, one might think that simply taking up to the third order would make it better. However, this is not necessary because it is difficult to obtain the third-order phase from the model, and in many cases, it is predicted from experiments that the second-order phase is almost sufficient. When a post or a dielectric is attached near the aperture, it should need a high-order expansion, but how to find it is a problem. A reflected wave is ignored as $\alpha = 0$, but this effect is usually small. When an aperture is small, or when a vswr is generally poor (2 or more), the calculation accuracy may be improved by considering reflected waves.

Now, by taking Fourier transformation to an aperture field distribution obtained in the previous section, the radiation pattern in the far-region can be obtained as a result. Here, let us start by replacing the electromagnetic field at an aperture surface with an equivalent electromagnetic current distribution. If this aperture current is applied, the vector potential can be calculated and the radiation field in the far-region can be also obtained, which is the same expression as Fourier transformation. From the equivalent principle of electromagnetic field, electro-magnetic currents and electromagnetic fields created by the current have a relationship as:

$$\boldsymbol{J} = \boldsymbol{i}_z \times \boldsymbol{H}_a, \qquad \boldsymbol{M} = -\boldsymbol{i}_z \times \boldsymbol{E}_a. \tag{20}$$

If these fields or currents present on an aperture, the electromagnetic field in the far-region can be determined by the above-described concept. Here, $\boldsymbol{J}$ and $\boldsymbol{M}$ are the electric and magnetic current across the aperture, respectively, $\boldsymbol{i}_z$ is the z-axis unit direction vector perpendicular to the aperture, and $\boldsymbol{E}_a$ and $\boldsymbol{M}_a$ are the electric and magnetic field at the aperture surface, respectively. If the aperture surface is a perfect conductor, either magnetically or electrically, the conductor can be removed assuming the electromagnetic current.

For $\nabla \cdot \boldsymbol{H} = 0$ which means that there is no magnetic charge, it is possible to specify magnetic field by a vector $\boldsymbol{A}$ expressed by $\boldsymbol{H} = \nabla \times \boldsymbol{A}$ because of the vector formula $\nabla \cdot \nabla \times \boldsymbol{A} = 0$. This vector $\boldsymbol{A}$ is called (magnetic) vector potential. On the other hand, expressing Ampère's law $\nabla \times \left(\boldsymbol{E} + j\omega\mu \boldsymbol{A}\right) = 0$ by using $\boldsymbol{A}$, it can be introduced scaler quantity $\boldsymbol{E} + j\omega\mu\boldsymbol{A} = \nabla \cdot \Phi$ because of the vector identity $\nabla \times \nabla \times \Phi = 0$. This Φ is called scalar potential. The two potentials $\boldsymbol{A}$ and Φ can be expressed by wave source $\boldsymbol{J}$ and electric charge ρ. Therefore, if these potentials are explicitly defined, required electromagnetic fields can be easily calculated. In order to express a relationship regarding $\boldsymbol{A}$ and Φ, Lorentz's gauge $\nabla \cdot \boldsymbol{A} = -j\omega\mu\Phi$ is selected, then

$$\nabla^2 \boldsymbol{A} + \omega^2 \mu\varepsilon \boldsymbol{A} = -\mu \boldsymbol{J},\ \nabla^2 \Phi + \omega^2 \mu\varepsilon\Phi = -\frac{\rho}{\varepsilon} \tag{21}$$

is finally calculated. The procedure for finding this solution is briefly described as follows. Expressing vector $\boldsymbol{A}$ by $\boldsymbol{A} = \boldsymbol{i}_x A_x + \boldsymbol{i}_y A_y + \boldsymbol{i}_z A_z$, equation (21) is reduced to a scaler wave equation as

$$\nabla^2 \psi + \omega^2 \mu\varepsilon\psi = -\delta(x)\delta(y)\delta(z), \tag{22}$$

where $\delta(\cdot)$ is Dirac's Delta function so that wave source $\boldsymbol{J}$ and the charge ρ are unit point sources. This wave equation is Helmholtz's equation $\nabla^2 \psi + \omega^2 \mu\varepsilon\psi = 0$ without the wave source. Solution of Helmholtz's equation is given by $\psi = \exp\left(-jkr\right)/\left(4\pi r\right)$ when only a diverging wave is taken. This wave is called three-dimensional Green's function in free space. The Green's function can be considered as a response function per unit. Taking the sum of all sources in whole space adding magnitudes, its response at a distance can be evaluated. Therefore, using Green's function, the vector potential is expressed as:

$$A = \int_V J(r') \frac{\exp(-jkr)}{4\pi r} dV'. \tag{23}$$

Strictly speaking, the standard method is obtained by expanding the Delta function to its series and taking it by contour integration.

Next, in order to obtain a desired electromagnetic field resulted from the vector potential, as shown in Figure 5, let us assume that the position vector of the wave source is denoted by $\boldsymbol{r}'$ and observation point is $\boldsymbol{r}$. Then, the vector potentials $\boldsymbol{A}$ and $\boldsymbol{F}$ are explicitly given as follows

$$\boldsymbol{A}(\boldsymbol{r}) = \frac{\exp(-jkr)}{4\pi r} \int_S \boldsymbol{J}(\boldsymbol{r}') \exp(-jk\boldsymbol{r}\cdot\boldsymbol{r}') dS', \tag{24a}$$

$$\boldsymbol{F}(\boldsymbol{r}) = \frac{\exp(-jkr)}{4\pi r} \int_S \boldsymbol{M}(\boldsymbol{r}') \exp(-jk\boldsymbol{r}\cdot\boldsymbol{r}') dS', \tag{24b}$$

where $\boldsymbol{r}' = \boldsymbol{i}_x x' + \boldsymbol{i}_y y'$, $dS' = dx'dy'$, and $\boldsymbol{F}$ is an electric vector potential due to the magnetic current. Origin of the coordinates is on the aperture, but it has to be noted that it is different from the coordinate system of a horn aperture that is a wave source. When the vector potential is obtained in this way, using the relationship between the vector potential and the radiation field in a far-region, they are given as

$$\boldsymbol{E} = -j\omega\mu\left(A_\theta \boldsymbol{i}_\theta + A_\phi \boldsymbol{i}_\phi\right), \qquad \boldsymbol{H} = j\omega\varepsilon\left(F_\theta \boldsymbol{i}_\theta + F_\phi \boldsymbol{i}_\phi\right), \tag{25}$$

*Figure 5. Aperture field method: Source in aperture and observation point in spherical coordinates (**r**, θ, ϕ) as radiation space*

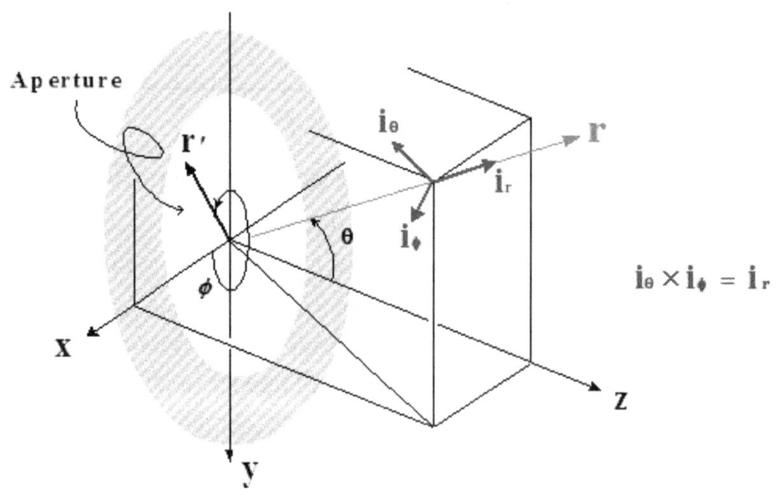

which are obtained by substituting $\boldsymbol{A}$ and $\boldsymbol{F}$ to the wave equation and neglecting the higher-order term. From these equations, the electric components are calculated as follows:

$$E_\theta = -j\omega\mu A_\theta - j\omega\varepsilon F_\phi, \qquad E_\phi = -j\omega\mu A_\phi - j\omega\varepsilon F_\theta. \tag{26}$$

Here, the relationship between the electric and the magnetic field in the far-region $\boldsymbol{E} = \sqrt{\mu / \varepsilon} \quad \boldsymbol{H} \times \boldsymbol{i}_r$ is applied, which means a far-region wave constitutes spherical TEM waves.

Referring to equation (24), Fourier transform of the aperture distribution $\boldsymbol{E}_a$ and $\boldsymbol{H}_a$ are calculated in advance as:

$$\boldsymbol{e}(\boldsymbol{r}) = \int_S \boldsymbol{E}_a(x', y') \exp(jk\boldsymbol{r} \cdot \boldsymbol{r}')\, dS' = \int_S \boldsymbol{E}_a(x', y') \mathrm{e}^{jk(ux'+vy')} dx'dy', \tag{27a}$$

$$\boldsymbol{h}(\boldsymbol{r}) = \int_S \boldsymbol{H}_a(x', y') \mathrm{e}^{jk(ux'+vy')} dx'dy'. \tag{27b}$$

These vectors are called the radiation vector. And, variables θ and ϕ are the zenith and the azimuth angle, respectively, have the relations with the rectangular coordinates as follows

$$\boldsymbol{r} = u\boldsymbol{i}_x + v\boldsymbol{i}_y + \cos\theta \boldsymbol{i}_z, \quad u = \sin\theta\cos\phi, \quad v = \sin\theta\sin\phi, \tag{28}$$

(see Figure 5). The radiation vector defined above takes the same form as equation (24). That is, Fourier transform of an aperture distribution gives a vector potential or a far-field. Relation between the radiation vector and the vector potential is obtained by using equation (20) as:

$$A(\boldsymbol{r}) = \frac{\exp(-jkr)}{4\pi r}\left(-h_y \boldsymbol{i}_x + h_x \boldsymbol{i}_y\right), \qquad \boldsymbol{F}(\boldsymbol{r}) = \frac{\exp(-jkr)}{4\pi r}\left(-e_x \boldsymbol{i}_x + e_y \boldsymbol{i}_y\right). \tag{29}$$

The relationship of the spherical and the rectangular coordinates are given as:

$$\boldsymbol{i}_x = u\boldsymbol{i}_r + \cos\theta\cos\phi \boldsymbol{i}_\theta - \sin\phi \boldsymbol{i}_\phi,$$

$$\boldsymbol{i}_y = v\boldsymbol{i}_r + \cos\theta\sin\phi \boldsymbol{i}_\theta + \cos\phi \boldsymbol{i}_\phi, \tag{30}$$

$$\boldsymbol{i}_z = \cos\theta \boldsymbol{i}_r - \sin\theta \boldsymbol{i}_{\theta,}.$$

so that equation (29) are modified as:

$$A(r)=\frac{\exp(-jkr)}{4\pi r}\begin{pmatrix} i_{\theta}\cos\theta\left(h_{x}\sin\phi-h_{y}\cos\phi\right) \\ i_{\phi}\left(h_{x}\cos\phi+h_{y}\sin\phi\right) \end{pmatrix}, \tag{31a}$$

$$F(r)=\frac{\exp(-jkr)}{4\pi r}\begin{pmatrix} i_{\theta}\cos\theta\left(e_{x}\sin\phi-e_{y}\cos\phi\right) \\ i_{\phi}\left(e_{x}\cos\phi+e_{y}\sin\phi\right) \end{pmatrix}. \tag{31b}$$

Finally, angle components of the radiation field is calculated using equation (26) as:

$$E_{\theta}=jk\frac{\exp(-jkr)}{4\pi r}\left\{e_{x}\cos\phi+e_{y}\sin\phi+\sqrt{\frac{\mu}{\varepsilon}}\cos\theta\left(h_{x}\cos\phi-h_{y}\sin\phi\right)\right\}, \tag{32a}$$

$$E_{\phi}=jk\frac{\exp(-jkr)}{4\pi r}\left\{\cos\theta\left(e_{y}\cos\phi-e_{x}\sin\phi\right)-\sqrt{\frac{\mu}{\varepsilon}}\left(h_{x}\cos\phi+h_{y}\sin\phi\right)\right\}. \tag{32b}$$

The $e_{x,y}$ and $h_{x,y}$ above indicate components in the subscript direction of a radiation vectors.

Thus, a radiation field can be calculated by Fourier transformed radiation vector of an aperture field distribution. Therefore, a typical antenna whose aperture is sufficiently larger than a wavelength has a high peak-value in the wave-number domain, which determines so-called beam-width. From Fourier theory, moreover, this fact means nothing but that all electromagnetic waves can be represented by the sum or the integration of plane wave spectrums with appropriate weight functions. When Fourier transformation of a wave source is calculated, the far-field distribution in an observation region can be found. In order to obtain antenna radiation far-field pattern, it is only necessary to perform Fourier transform to electromagnetic field or current on the antenna as a source. If this electromagnetic current is exactly obtained, the radiation pattern in far-region becomes to be accurate. In this chapter, the radiation pattern of the pyramidal horn antenna is formulated by using the aperture field method (AFM), which can be generalized to physical optics (PO). In radiation and scattering problems of electromagnetic theory, the PO method is a typical and powerful high-frequency technology (Kobayashi, 2011).

Far-Field of Sectoral Horn

At first, let us consider an H-plane sectoral horn. The aperture distribution of an H-plane sectoral horn is given by equation (18). The amplitude behaves as a cosine wave and reaches 0 on the wall. The phase proceeds in an arc in the *z*-direction, and the second order term is taken into account according to the flat aperture shape. This quadratic phase relationship leads to the introduction of Fresnel's integration, which appears later. Since only the E_x component of aperture distribution exists, the radiation vector of equation (27) is given as follows:

$$e_x(u,v)=\int_S E_x(x',y')\exp\{jk(ux'+vy')\}dx'dy'. \tag{33}$$

On the other hand, the relationship between an electric and a magnetic field in the far-region is $\boldsymbol{E}=-\sqrt{\mu/\varepsilon}\quad \boldsymbol{i}_r\times\boldsymbol{H}$, so the radiation vector has also the relation in $h_y=\sqrt{\varepsilon/\mu}\quad e_x$, $h_x=\sqrt{\varepsilon/\mu}\quad e_y=0$ in the far-region. Therefore, final equation (32) are calculated as:

$$E_\theta=jk\frac{\exp(-jkr)}{4\pi r}(1+\cos\theta)\cos\phi\cdot e_x, \tag{34a}$$

$$E_\phi=jk\frac{\exp(-jkr)}{4\pi r}(1+\cos\theta)\sin\phi\cdot e_x. \tag{34b}$$

Their vector expression is given as:

$$\boldsymbol{E}_H(\theta,\phi)=jk\frac{\exp(-jkr)}{4\pi r}(1+\cos\theta)\{\cos\phi\boldsymbol{i}_\theta+\sin\phi\boldsymbol{i}_\phi\}\cdot e_x. \tag{35}$$

The remaining problem is the integration in equation (34). Speaking from a conclusion, this integration can be given by Fresnel's integral defined by the following equation:

$$F(x)=C(x)-jS(x)=\int_0^x\exp\left(-j\frac{\pi}{2}t^2\right)dt. \tag{36}$$

The aperture distribution in equation (18) is substituted into the radiation integral in equation (33). Since there is no variation in the x-direction, its calculation is easy and the result is given as follows

$$e_x(u,v)=K_h(u)\int_{-A/2}^{A/2}\cos\frac{\pi y}{A}\exp\left\{-j\frac{(ky)^2}{2k\rho_1}\right\}\exp(jkvy)dy, \tag{37a}$$

where, $C_1=1$ and

$$K_h(u)=\int_{-B/2}^{B/2}\exp(jkux)dx=B\frac{\sin(kBu/2)}{kBu/2}. \tag{37b}$$

The variables are defined by $u = \sin\theta\cos\phi,\ \ v = \sin\theta\sin\phi$. The function $K_h(u)$ presents the pattern function due to uniform distribution and is called a sampling function, whose symbol is usually $(\sin x)/x = \mathrm{sinc}(x)$. When $\theta = 0$ or $\phi = \pi/2$, $K_h(u = 0) = B$. From this, it is found that the radiation in the plane $\phi = 0$ is completely dominated by this function, and $\phi = \pi/2$ is not involved at all.

Next, let us integrate equation (37) with respect to dy. In order to perform this, using Euler's formula to convert a trigonometric to an exponential function and separate the integral into two, then we can obtain as:

$$e_x(u,v) = \frac{K_h(u)}{2} \tag{38}$$

$$\bullet\left[\int_{-\frac{A}{2}}^{\frac{A}{2}} \exp\left\{-jk\left(\frac{y^2}{k\rho_1} - \frac{\pi y}{A} - vy\right)\right\}dy + \int_{-\frac{A}{2}}^{\frac{A}{2}} \exp\left\{-jk\left(\frac{y^2}{k\rho_1} - \frac{\pi y}{A} - vy\right)\right\}dy\right].$$

This can be done by solving the following integral for both terms

$$f(p,q,s) = \int_{-s}^{s} \exp\left(-jpt^2 + jqt\right)dt, \tag{39a}$$

which is similar to Fresnel's integral, so modifying as

$$f(p,q,s) = \exp\left(\frac{jq^2}{4p}\right)\int_{-s}^{s} \exp\left[-jp\left(t - \frac{q}{2p}\right)^2\right]dt, \tag{39b}$$

by variables transformation, $\sqrt{p}\left(t - q/2p\right) = \sqrt{\pi/2t}$, and changing the integration path as $\int_{-x'}^{x} = \int_{0}^{x} - \int_{0}^{-x'}$. Then, the integration $f(p, q, s)$ is derived as:

$$f(p,q,s) = \sqrt{\frac{\pi}{2p}}\exp\left(\frac{jq^2}{4p}\right)\left\{F\left[\sqrt{\frac{2p}{\pi}}\left(s - \frac{q}{2p}\right)\right] - F\left[\sqrt{\frac{2p}{\pi}}\left(-s - \frac{q}{2p}\right)\right]\right\}. \tag{39c}$$

It is found that the far-field radiation pattern can be expressed by Fresnel's integration, which is a well-defined function and its numerical code is found in many books and web. All that remains is to organize these expressions. This is summarized below, for convenience of numerical calculation, the function f has been redefined with the function g.

$$\boldsymbol{E}_H(\theta,\phi) = j\frac{\exp(-jkr)}{4\pi r}\frac{1+\cos\theta}{2}\{\cos\phi\boldsymbol{i}_\theta + \sin\phi\boldsymbol{i}_\phi\}\cdot K_h(\theta,\phi)$$

$$\cdot\left\{g\left(\frac{1}{2k\rho_1},\frac{\pi}{kA}+\sin\theta\sin\phi,\frac{kA}{2}\right)+g\left(\frac{1}{2k\rho_1},-\frac{\pi}{kA}+\sin\theta\sin\phi,\frac{kA}{2}\right)\right\}, \tag{40}$$

where all dimensions are normalized by the wave-number k.

Regarding the E-plane sectoral horn, the only difference between the E-plane and the H-plane horn is that the phase variation on the aperture occurs in the uniform direction of the electric field. The aperture distribution on the E-plane is shown by equation (19). From viewpoint of the shape, it can be calculated mechanically by imitating the H-plane horn. The component of radiation vector e_x is expressed as:

$$e_x(\theta,\phi) = \int_{-\frac{A}{2}}^{\frac{A}{2}}\int_{-\frac{B}{2}}^{\frac{B}{2}}\cos\frac{\pi y}{A}\exp\left[-jk\frac{(kx)^2}{2k\rho_2}\right]\exp\left[jk\sin\theta(x\cos\phi + y\sin\phi)\right]dxdy. \tag{41}$$

If the integral for dy is set to K_e corresponding to K_h in the H-plane, the following equation can be easily obtained:

$$K_e(\theta,\phi) = \frac{kA}{2}\left\{\frac{\sin\left(\frac{\pi}{2}+\frac{kA}{2}\sin\theta\cos\phi\right)}{\frac{\pi}{2}+\frac{kA}{2}\sin\theta\cos\phi}+\frac{\sin\left(-\frac{\pi}{2}+\frac{kA}{2}\sin\theta cos\phi\right)}{-\frac{\pi}{2}+\frac{kA}{2}\sin\theta\cos\phi}\right\}. \tag{42}$$

On the other hand, the integration for dx is as follows using Fresnel's integral:

$$e_x(\theta,\phi) = K_e(\theta,\phi)\int_{-\frac{B}{2}}^{\frac{B}{2}}\exp\left[-jk\left(\frac{x^2}{2k\rho_2}-x\sin\theta\cos\phi\right)\right]dx$$

$$= K_e(\theta,\phi)\,g\left(\frac{1}{2k\rho_2},\sin\theta\sin\phi,\frac{kB}{2}\right). \tag{43}$$

The vector expression for the E-plane sectoral horn is given as:

$$\boldsymbol{E}_E(\theta,\phi) = j\frac{\exp(-jkr)}{4\pi r}\frac{1+\cos\theta}{2}\{\cos\phi\boldsymbol{i}_\theta + \sin\phi\boldsymbol{i}_\phi\}K_e(\theta,\phi)\,g\left(\frac{1}{2k\rho_2},\sin\theta\cos\phi,\frac{kB}{2}\right). \tag{44}$$

$\boldsymbol{E}_E$ is not much different from $\boldsymbol{E}_H$, but the main-lobe is larger than that when their aperture dimensions are same. Therefore, in order to make a pyramidal horn with the same beam-width in both the vertical and the horizontal plane, the aperture length of the E-plane is usually shorter than that of the H-plane.

Now, let us consider the secondary phase distribution introduced when determining the aperture distribution of a finite horn, which is given by $\Delta y = -(ky)^2/(2k\rho_1)$ for the H-plane. This maximum value is given as:

$$\Delta y_{\max} = -\frac{1}{8}\frac{(kA)^2}{k\rho_1} = -\frac{1}{8}t_h \qquad \text{for} \qquad y = \frac{A}{2}. \tag{45}$$

Therefore, if the parameter is changed slightly to, the phase error of an aperture distribution can be understandable (Balanis, 1982). In addition, the approximation for the quadrature phase is applicable to the radiation pattern analysis such as a curved patch antenna (Kobayashi, 2018).

Far-Field Radiation Pattern for Pyramidal Horn

A pyramidal horn antenna has both E- and H-planes protruding from a waveguide aperture as shown in Figure 1. A sectoral horn has a sharp beam in the protruding plane, but the other plane remains to be broad. Therefore, the pyramidal horn is designed when it is desired to use a sharp beam on the both planes and a high gain antenna. Since E- and H-fields are independent each other, namely orthogonal, the pyramidal horn can be obtained by incorporating two sectoral horns. Some pyramidal horns have tapered inside-walls to reduce unwanted modes due to the discontinuous junction with the waveguide, however here let us continue to analyze a conventional straight- wall horn.

Geometric parameters of the pyramidal horn are $\rho_{h,e}$: flare lengths of the H- and E-planes, A, B, a, b: aperture lengths of H- and E-planes and the feeding waveguide, respectively. If five of these six parameters are determined geometrically, the rest can be obtained. The relation is given as:

$$\frac{A-a}{B-b} = \frac{A}{B}\frac{\rho_h^2 - (B/2)^2}{\rho_e^2 - (A/2)^2}. \tag{46}$$

For horn shapes with the same beam-width on E- and H-planes, the E-plane is usually wider if the aperture lengths are same, so that firstly we should determine the parameter A, ρ_e and then determine B. There is a special case where vertices of E and H-planes overlap, and it is found $A/B = a/b$. The vertex includes exactly a straight line, and in differential geometry it is called center of curvature and corresponds to a focal line (caustic). The so-called "focus" with a lens refers to this special case. The focal line of a horn antenna behaves like a wave source, and is a phase center of the radiation field. This phase center depends on distance from an observation point. When the observation point is sufficiently far from the aperture, the center of a geometric aperture becomes to be this phase center. Therefore, when used as a primary feeding antenna for a reflector antenna, the center of the physical aperture is generally placed at the focal point of the reflector. In addition, when a dielectric lens is attached to a horn aperture to improve antenna characteristics, the lens should be focused at the phase center of $\rho = 0$. And, when A

$= a$, $B = b$ in equation (46). This means that there is no overhang, that is, the horn aperture is the same to a feeding waveguide.

Now, the aperture distribution of the pyramidal horn can be assumed as follows, with two sectoral horns superimposed:

$$E_x(x,y) = C_P \cos\frac{\pi y}{A}\exp(-jk\Phi), \qquad H_\phi(x,y) = \sqrt{\frac{\mu}{\varepsilon}}E_x(x,y), \tag{47}$$

where Φ indicates the relative phase difference of each point on the aperture, which is given as:

$$\Phi = \frac{y^2}{\rho_1} + \frac{x^2}{\rho_2}. \tag{48}$$

Applying the above aperture distribution to equation (32), the following expression is derived as:

$$\boldsymbol{E}_P(\theta,\phi) = j\frac{\exp(-jkr)}{4\pi r}(1+\cos\theta)\{\cos\phi\boldsymbol{i}_\theta + \sin\phi\boldsymbol{i}_\phi\}e_x, \tag{49a}$$

$$e_x(\theta,\phi) = C_P\int_{-\frac{A}{2}}^{\frac{A}{2}}\int_{-\frac{B}{2}}^{\frac{B}{2}} k\cos\frac{\pi y}{A}\exp(-jk\Phi)\exp\left[jk\sin\theta(x\cos\phi + y\sin\phi)\right]dxdy. \tag{49b}$$

Integration of e_x is separated as follows, with $C_P = 1$,

$$e_x(\theta,\phi) = k\int_{-\frac{B}{2}}^{\frac{B}{2}}\exp\left(\frac{-jkx^2}{2\rho_e}\right)\exp(jkx\sin\theta\cos\phi)dx$$

$$\bullet\int_{-\frac{A}{2}}^{\frac{A}{2}}\cos\frac{\pi y}{A}\exp\left(\frac{-jky^2}{2\rho_h}\right)\exp(jky\sin\theta\sin\phi)dy. \tag{50}$$

This expression is exactly the same as $\boldsymbol{E}_H$ and $\boldsymbol{E}_E$. Therefore, using function $g(p, q, s)$, $\boldsymbol{E}_P$ in equation (49) is easily given as:

$$\boldsymbol{E}_P(\theta,\phi) = j\frac{\exp(-jkr)}{4\pi r}\frac{1+\cos\theta}{2}\{\cos\phi\boldsymbol{i}_\theta + \sin\phi\boldsymbol{i}_\phi\}g\left(\frac{1}{2k\rho_2}, \sin\theta\cos\phi, \frac{kB}{2}\right)$$

Figure 6. Pyramidal horn pattern

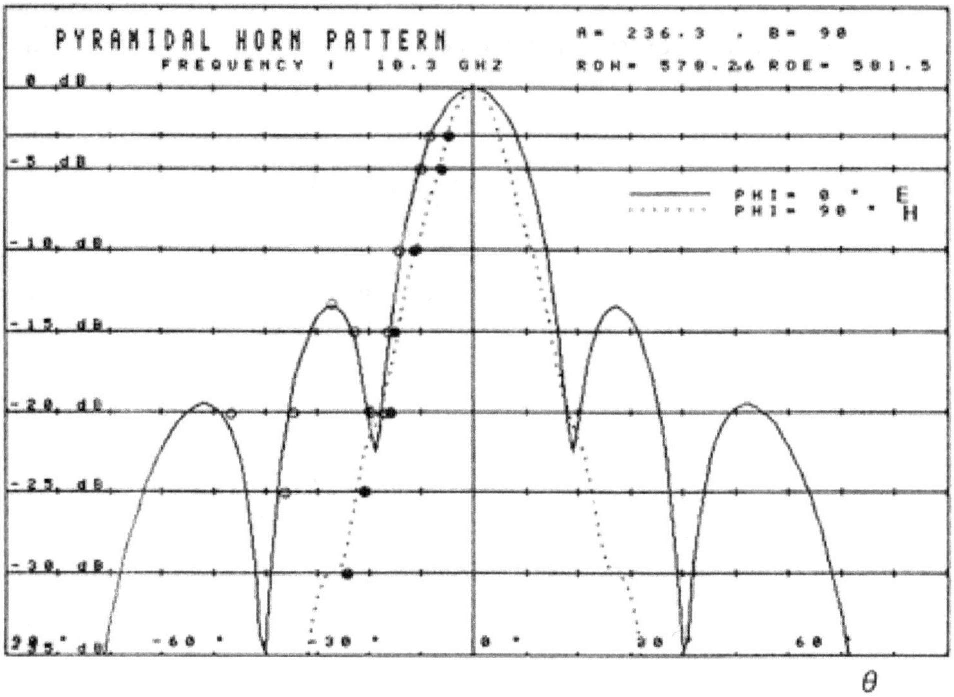

$$\bullet\left\{ g\left(\frac{1}{2k\rho_1},\frac{\pi}{kA}+\sin\theta\sin\phi,\frac{kA}{2}\right)+g\left(\frac{1}{2k\rho_1},-\frac{\pi}{kA}+\sin\theta\sin\phi,\frac{kA}{2}\right)\right\}. \tag{51}$$

After all, the numerical calculation for equation (51) is based on Fresnel's integration. The calculation algorithm is not so difficult by using Taylor's expansion and more convergent asymptotic expansion. Let us examine how much these formula here coincides with measurement. Figure 6 compares calculated and measured values for a relatively large pyramidal horn antenna. The shape parameters are A = 236.3, B = 98.0, $\rho_h = 578.3$, $\rho_h = 581.5$ [mm], and measuring frequency is 10.3 [GHz]. In the figure, it is recognized that the angle is slightly shifted in the second side-lobe in the E-plane pattern. For small horn antennas whose aperture is less than 2 wavelengths, calculated and measured results almost agree, and especially at their main-lobe region.

The calculation of the pyramidal horn can be simply obtained from equation (51), however, the most important thing here is understanding the radiation mechanism and the mathematical derivation under the electromagnetism.

Diagonal Horn and Double-Ridged Small Aperture Horn

The diagonal horn antenna is a modified type of pyramidal horn with an electric field direction at ϕ = 45° as shown in Figure 7. This is used to improve the symmetry of the beam, but as will be understood later, there are some inferior points in the productivity of the feeding system, and there are few applications. However, the calculation can be derived immediately from the pyramidal horn. The diagonal

Figure 7. Outline of diagonal horn antenna with square aperture

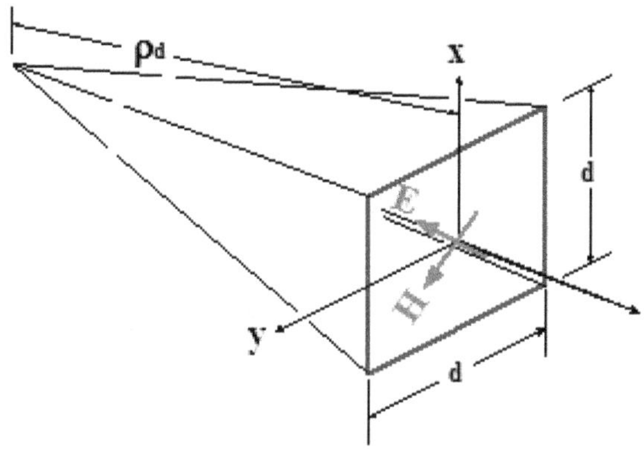

horn excites an electric field along one diagonal direction. Therefore, it can be expected that the power feeding system is slightly complicated.

The aperture distribution is determined in the following manner. Here, the diagonal aperture shape with a relatively smooth overhang angle is a square, and the aperture distribution is given by the vector combination of TE_{10} and TE_{01}modes shown in Figure 8. Since the boundary condition is $x = \pm d/2, y = \pm d/2$ on the waveguide wall, it can be given as

$$\boldsymbol{E}_a = C_d \left\{ \cos\frac{\pi y}{d}\boldsymbol{i}_x + \cos\frac{\pi x}{d}\boldsymbol{i}_y \right\} \qquad \exp(-jk\Phi), \tag{52}$$

where C_d is an amplitude of the propagating wave, and Φ, which indicates the phase displacement at each point of the aperture, should be assumed as:

$$\Phi = -\frac{x^2 + y^2}{2\rho_d}. \tag{53}$$

Figure 8. Aperture field of diagonal horn antenna with square aperture

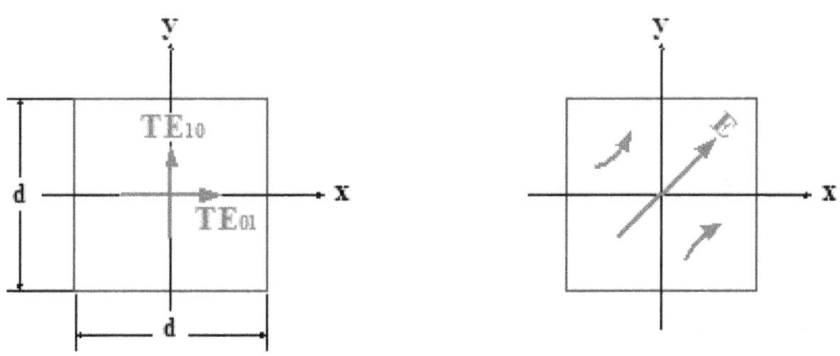

Figure 9. Feeding system for diagonal horn antenna

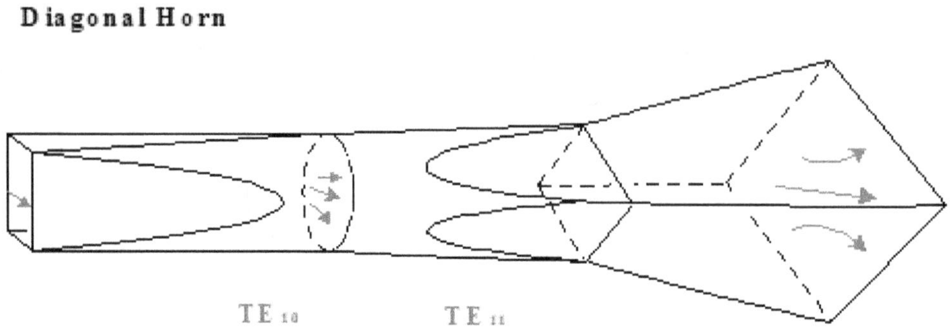

The vector magnitude is $\left|\boldsymbol{E}_a\right| = C_d\sqrt{\cos^2(\pi x/d) + \cos^2(\pi y/d)}$. From this, the magnitude of the electric field distribution can be grasped to some extent, and it can be expected to be similar to TE_{11} mode of a circular waveguide. Therefore, as shown in Figure 9, the feeding system of this antenna is changed to TE_{11} of the circular waveguide through TE_{10} of the rectangular waveguide. It is quite difficult to physically realize this conversion, and this is the biggest reason why this antenna is not widespread.

As expected, since the vector that gives the diagonal aperture distribution is generated from orthogonal TE_{10} and TE_{01} modes, the calculation can be considered separately. In other words, it was sufficient to consider only the component e_x for the radiation vector corresponding to TE_{10} mode, and it is found that only the variables need to be changed due to symmetry considering the component e_y in this time. In the pyramid horn expression, $\phi = \pi/4$ is considered to be the diagonal E-plane and $\phi = 3\pi/4$ to be H-plane. Substituting these values, the far-fields of the diagonal horn are calculated as follows:

$$\text{H-plane} \qquad E_\theta = \frac{jk\exp(-jkr)}{4\pi r}\frac{1+\cos\theta}{\sqrt{2}}\left(-e_x + e_y\right) = 0, \tag{54a}$$

$$E_\phi = \frac{jk\exp(-jkr)}{4\pi r}\frac{-2(1+\cos\theta)}{\sqrt{2}}e_x\left(\phi = \frac{3\pi}{4}\right), \tag{54b}$$

$$\text{E-plane} \qquad E_\theta = \frac{jk\exp(-jkr)}{4\pi r}\frac{2(1+\cos\theta)}{\sqrt{2}}e_x\left(\phi = \frac{\pi}{4}\right), \tag{54c}$$

$$E_\phi = \frac{jk\exp(-jkr)}{4\pi r}\frac{1+\cos\theta}{\sqrt{2}}\left(-e_x + e_y\right) = 0. \tag{54d}$$

Therefore, the vector radiation field is as follows:

Figure 10. Ultra-broadband double ridged horn antenna with small aperture, developed for active phased array antenna

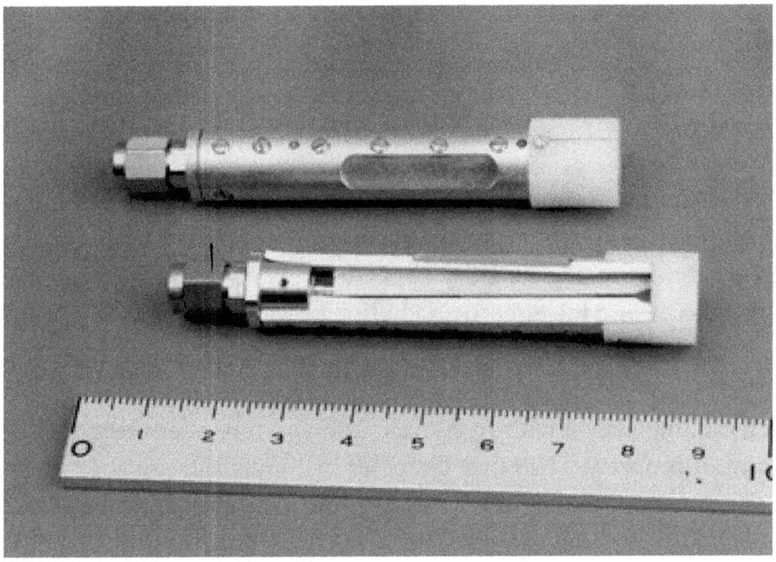

$$\boldsymbol{E}_D(\theta,\phi) = \boldsymbol{i}_\theta \sin\left(\phi+\frac{\pi}{4}\right)E_\theta\left(\phi+\frac{\pi}{4}\right)+\boldsymbol{i}_\phi \sin\left(\phi-\frac{\pi}{4}\right)E_\theta\left(\phi-\frac{\pi}{4}\right). \tag{55}$$

Furthermore, a little simpler expression is obtained by the first-order approximation of the phase distribution. Naturally, the approximation accuracy deteriorates. The function $g(p, q, s)$ is set as $p = 0$, which is same to ignore the phase variation in equation (53). At this time, the related expressions are simplified as follows:

$$g(p=0,q,s) = \int_{-d/2}^{d/2} \exp(jqy)dy = \frac{2\sin\left(\frac{dq}{2}\right)}{q}, \tag{56}$$

so that we can obtain the follows:

$$e_x = \frac{C_d}{2}\mathrm{sinc}\left(\frac{1}{2}kd\sin\theta\cos\phi\right)\left\{\mathrm{sinc}\left(\frac{\pi}{2}+\frac{1}{2}kd\sin\theta\sin\phi\right)+\mathrm{sinc}\left(-\frac{\pi}{2}+\frac{1}{2}kd\sin\theta\sin\phi\right)\right\},$$

$$e_y = e_x\begin{Bmatrix}\cos\phi \to \sin\phi \\ \sin\phi \to \cos\phi\end{Bmatrix}. \tag{57}$$

This first order approximation can naturally be applied to a pyramidal horn. However, since the numerical calculation of Fresnel's integral is not so difficult with the current computer, the second order approximation is recommended as it is.

Next, let us show a practical application example. Figure 10 shows a development of a small size and ultra-broadband horn antenna with the double ridged waveguide inside (Kobayashi, 1998). This antenna has many features as follows:

1. Broadband used at full frequency band of X-Ku band: specific band-width 77%.
2. Ultra small size for active phased array applications: aperture diameter 12mm.
3. Interchangeability for active phased array applications: end-launcher connecting.
4. High antenna gain to ensure the effective radiation power: $G_e = 1.0 \sim 2.0[\mathrm{dB_i}]$.
5. Excellent power durability against high microwave power: more than 2 [W].

In order to match the impedance with free space, the aperture is covered with a dielectric cap that also serves as waterproof property. The outer diameter including the cap is 12 mm, and the waveguide portion is 10 mm. When normalized by the wavelength of the lower limit frequency of 8 [GHz], 0.25 wavelength. This is a condition to suppress grating-lobes for beam scanning of around 50° in the phased array antenna. Furthermore, since the transmission of this phased array is an active type in which microwave amplifiers are distributed and arranged for each element, it is an end launcher type that is connected in the axial direction of the antenna. For this end launcher, the coaxial center conductor is connected to one of the waveguide ridges to form a cavity resonance room, and the mode conversion from the coaxial waveguide is performed. The ridge is a fin-like plate provided on the electric field surface of the waveguide, and there are a single and a double ridge type. The purpose of this ridge equipping is to lower the cutoff frequency of the waveguide, and the size of the waveguide can be reduced equivalently. When there is no aperture cap, which is composed with a low and high density dielectrics, the antenna gain is $0[\mathrm{dB_i}]$ or less in the entire band, and a large ripple occurs in the radiation pattern.

Approximating the aperture of the double ridged horn to a pyramidal horn without the ridges, the calculation results are relatively consistent with the overall radiation pattern compared to the measured one. The calculation naturally ignores the effect of the ridge near the aperture. Figure 11 shows the measured antenna patterns when the cap is attached. In the figure, the patterns for the E-plane, the 45° diagonal plane, and the H-plane from the widest beam-width among the three patterns. It can be found that the pattern in a 45° plane is roughly the midpoint between the E and H -planes. The electromagnetic field in the waveguide with the ridge is in a hybrid mode, and it is necessary to take a procedure such as expansion by eigenfunction. So, for only antenna designing, it is better and easy to employ the current electromagnetic simulators.

Pyramidal and Corrugated Conical Horn Fabricated by 3D-Printer

In contrast to an ordinary printer that prints on paper, a 3D-printer is a device that produces solid objects directly from CAD and CG data. An aperture antenna such as the pyramidal or the corrugated horn can be manufactured using 3D-printer which is a hot melt lamination method by ABS resin as a material. Since the surface of the fabricated antenna is still dielectric resin, some conductive paint is applied or sprayed to ensure the reflectivity at an operating frequency. After all, the electrical performance of the antenna

Figure 11. Measured Radiation patterns by ultra-broadband double ridged horn antenna with small aperture in Figure 10, from top, 8, 13, 18 [GHz]

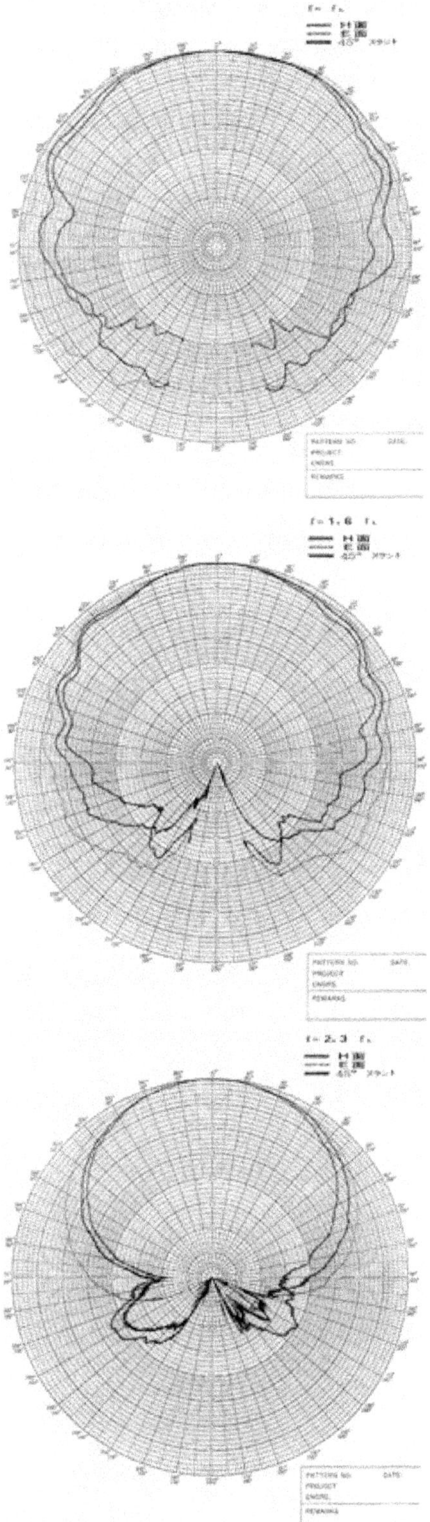

by 3D-printer is based on the reflectivity by this conductive paint. In order to confirm the reflectivity by the resin with the metal coating, the aluminum flat metal is attached to the waveguide aperture of the coaxial/waveguide adaptor, and then, which is replaced with the painted resin plate. The difference in return loss is measured. As a result, the difference was within 0.1 [dB], which ensures that it exhibits the normal metal characteristics (Kobayashi, 2014). In recent years, it has become possible to produce a metal object by 3D-printer, and in this case, except for special antennas that require high dimensional accuracy, almost relatively small aperture antennas can be fabricated by 3D-printer without no painting. The proposal of a low-loss resin material in microwave-band is also expected. In the near future, the patch antennas currently produced by etching on a printed circuit boards will be manufactured at low cost in a short time by simultaneous processing with the previous mentioned metal material 3D-printer.

Now, theoretical radiation patterns of a pyramidal horn is shown in Figure 12, whose dimensions are A = 77.0, B = 62.0, ρ_h = 150.0, ρ_e = 145.0 [mm] and calculated by equation (51). The rectangular waveguide for feeding is assumed to be a standard product (WR-90), and the frequency in the figure is 8, 10, and 12 [GHz], which are the lower, center, and upper frequency of X-band. Figure 13 shows an

Figure 12. Theoretical radiation patterns of pyramidal horn, frequency 8, 10, 12 [GHz] from upper

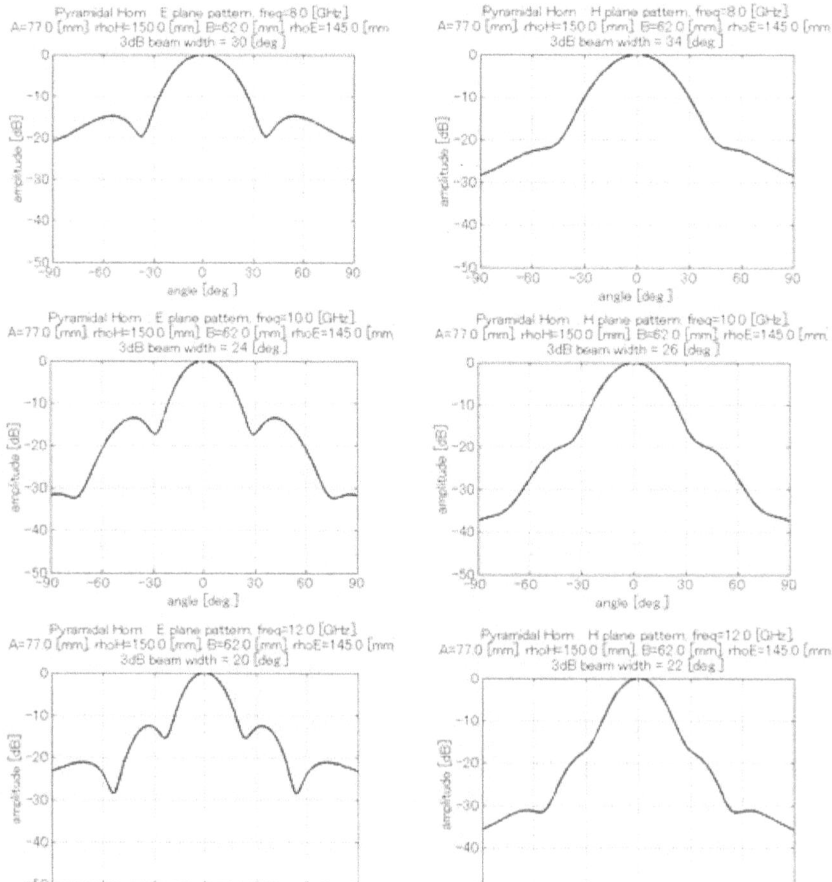

Figure 13. Photos of resin pyramidal horn fabricated by 3D-PRT (left: CAD model, centre: before painting, right: after painting)}

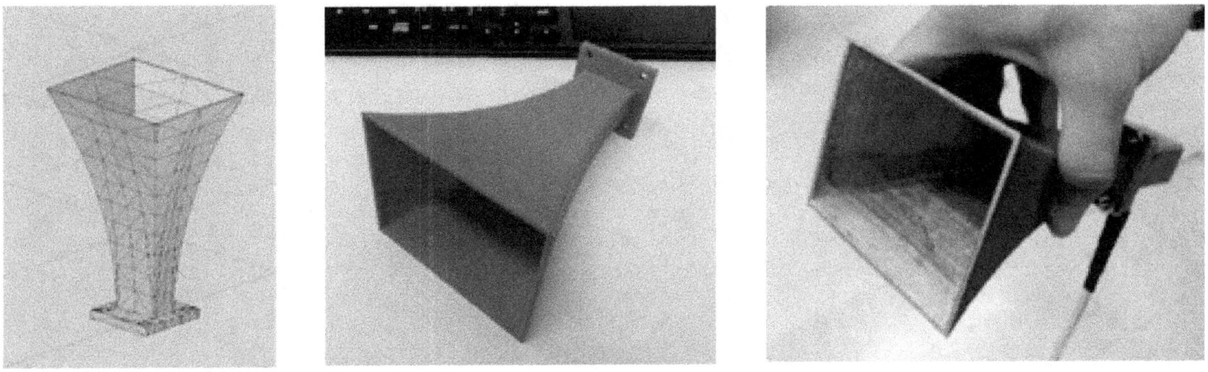

external view of the prototype pyramid horn with the above dimensions. In the figure, it refers to the outline of the CAD model (left), before applying conductive paint (center), and after applying (right). The wall of the horn has a smooth curved surface inside. This equation is given as:

E-plane: $$x = 0.0017z^2 - 0.0191z + 5.1, \tag{58a}$$

H-plane: $$x = 0.0013z^2 + 0.0388z + 11.4. \tag{58b}$$

Although the straight wall employs a smooth quadratic surface to improve antenna vswr, it has been confirmed by the numerical matrix simulators that there is no significant effect on the radiation pattern other than the side-lobes in the wide angle region compared with a flat surface. Figure 14 is the measured far-field patterns corresponding to the theoretical radiation field of Figure 13. The left side of the figure is the E-plane, and the right is the H-plane pattern and cross- polarization characteristics. The measurement frequency is 8, 10, 12 [GHz] from the top. In all patterns, the maximum value is normalized to 0 [dB]. Although there is a difference in the side-lobe etc. compared to the theoretical value, the half power beam-width etc. are almost the same. The H-plane pattern also shows the characteristics of cross-polarization with respect to co-polarization. The levels of co-polarization are about -20 to -30 [dB] from the co-polarization, which is considered to be same to a metal horn.

The corrugated horn, which is a typical hybrid-mode antenna, was originally developed with the aim to reduce the spillover of a reflector antenna. It is a high-grade antenna with many advantages such as low cross-polarization component, axisymmetric beam, and wider band-width characteristics (Lo, 1988).

This hybrid-mode is realized by loading a corrugation on the inner wall. Typical propagation mode is HE_{11} (TE_{11} + TM_{11} + others), so that a cross-polarization component of the aperture distribution can be almost eliminated. When the corrugation height is approximately 1/4 wavelength, HE_{11} mode has a Gaussian's distribution in the radial direction and an axisymmetric shape with no variation in the circumferential direction [Rudge, 1986]. Figure 15 illustrates this mode. Since Fourier transform of Gaussian's distribution results in Gaussian's one, the far-field pattern can be expected to exhibit low side-lobe characteristics, and the radiation beam which is symmetric with respect to the antenna boresight-axis and insignificant cross-polarization is realized.

Figure 14. Measured radiation patterns of pyramidal horn in Figure 13, frequency 8, 10, 12 [GHz] from upper

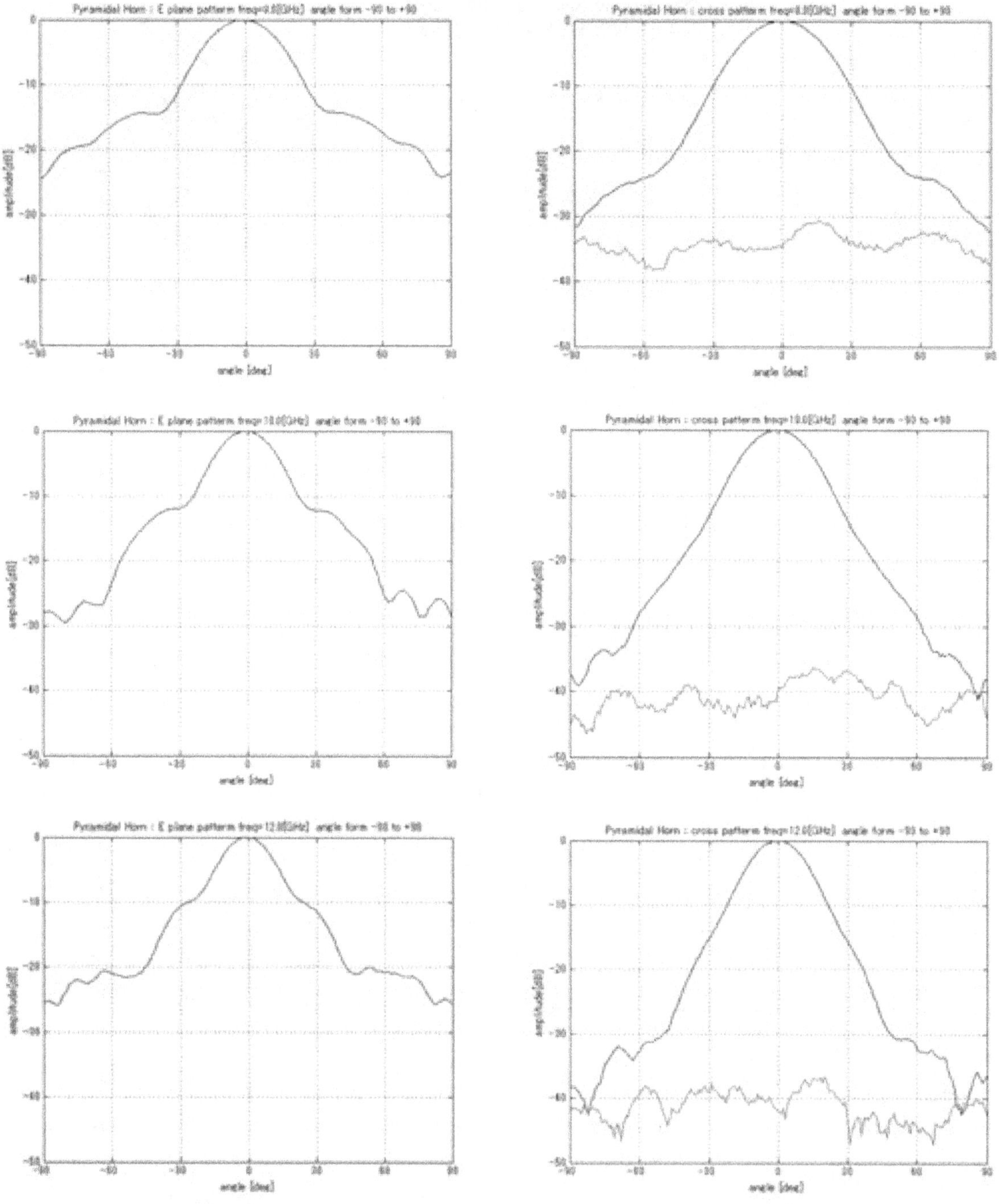

Figure 15. Hybrid HE_{11} mode generation

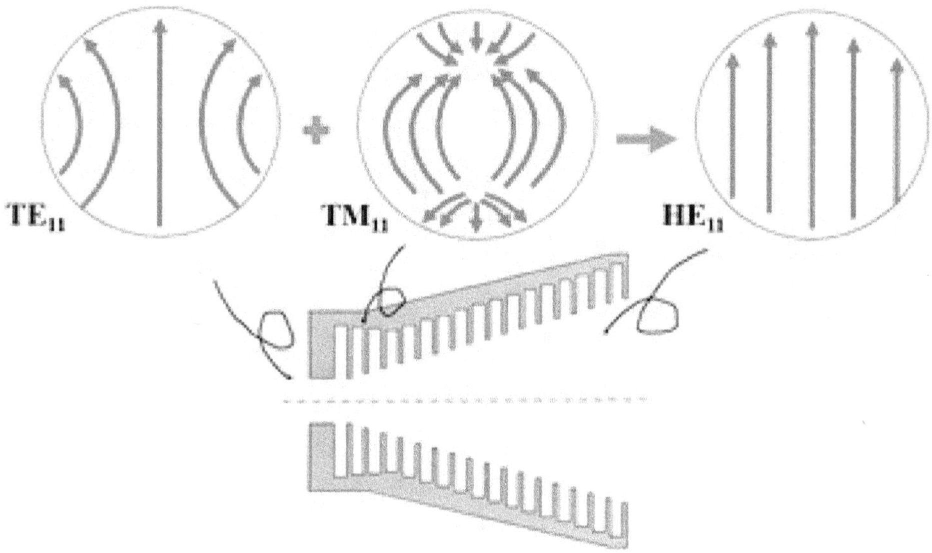

Here, using the design formula described in (Balanis, 2008), a prototype of corrugated conical horn is fabricated by 3D-printer. As shown in Figure 16, there are two types of corrugated conical horns, one with a corrugated slot in the radial direction along the horn wall described above: Figure 16(a), the other, slots are provided in a ring shape with respect to the central axis of the horn: Figure 16(b). The former is a relatively high gain antenna, and the latter is a choke-type circuit applied to the antenna aperture which is suitable for low and medium gain antenna. If the antenna operating center wavelength is λ and the antenna gain is $G_{\text{in dB}_\text{i}}$, the following expression can be obtained.

(a) Hybrid-mode type: $13 \le G_{\text{in dB}_\text{i}} \le 22$ (59)

$$a_i = \frac{3\lambda}{2\pi}, p = \frac{\lambda}{8}, w = \frac{4p}{5}, L = pN_{\text{slots}}, N_{\text{MC}} = 5,$$

$$N_{\text{slots}} = \text{nearest integer of } \frac{4a_0}{p},$$

$$\alpha = 192.351 - 17.7364G_{\text{in dB}_\text{i}} + 0.61362\left(G_{\text{in dB}_\text{i}}\right)^2 - 0.007712\left(G_{\text{in dB}_\text{i}}\right)^3,$$

$$a_0 = \left(8.72704 - 0.740515\alpha + 0.0295435\alpha^2 - 0.00055165\alpha^3 + 0.00000387765\alpha^4\right)\lambda$$

Figure 16. Corrugated conical horn with linear corrugation profile

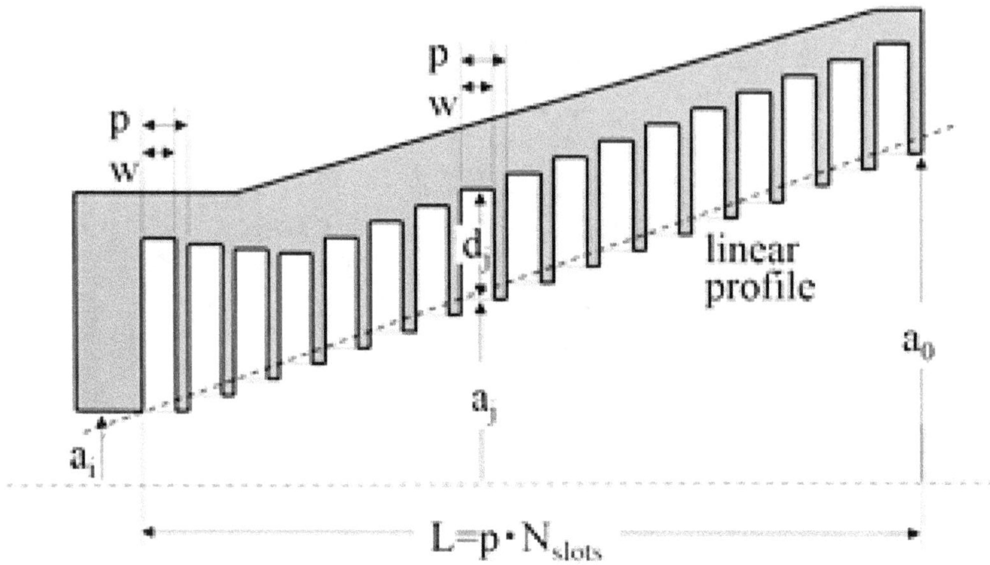

$$a_j = a_i + (j-1)\frac{a_0 - a_i}{N_{\text{slots}} - 1} \qquad \text{for} \qquad 1 \le j \le N_{\text{slots}}$$

$$d_j = \left\{0.42 - \left(0.42 - \frac{1}{4}\exp\left[\frac{1}{2.114(2\pi a_j/\lambda)^{1.134}}\right]\right)\frac{j-1}{N_{\text{MC}}}\right\}\lambda \quad \text{for} \qquad 1 \le j \le N_{\text{MC}}$$

$$d_j = \frac{\lambda}{4}\exp\left[\frac{1}{2.114(2\pi a_j/\lambda)^{1.134}}\right] \qquad \text{for} \qquad N_{\text{MC}} \le j \le N_{\text{slots}}.$$

(b) Choke type: $10.5 \le G_{\text{in dB}_i} \le 14.5$ (60)

$$a_i = \frac{3\lambda}{2\pi}, p = \frac{\lambda}{8}, w = \frac{4p}{5}, L = pN_{\text{slots}}, \theta = 45^\circ,$$

N_{slots} = nearest integer of

$$-343.325 + 84.7229 G_{\text{in dB}_i} - 6.99153\left(G_{\text{in dB}_i}\right)^2 + 0.194452\left(G_{\text{in dB}_i}\right)^3,$$

Figure 17. CAD model and dimensions of corrugated conical resin horn: type (a)

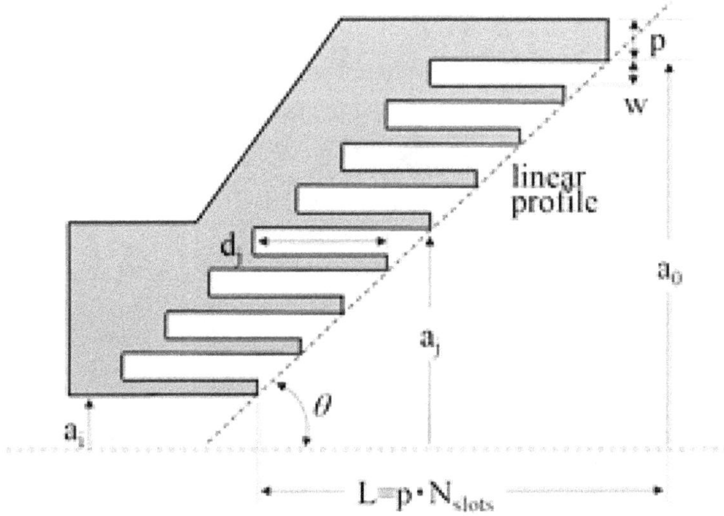

$$a_j = a_i + jp \quad \text{for} \quad 1 \le j \le N_{\text{slots}},$$

$$d_j = \frac{\lambda}{4}\exp\left[\frac{1}{2.114\left(2\pi a_j/\lambda\right)^{1.134}}\right] \quad \text{for} \quad N_{\text{MC}} \le j \le N_{\text{slots}}.$$

(a) Hybrid-mode type with radial corrugated slots
(b) Choke type with axially corrugated slots

Figure 17 shows a prototype CAD model and dimensions of a conical horn with a hybrid-mode type corrugated: type (a). The number of stages is $N = 10$ and the central operating frequency is 10 [GHz]. The shape of the corrugation is a linear profile, and the aperture and axial length are 80 and 125 [mm], respectively. Figure 18 shows an external photograph of the corrugated conical horn fabricated by 3D-printer together with an adaptor from the rectangular to the circular waveguide. The measurement result of the input impedance shows a good characteristic of -15 [dB] or less in the entire 8-12 [GHz] range. On the other hand, if the insertion loss is simply evaluated by closing the aperture with a metal flat plate, it is large in the low-range frequency. In fact, it is found later that the antenna pattern also has a broken beam, which indicates that the proper transmission mode has not been generated.

Figure 19 shows the measured far-field pattern of the above-described corrugated conical horn. The left side is an E-plane, the right side is H-plane patterns, and the cross-polarization characteristics are also shown on the H-plane patterns. The rectangular waveguide at the input port is WR-90, whose guaranteed band-width for TE_{10} mode only is 8-12 [GHz]. However, since the generation of higher-order modes is considered to be smaller than that of rectangular waveguide due to the conical waveguide with

Figure 18. Outline after painted of corrugated conical resin horn: type (a)

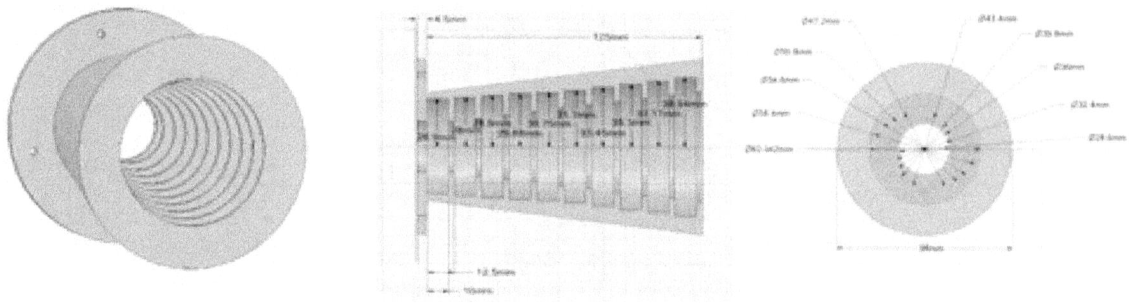

Figure 19. Measured radiation field of corrugated conical horn in Fig.18, frequency 10, 13, 15 [GHz] from upper

corrugation, the measurement frequencies are 10, 13, and 15 [GHz]. The E- and the H-plane patterns are normalized with the maximum value being 0 [dB]. It can be found that the side-lobe characteristics of the E-plane, which is remarkable in the pyramid horn, are greatly improved. It can also be seen that the beam symmetry in both planes and the frequency characteristics of the beam-width exhibit the corrugated antenna. On the other hand, the cross-polarization characteristics are approximately -25 [dB] from the co-polarization, leaving room for further improvement in the formation of the EH_{11} mode.

Next, Figure 20 shows a prototype CAD model of the same conical horn: type (b), which is a choke circuit corrugated horn. As with the hybrid-mode type described above, the center operating frequency is 10 [GHz] and the number of stages is $N = 11$. The aperture and axial length are 112.5 and 60.5 [mm], respectively. The input impedance is less than -15 [dB] over the entire 8-12 GHz range and exhibits good characteristics. Figure 21 is a photograph showing a pattern measurement scene of the resin antenna. Figure 22 shows the measured far-field pattern of the above choke type corrugated resin horn. The left side of the figure is the E-plane, the right is the H-plane pattern, and the cross-polarization characteristics are also shown on the H-plane patterns. The rectangular waveguide at the input end is WR-90

Figure 20. CAD model and dimensions of corrugated conical resin horn: type (b)

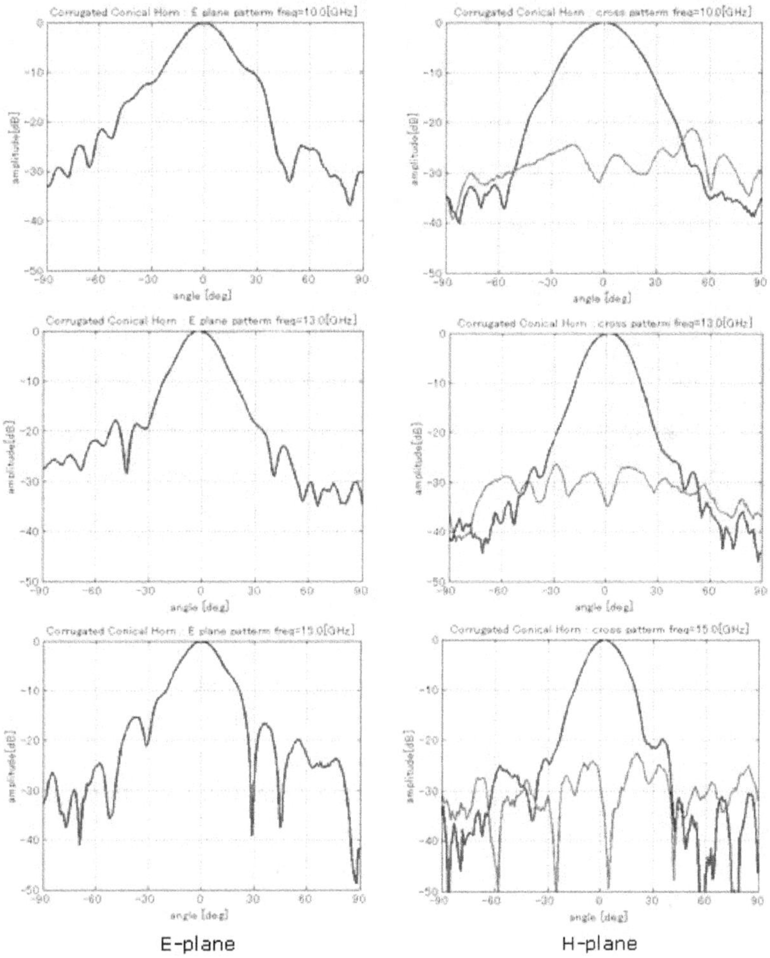

Figure 21. Outline after painted of corrugated conical resin horn: type (b)

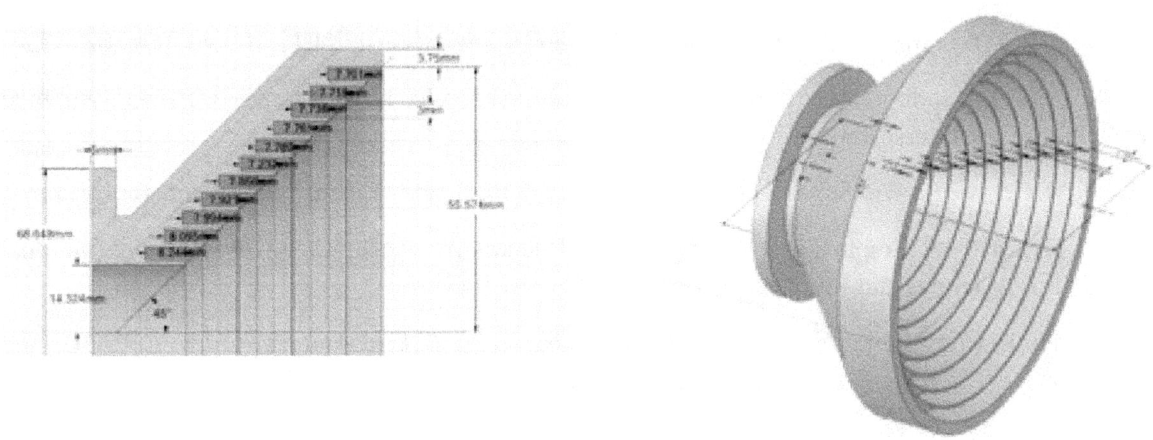

Figure 22. Measured radiation field of corrugated conical horn in Fig.21, frequency 8, 10, 12 [GHz]

standard, and the frequency band is 8-12 GHz. The E- and H-plane is normalized with the maximum value being 0 [dB]. Although not shown in the figure, the input impedance is approximately -15 [dB] or less. Compared to the hybrid-mode type described earlier, the choke type corrugated horn can reliably apply the paint at the root of the corrugated fin. The side-lobe characteristics of the E-plane unique to the pyramid horn are greatly improved in the same way as the hybrid-mode type, and the symmetry of the beam on both sides and the frequency characteristics of the beam-width also exhibit the features of the corrugated antenna. On the other hand, the cross-polarization characteristics are about -20 dB from the co-polarization, which is not better than the pyramid horn. One of this reason is presumed that the conductive paint is sufficiently applied or not.

REFERENCES

Balanis, C. A. (1982). *Antenna Theory -Analysis and Design*. NY: Harper & Row.

Balanis, C. A. (2008). *Modern Antenna Handbook*. John Wiley & Sons. doi:10.1002/9780470294154

Chang, K. (1989). Handbook of Microwave and Components. Vol. I, Microwave Passive & Antenna Components, Wiley Interscience.

Kobayashi, H. (2011). *Electromagnetic Wave in Space-Domain*. Tokyo, Japan: Sakura-tech. (in Japanese)

Kobayashi, H. (2018). A Simple Expression for Curved Rectangular Patch Antenna Pattern. International Symposium on Antennas and Propagation (ISAP2018), A8_1007.

Kobayashi, H., Kawamura, R., Yamada, Y. (2014). Fabrication of Horn Antenna using 3D Printer and its Evaluation. *IEICE Tech. Rep., 114*(193), AP2014-72, pp. 1-6. (in Japanese)

Kobayashi, H., & Tokumitsu, Y. (1998). The Element Antenna, Phase Shifter and Amplifier for Wide-band Systems. In *Proceedings of 1998 China-Japan Joint Meeting on Microwaves (CJMW'98)*, pp. 283-286.

Lo, Y. T., & Lee, S. W. (1988). *Antenna Handbook*. New York: Van Nostrand Reinhold. doi:10.1007/978-1-4615-6459-1

Rudge, A. W., Milne, K., Olver, A. D., & Knight, P. (1986). *The handbook of antenna design*. IEE UK.

Silver, S. (1949). *Microwave Antenna Theory and Design. Radiation Laboratories Series*. McGraw Hill.

Chapter 9
Microstrip Patch Antenna

Kim Huat Yeap
Westlake International School, Malaysia

ABSTRACT

This chapter elaborates in detail on the microstrip patch antenna, which is widely utilized in the receivers of radio telescopes, as well as in the wireless communication industry today. Several models have been developed to analyze and design the patch antennas. The three most common ones are the transmission line model, the cavity model, and the Method of Moments model. Apart from this, the important parameters used in characterizing the patch antenna are also covered, which are its gain, efficiency, directivity, radiation pattern, return loss, bandwidth, and polarization. This is followed by the introduction of the radiation regions, which are basically classified as the Fresnel region and the Fraunhofer region. Finally, the dual-frequency microstrip patch antenna is introduced. Three popular approaches adopted for the design are orthogonal-mode polarization, multi-layer patching, and reactive loading.

INTRODUCTION

Be it leading corporate or mere civilian, the fact that many could hardly keep track with the concatenated release of wireless gadgets is a sheer witness of how rampancy technological advancement is altering in the telecommunication industry today. Indeed, such occurrence is attributed to the rapid evolution of the microstrip patch antenna (or simply, patch antenna).

Though its debut could be traced back as early as the 1950s (Deschamps, 1953; Gutton, 1955), the patch antenna only received considerable attention by the 1970s (Howell, 1975; Derneryd, 1976, 1978, 1979a, 1979b; Shen *et al.*, 1977; Agrawal & Bailey, 1977; Long & Walton, 1979; Uzunoglu *et al.*, 1979; Bailey, 1979), as transition to wireless mediums commenced. It became the prime subject of study since then. Extensive development has been conducted and designs of various geometries have been produced (Wong, 2002; Lee & Chen, 1997).

Among the shapes most popularly adopted for the patch antenna design is the rectangular patch, as shown in Figure 1. This is due to its ease of analysis and fabrication. It is a narrowband, wide-beam antenna fabricated by photo-etching a thin conductive patch onto the top layer of an insulating dielectric substrate (Richards, 1988). Meanwhile, the bottom side is continuously electroplated with a conductive

DOI: 10.4018/978-1-7998-2381-0.ch009

platform to act as its ground plane. Consequently, the substrate is embraced in a sandwich-like structure. A feed, usually in the form of a microstrip or a coaxial probe, is also connected to the patch as a transmission channel to the external circuitries. The conductive layers are usually made of copper or gold material while the substrates are commonly made of the Flame-Retardant level 4 (FR4), Duroid, or polytetrafluoral ethylene (PTFE) material.

A rectangular patch antenna could be viewed as an open-ended transmission line which is significantly widened at its end, such that when electromagnetic (EM) wave propagates along it, the energy is primarily radiated into the surrounding at the non-homogeneities of the line. Deliberately sized and shaped, such antenna is geometrically corresponsive to the maximum radiation of the desired wavelength. Its radiation is an effect of a strong horizontal component of electric field intensity vector at its edges due to the current distribution source. With the ground plane acting as a reflector with null potential, the rearward radiation is limited and the gain is increased in the forward direction. Thereon, the maximum radiation pattern is normal to the patch and a broadside radiator is designed.

Advantages

The overwhelming popularity the patch antenna received is absolutely non-coincidental. It is in fact, the many advantages incorporated in its appearance which have placed it under the core of research. Namely, the patch antenna is small sized and light weight (Yang *et al.*, 2008), offering a low profile well-suited for high-performance airframes. Its versatility and flexibility enable it to produce a wide variety of patterns and polarizations, depending on the mode of excitation and the shape of the patch (Boualleg & Merabtine, 2005). It is also conformable to non-planar surfaces where its curved implementations can be made to conform to aircraft hulls (Wong, 1999). Mechanically robust but inexpensive, the patch antenna is well producible (Stutzman & Thiele, 1998), since it is manufactured the same way as a printed circuit board (PCB).

Further still, by adding appropriately-loaded elements, such as pin/varactor diodes between the patch and its ground plane (Richards & Long, 1986a), as well as by changing the bias voltage on the diodes (Richards & Long, 1986b), the patch antenna can easily vary its resonant frequency, polarization, impedance, and other parameters accordingly (Richards & Lo, 1983; Richards *et al.*, 1981; Bhartia & Bahl, 1982; Purchine *et al.*, 1994). This subsequently, offers an ease of installation and integration into microwave circuitries, avoiding the additional need of transmission lines, connectors, and symmetrization circuits. Such planar technology is well-compatible with Monolithic Microwave Intergrated Circuit (MMIC) and

Figure 1. Rectangular microstrip patch antenna

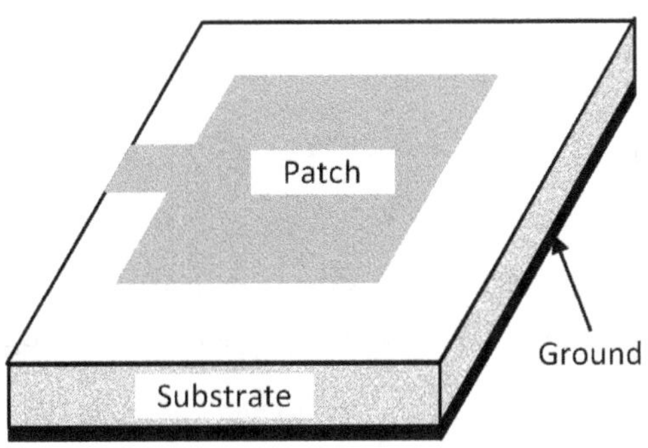

is thus useful in modern communication systems and satellite links (Kaur & Khanna, 2013). All these conclusively sums into its irresistible usages.

Applications

Due to its lightweight and many other advantages, the microstrip patch antenna is widely used in mobile systems, such as handsets, Global Positioning System (GPS), and other wireless devices. With modern computer-aided design (CAD) tools further enhancing its conformability to follow sophisticated curving contours, it can now be flexibly mounted on walls, vehicles, aircrafts, satellites, and even military systems like radars and missile guides. In addition, adaptive technology that can vary both phase shift and power to each element allows the patch antenna electronically steerable and thus enables its system to operate with higher capacity and more reliable satellite links. As such, airborne satellite communication antenna radars (ASCAR) are installed on most commercial passenger aircrafts today, replacing the former shortwave radio communication used on trans-oceanic flights (Breed, 2009).

METHODS OF ANALYSIS

The inherent principle of a microstrip patch antenna is operationally complex and structurally abstruse. In analyzing its operating mechanism, several analytical methods have been developed. The main objective of these methods is to provide a methodological platform in gaining a deeper insight of its pragmatic nature as well as visualizing its input impedance and radiation characteristics. This, in parallel, ascertains its limitations and thus offers a comprehensive yet robust model to the design of the antenna itself as well as its relevant circuitries. Among the three most popular models in this aspect are the transmission line model, the cavity model, and the method of moments (MoM) model.

Transmission Line Model

In the transmission line model, the rectangular microstrip patch antenna is visualized as two radiating slots separated by a transmission line (Munson, 1974). As shown in Figure 1, the electric field propagates in a nonhomogeneous structure, whereby the major part of it is concentrated in the substrate while the remaining part resides in open space, normally air. As a result, the transmission line does not operate in pure transverse EM (TEM) mode due to the phase difference in air and the substrate, rather the dominant mode of propagation is quasi-TEM. Thereby, an effective dielectric constant (ε_{reff}) is introduced to account for the fringing fields and wave propagation in the line,

$$\varepsilon_{reff} = \frac{\varepsilon_r + 1}{2} + \frac{\varepsilon_r - 1}{2}\left[1 + 12\frac{h}{W}\right]^{-\frac{1}{2}}, \tag{1}$$

where ε_r is the dielectric permittivity of the substrate, h is the height of the substrate, and W is the width of the patch. ε_{reff} is found slightly lower than ε_r as the fringing fields around the periphery of the patch are not confined within the dielectric substrate alone but also spreaded into the air surroundings.

To operate in the fundamental transverse magnetic (TM) mode and by accounting in the fringing effects, the length of the patch is longer electrically than its physical dimension (l). This is calculated with (Zurcher, 1995; Hammerstad, 1975)

$$l = \frac{v_o}{2f_r\sqrt{\varepsilon_{reff}}} - 2\Delta l , \tag{2}$$

$$\Delta l = 0.412h\frac{\left(\varepsilon_{reff}+0.3\right)\left(\frac{W}{h}+0.264\right)}{\left(\varepsilon_{reff}-0.258\right)\left(\frac{W}{h}+0.8\right)}, \tag{3}$$

where Δl is the fringing length at one edge of the patch, v_o is the speed of light in free space, and f_r is the resonant frequency of the antenna. l is slightly shorter than half the wavelength of the resonating field within the substrate while its corresponding width (W) is expressed as (Kumar & Gupta, 1985),

$$W = \frac{v_o}{2f_r}\sqrt{\frac{2}{\varepsilon_r+1}} . \tag{4}$$

The basic form of the transmission line model can be observed from a simple rectangular patch antenna as shown in Figure 2. The two radiating slots along the width of the patch is represented by two identical lines of equivalent admittance, Y_1 and Y_2 respectively. Given (Harrington, 1961)

$$Y_1 = G_1 + jB_1 , \tag{5a}$$

$$G_1 = \frac{d}{120\lambda_o}\left[1-\frac{1}{24}\left(k_oh\right)^2\right], \text{ for } \frac{h}{\lambda_o}<\frac{1}{10}, \tag{5b}$$

$$B_1 = \frac{d}{120\lambda_o}\left[1-0.636ln\left(k_oh\right)\right], \text{ for } \frac{h}{\lambda_o}<\frac{1}{10}, \tag{5c}$$

where G is the conductance and B is the susceptance of the radiating slot. d is the dimension of the slot, that is W in this case. k_o is the wavenumber in free space,

$$k_o = \omega\sqrt{\mu_o\varepsilon_o} , \tag{5d}$$

where μ_o and ε_o are the respective permeability and permittivity of free space. Since,

$$Y_1 = Y_2, \tag{6a}$$

$$G_1 = G_2 \text{ and } B_1 = B_2, \tag{6b}$$

both radiating slots are separated by a low-impedance (Z_c) transmission line of length *l*, expressed as (Balanis, 1989)

$$Y_c = \frac{1}{Z_c}, \tag{7a}$$

$$Z_c = \begin{cases} \dfrac{60}{\sqrt{\varepsilon_{reff}}} ln\left[\dfrac{8h}{W_o} + \dfrac{W_o}{4h}\right], \dfrac{W_o}{h} \le 1 \\ \dfrac{120\pi}{\sqrt{\varepsilon_{reff}}\left[\dfrac{W_o}{h} + 1.393 + 0.667 ln\left(\dfrac{W_o}{h} + 1.444\right)\right]}, \dfrac{W_o}{h} > 1 \end{cases}, \tag{7b}$$

where Y_c is the admittance of the transmission line and W_o is the width of microstrip feed connected to the patch.

Despite with less computation, the transmission line model is able to provide reasonable results for the radiating fields and input impedance. Unfortunately, it is only applicable for rectangular patch since it does not account for field variations in the orthogonal direction of propagation (Kumar & Ray, 2003). It is also difficult to model its coupling effect (Chang *et al.*, 2002).

Cavity Model

The cavity model conceives the region between the patch and the ground plane of the microstrip patch antenna as an EM cavity/resonator that is bounded by magnetic walls around the periphery as well as by electric walls on the top and bottom sides (Lo *et al.*, 1979). Its effective dimension is larger than its physical size due to the extension of its boundaries to account for the fringing effects. The basis of this model is subjected to the properties of its very thin substrate that postulates several plausible observations. This could be well explicated by studying the charge distribution on the top and bottom surfaces of the patch.

When the patch is energized, there exist two different reactions implicated by the charge distribution, the attractive and repulsive tendencies (Richards, 1988). As shown in Figure 3, the attractive tendency keeps the charges concentrated on the bottom surface due to its interaction with the opposite charges on the ground plane, while the repulsive tendency pushes the like charges on the bottom surface along the edge to the top. This creates the current densities $\boldsymbol{J}_t$ and $\boldsymbol{J}_b$ on its respective top and bottom surfaces.

Figure 2. Rectangular patch antenna and its transmission line model

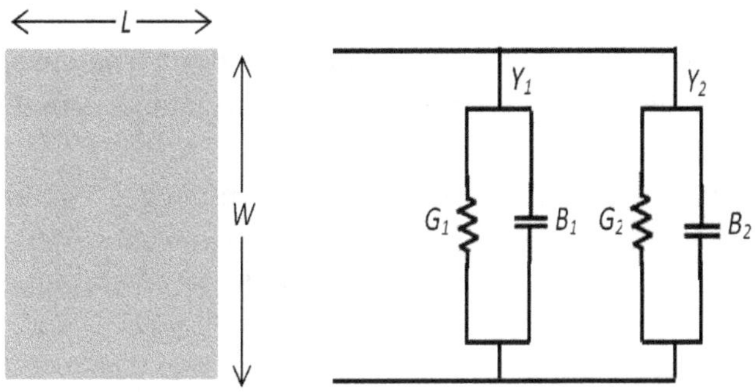

For very thin substrate ($h \ll \lambda$), the attractive tendency dominates where most of the charges concentrate on the bottom of the patch. The current flows along the edges are very small and ultimately assumed zero. Consequently, the tangential component of the magnetic field (H_τ) along the edge is assumed zero as well. Hence the side walls are modeled as the perfect magnetic conductors (PMC). Since the substrate is very thin and also assuming that the patch is perfectly conductive, the electric field bounded by the magnetic walls renders no tangential component (E_τ) and is uniform with no variation in the normal direction. Thereby, the electric field is only normal directed to the patch while the magnetic field has only the transverse components in the region bounded by the patch metallization and the ground plane. This creates the electric walls on the top and bottom of the cavity. Thus, only the TM configuration is considered within the cavity.

Manipulating the Maxwell equations with the field equivalence principle (Huygens, 1690), the current densities along the periphery of the cavity are found as

$$\boldsymbol{J}_s = \hat{n} \times \boldsymbol{H}_a , \tag{8a}$$

$$\boldsymbol{M}_s = -\hat{n} \times \boldsymbol{E}_a , \tag{8b}$$

where $\boldsymbol{J}_s$, $\boldsymbol{M}_s$ is the respective electric and magnetic current density while $\boldsymbol{E}_a$, $\boldsymbol{H}_a$ is the respective electric and magnetic field. $\boldsymbol{J}_s$, $\boldsymbol{J}_t$ are zero due to the thin substrate assumption and from image theory,

$$\boldsymbol{M}_s = -2\hat{n} \times \boldsymbol{E}_a , \tag{9}$$

due to the presence of the ground plane.

A perfectly bounded cavity is purely reactive. In practicality however, the imperfections within the cavity contribute to its resistivity, hence account for its radiation.

MoM Model

The MoM model employs full-wave analysis onto the microstrip patch antenna. In this model, the surface current is evaluated to model the patch while the volume polarization current is calculated to model the fields in the dielectric structure.

It has been shown in Newman and Tulyathan (1981) that an integral equation is formulated for these unknown currents and using MoM, these field integral equations are converted into its matrix equivalence which are then resolved algebraically. In its succinct form, the electric field ($\boldsymbol{E}$) is expressed in terms of current density ($\boldsymbol{J}$) as

$$\boldsymbol{E} = f_e(\boldsymbol{J}), \tag{10}$$

where f_e is a known linear operator. $\boldsymbol{J}$ can be expanded as a linear combination of N terms, giving

$$\boldsymbol{J} = \sum_{n=1}^{N} a_n \boldsymbol{J}_n = a_1\boldsymbol{J}_1 + a_2\boldsymbol{J}_2 + a_3\boldsymbol{J}_3 + \cdots + a_N\boldsymbol{J}_N, \tag{11}$$

where a_n is an unknown constant and $\boldsymbol{J}_n$ is the basis vector function. Substituting (11) into (10) and applying its linear property,

$$\sum_{n=1}^{N} a_n f_e(\boldsymbol{J}_n) = \boldsymbol{E}. \tag{12}$$

a_n is found using the weighted residual method whereby a set of trial solutions is established with several variable parameters. The residuals are a measure of the difference between the trial and actual solutions. The variable parameters are then sensibly selected to assert the pertinent trial functions based on the minimization of the residuals. This is done through a set of weighted functions w_m in the domain of the operator (f_e). Taking the inner product of these functions, (12) becomes

Figure 3. Charge distribution in a microstrip patch antenna

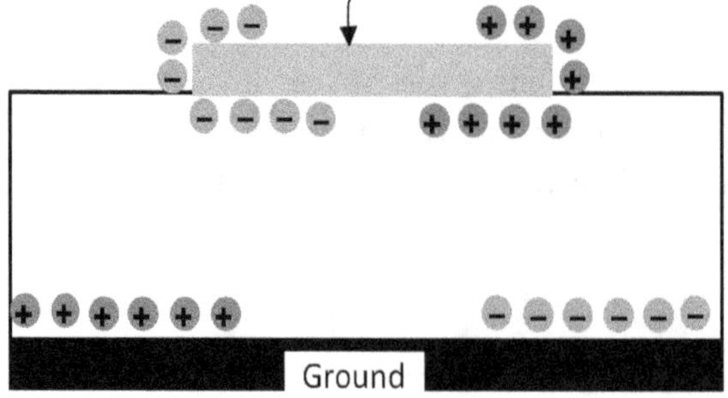

$$\sum_{n=1}^{N} w_m, f_e(\boldsymbol{J}_n, a_n) = w_m, \boldsymbol{E}, \; m = 1,2,3,\cdots,N, \tag{13}$$

and is expressed in matrix form as

$$[Z_{mn}][\boldsymbol{J}] = [\boldsymbol{E}_m], \tag{14a}$$

$$[Z_{mn}] = w_m, f_e(a_n) = \begin{bmatrix} w_1, f_e(a_1) w_1, f_e(a_2) \cdots \\ w_2, f_e(a_1) w_2, f_e(a_2) \cdots \\ \vdots \\ \vdots \end{bmatrix}, \tag{14b}$$

$$[\boldsymbol{E}_m] = w_m, \boldsymbol{E} = \begin{bmatrix} w_1, \boldsymbol{E} \\ w_2, \boldsymbol{E} \\ \vdots \\ w_N, \boldsymbol{E} \end{bmatrix}, \tag{14c}$$

where $\boldsymbol{E}$ corresponds to the known incident field quantities and the terms in the Z_{mn} matrix are the functions of geometry. The unknown coefficients of the induced current density are associated with the terms of the $\boldsymbol{J}$ vector. a_n can be solved using algebraic techniques such as Gaussian elimination, while the appropriate selection of the weighting functions is necessary so that the elements of w_m are not only linearly independent, but also minimize the computations for the inner products. One legitimate way to facilitate this is by setting $w_n = \boldsymbol{J}_n$ as suggested by the Galerkin's Method (Galerkin, 1915). Solving these matrices yield the desired current density $\boldsymbol{J}$. Subsequently, the scattered electric and magnetic fields can be derived.

It is readily seen that this model involves extensive numerical analysis but corresponds to a highly precise parametric calculation.

TEST PARAMETERS

This Section details some of the important parameters used to characterize the performances of an antenna.

Gain

Gain (G) is a measure of the ability of the antenna to direct the input power (P_{in}) into radiation in a particular direction and is measured at the peak radiation intensity. It is defined as the ratio of the radiation intensity in a given direction ($U(\theta,\varnothing)$), to the radiation intensity that would be obtained if the power accepted by the antenna were radiated isotropically (Balanis, 2005).

$$G = 4\pi \frac{U(\theta, \varnothing)}{P_{in}}, \tag{15}$$

where θ and $\varnothing$ is the respective elevation and azimuth angle of the spherical coordinate system.

The gain of an antenna is obtained by comparing it to that of an isotropic source, measured in dBi. An isotropic source is a non-existent ideal radiator, which is considered to be 100% efficient and has an omnidirectional radiation characteristic in the E- and H-plane. Alternatively, the gain can also be obtained by comparing it to a resonant dipole, measured in dBd. There is a 2.15 dB difference in gain between both methods of comparison, as a dipole possesses a superior gain of 2.15 dBi.

Even though some antennas can supply higher gain than an isotropic or dipole structure, they are not able to amplify the signal in the sense that a transistor can, because they are passive structures. In reality, one does not achieve an increment in energy via antenna gain. An antenna at best, can only exploit the RF energy supplied to their input port and then passes it out into space in a particular direction with a certain gain over that of the reference antenna. The total radiated power will never be greater than the energy fed into its input. In fact, there will always be a certain loss in total power within the antenna, since the radiation efficiency can never be 100% practically. This is due to the losses induced within the structure itself, its matching network, its substrate, its dielectric housing, and its surrounding structures.

Efficiency

For a microstrip patch antenna, efficiency (e) is defined as the power radiated from the microstrip element (P_{rad}) divided by the power received at its input (P_{rec}), given as

$$e = \frac{P_{rad}}{P_{rec}}. \tag{16}$$

The efficiency of the antenna takes into account the losses at the input terminals and within its structure. The major factors that affect the efficiency of an antenna include the efficiency of the dielectric (e_d), the conductor (e_c), and the reflected power (e_r), such that

$$e = e_r e_c e_d. \tag{17}$$

Directivity

Directivity (D) is the ability of an antenna to focus energy in a particular direction and could be easily estimated through radiation patterns. It is defined as the ratio of the radiation intensity in a given direction ($U(\theta, \varnothing)$) from the antenna to the radiation intensity averaged over all directions (U_o). The averaged radiation intensity (U_o) is the total power radiated (P_{rad}) by the antenna divided by 4π (Balanis, 2005).

$$D = \frac{U(\theta,\varnothing)}{U_0} = \frac{4\pi U(\theta,\varnothing)}{P_{rad}}. \tag{18}$$

The integrated average of the directivity pattern over an entire sphere has to be 0 dBi, that means creating directivity in a certain direction reduces it in other directions.

Radiation Pattern

The radiation pattern of an antenna is defined as a mathematical function or a graphical representation of the radiation properties of the antenna, expressed in space coordinates (Balanis, 2005). It is a plot of the far-field radiation properties of an antenna as a function of the spherical coordinates, θ and $\varnothing$, for a constant radial distance and frequency. It describes how it radiates energy out to space or how it receives energy.

Radiation pattern can be represented by a three-dimensional (3D) or two-dimensional (2D) spatial distribution plot of power flux density, radiation intensity, or field strength. In a 2D polar plot, the radiation pattern is displayed in the E- and H-plane, where its field and power patterns are normalized with respect to its maximum value. It is plotted with either of the angles held fixed while the other is varied.

Return Loss

Return loss (RL) is defined by the ratio of the incident power (P_f) of the antenna to the power reflected back from the source (P_r).

$$RL = 10\log\frac{P_f}{P_r}. \tag{19a}$$

It is the efficiency of power transmitted to an antenna and is related to impedance matching. Factors that affect the RL of an antenna include its geometry, its method of excitation, and its proximity to surrounding objects. RL can also be expressed in terms of voltage standing wave ratio (*VSWR*) as given in (Bird, 2009; Pozar, 1990)

$$RL = 20\log\left|\frac{VSWR+1}{VSWR-1}\right|, \tag{19b}$$

or as the complex input reflection coefficient (Γ),

$$RL = 10\log\left|\frac{1}{\Gamma^2}\right| = -20\log|\Gamma|. \tag{19c}$$

From (19c), it could be seen that RL is the negation of Γ. For perfect matching between the feeding system and the antenna, $\Gamma = 0$ and from (19c), $RL = \infty$. In this case, no power is reflected back. Similarly, at $\Gamma = 1$, $RL = 0dB$ and all power is reflected back. This implies that for good matching, RL is high while Γ is low, and vice-versa. For practical antenna applications, $VSWR = 2$ is required and this corresponds to $RL = 9.542dB$. However, RL ranging from 10 dB to 12 dB are acceptable.

In the case of a patch antenna, the input impedance with the source impedance is used as an intermediate parameter for determining Γ in terms of the S11 parameter expressed in dB, assuming Port 1 is used for the measurement.

Bandwidth

Most microstrip patch antennas have narrow bandwidths. It is a measure of the range of frequencies where the antenna parameter, such as input impedance, radiation pattern, radiation efficiency, beamwidth, polarization, side-lobe level, and gain, is within an acceptable value from those at the center frequency.

The bandwidth (BW) for broadband antenna is usually expressed as the ratio of the upper-to-lower frequency of operation,

$$BW = \frac{f_u}{f_l}, \tag{20a}$$

whereas the bandwidth of narrowband antenna is often expressed as a percentage of the frequency difference (Balanis, 2005),

$$BW = \frac{f_u - f_l}{\sqrt{f_u f_l}} \times 100\%, \tag{20b}$$

where f_h and f_l is the respective upper and lower frequency that coincides with the 9.542 dB RL value.

Polarization

The polarization of an EM field describes the magnitude and direction of the electric field vector as a function of time. In other words, it is the orientation of the electric field for a given position in space. The polarization of an antenna is the polarization of the wave radiated from the antenna.

Two common types of polarization are linear and circular. In linear polarization, the antenna radiates power in only one plane of propagation and the electric field varies in only one direction. Linear polarization can be either vertical or horizontal, depending on the orientation of the patch antenna. The antenna is vertically linear polarized when the electric field is perpendicular to the earth surface and horizontally linear polarized when the electric field is parallel to the earth surface. Linear polarization is mostly used in application like channel broadcasts.

Circular polarization antenna radiates power in all planes with the same magnitude in the direction of propagation. The plane of propagation rotates in circle making one complete cycle in one period of wave. A circularly polarized antenna can either be right-hand circularly polarized (RHCP) or left-hand circularly polarized (LHCP). The antenna is RHCP when the phases are $0°$ and $-90°$, and it is LHCP when the phases are $0°$ and $+90°$. Circular polarization is widely used in satellite communications since it is not sensitive to antenna orientation, unlike linear polarization does.

Considering Figure 4 and assuming the EM wave radiated by an antenna has an electric field ($\boldsymbol{E}$) with two components E_x and E_y,

$$E_x = |E_x|\cos(\omega t - \beta z), \tag{21a}$$

$$E_y = |E_y|\cos(\omega t - \beta z + \varnothing), \tag{21b}$$

where $|E_x|$ and $|E_y|$ are the amplitude of the field components in the x and y direction respectively. The wave is linearly polarized when either $|E_x| = 0$ or $|E_y| = 0$. On the other hand, it is circularly polarized when $|E_x| = |E_y| \neq 0$.

RADIATION REGIONS

Basically, the antenna radiation zones are classified into two principal regions, the Fresnel region and the Fraunhofer region. The Fresnel region covers radiation in the near field and its boundaries are given by

$$0.62\sqrt{\frac{D^3}{\lambda}} \leq R \leq \frac{2D^2}{\lambda}, \tag{22}$$

where R is the distance from the antenna and D is the largest dimension of the antenna. Most of the energy in this region is reactive and non-propagating. Its angular field distribution is dependent upon the distance from the antenna. The Fresnel region is also sometimes referred to as the induction region because the energy in this region is stored and returned to the antenna during each cycle, similar to the way in which an inductor stores and releases energy.

Figure 4. Polarization of EM wave

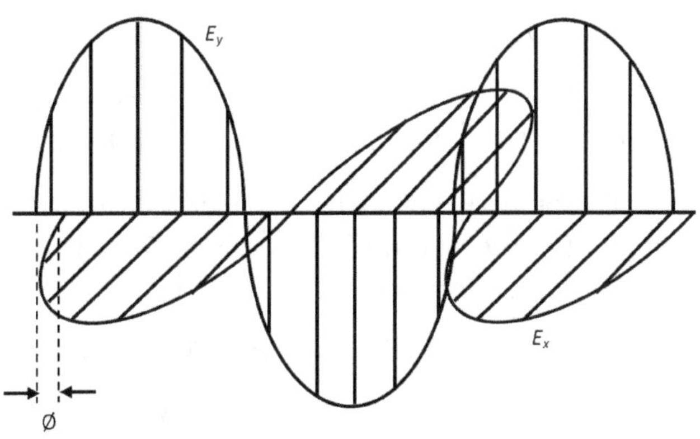

On the other hand, the Fraunhofer region covers radiation in the far field and its boundary is given by (Lehpamer, 2008)

$$R > \frac{2D^2}{\lambda}. \tag{23}$$

This is the region away from the antenna where the radiated wave takes the form of a plane wave. Most of the energy in this region is propagating. This means that the power that reaches this region continues to radiate outward and does not return to the antenna. Therefore, its angular field distribution is essentially independent of the distance from the antenna.

The boundaries given in (22) and (23) are only defined in approximation for the convenience of analysis. In reality, the boundary between the Fresnel and Fraunhofer regions is not delineated exactly due to the gradual changes that occur in the EM field. There is no abrupt change in the field configuration during which the boundaries are crossed. In fact, as the distance between the antenna and the radiating source increases, the energy is continually converted from reactive to radiating.

SUMMARY

This chapter elaborates on the microstrip patch antenna, which is widely utilized in the wireless industry today due to its many advantages, like low profile, light weight, high conformability, low cost, easy to manufacture, etc. Several models have been developed to analyze and design the patch antennas. Among them, the three most common ones are the transmission line model, the cavity model, and the MoM model. The transmission line model is the most straightforward and it involves less cumbersome calculations. However, it is only restricted to rectangular patch design. The cavity model, on the other hand, presents a more in depth view of the patch operation through the study of its current distribution. The MoM model is the most precise among all as it performs a full-wave analysis onto the patch. Nevertheless, it also consumes much memory space and time due to its extensive calculations. Apart from this, the important parameters in characterizing the patch antenna are also covered, namely its gain, efficiency, directivity, radiation pattern, RL, BW, and polarization. This is followed by the introduction of the radiation regions, which are basically classified as the Fresnel region and the Fraunhofer region.

REFERENCES

Agrawal, P. K., & Bailey, M. C. (1977). An analysis technique for microstrip antennas. *IEEE Transactions on Antennas and Propagation*, *AP-25*(6), 756–759. doi:10.1109/TAP.1977.1141706

Bailey, M. C. (1979) Analysis of the properties of microstrip antennas using strips embedded in a grounded dielectric slab. In *Proceedings IEEE Antennas and Propagation Society International Symposium*, pp. 370-373. June 1979, Seattle. IEEE. 10.1109/APS.1979.1148143

Balanis, C. A. (1989). *Advanced Engineering Electromagnetics*. New York: John Wiley & Sons.

Balanis, C. A. (2005). *Antenna Theory, Analysis and Design*. New York: John Wiley & Sons.

Bhartia, P., & Bahl, I. J. (1982) A frequency agile microstrip antenna. In *Proceedings IEEE Antennas and Propagation Society International Symposium*, pp. 304-307. May 1982, Albuquerque. IEEE.

Bird, T. S. (2009). Definition and misuse of return loss. *IEEE Antennas & Propagation Magazine*, *51*(2), 166–167. doi:10.1109/MAP.2009.5162049

Boualleg, A., & Merabtine, N. (2005). Analysis of radiation patterns of rectangular microstrip antennas with uniform substrate. Semiconductor Physics. *Quantum Electronics & Optoelectronics.*, *8*(3), 88–91. doi:10.15407pqeo8.03.088

Breed, G. (2009). *The fundamentals of patch antenna design and performance* (pp. 48–51). High Frequency Electronics.

Chang, K., Bahl, I., & Nair, V. (2002). *RF and Microwave Circuit and Component Design for Wireless Applications*. New York: John Wiley & Sons.

Derneryd, A. G. (1976). Linearly polarized microstrip antennas. *IEEE Transactions on Antennas and Propagation*, *AP-24*(6), 846–851. doi:10.1109/TAP.1976.1141445

Derneryd, A. G. (1978). A theoretical investigation of the rectangular microstrip antenna element. *IEEE Transactions on Antennas and Propagation*, *AP-26*(4), 532–535. doi:10.1109/TAP.1978.1141890

Derneryd, A. G. (1979a). Analysis of the microstrip disk antenna element. *IEEE Transactions on Antennas and Propagation*, *AP-27*(5), 660–664. doi:10.1109/TAP.1979.1142159

Derneryd, A. G., & Lind, A. (1979b). Extended analysis of rectangular microstrip resonator antennas. *IEEE Transactions on Antennas and Propagation*, *AP-27*(6), 846–849. doi:10.1109/TAP.1979.1142206

Deschamps, G. A. (1953). Microstrip microwave antennas. In *Proceedings of the Third Symposium on the USAF Antenna Research and Development Program, Oct* (pp. 18-22).

Galerkin, B. G. (1915). On electrical circuits for the approximate solution of the Laplace equation. *Vestnik Inzh.*, *19*, 897–908.

Gutton, H., & Baissinot, G. (1955) Flat aerial for ultra-high frequencies. French Patent No. 703 113.

Hammerstad, E. O. (1975) Equations for microstrip circuit design. In *Proceedings of the 5th European Microwave Conference*, pp. 268-272. September 1975, Hamburg. IEEE.

Harrington, R. F. (1961). *Time-Harmonic Electromagnetic Fields* (p. 183). New York: McGraw-Hill.

Howell, J. Q. (1975). Microstrip antennas. *IEEE Transactions on Antennas and Propagation*, *AP-23*(1), 90–93. doi:10.1109/TAP.1975.1141009

Huygens, C. (1690). Leyden: Traite de la Lumiere.

Kaur, J., & Khanna, R. (2013). Co-axial fed rectangular microstrip patch antenna for 5.2 GHz WLAN application. *Universal Journal of Electrical and Electronic Engineering*, *1*(3), 94–98. doi:10.13189/ujeee.2013.010306

Kumar, G., & Gupta, K. C. (1985). Non-radiating edges and four edges gap coupled multiple resonator broadband microstrip antennas. *IEEE Transactions on Antennas and Propagation*, *AP-33*(2), 173–178. doi:10.1109/TAP.1985.1143563

Kumar, G., & Ray, K. P. (2003). *Broadband Microstrip Antennas* (p. 8). Norwood: Artech House.

Lee, K. F., & Chen, W. (1997). *Advances in Microstrip and Printed Antennas* (pp. 53–62). New York: John Wiley & Sons.

Lehpamer, H. (2008). *RFID Design Principles*. Boston, MA: Artech House.

Lo, Y. T., Solomon, D., & Richards, W. F. (1979). Theory and experiment on microstrip antennas. *IEEE Transactions on Antennas and Propagation*, *AP-27*(2), 137–145. doi:10.1109/TAP.1979.1142057

Long, S. A., & Walton, M. D. (1979). A dual-frequency stacked circular-disc antenna. *IEEE Transactions on Antennas and Propagation*, *AP-27*(2), 270–273. doi:10.1109/TAP.1979.1142078

Munson, R. E. (1974). Conformal microstrip antennas and microstrip phased arrays. *IEEE Transactions on Antennas and Propagation*, *AP-22*(1), 74–78. doi:10.1109/TAP.1974.1140723

Newman, E. H., & Tulyathan, P. (1981). Analysis of microstrip antennas using Moment Methods. *IEEE Transactions on Antennas and Propagation*, *AP-29*(1), 47–53. doi:10.1109/TAP.1981.1142532

Pozar, D. M. (1990). *Microwave Engineering*. New York: Addison-Wesley.

Purchine, M. P., & Aberle, J. T. (1994). A tunable L-band circular microstrip patch antenna. *Microwave Journal*, 80–88.

Richards, W. F. (1988). Microstrip Antennas, Chapter 7 in Antennas Handbook – Theory, Applications, and Design. New York: Van Nostrand Reinhold Co.

Richards, W. F., & Lo, Y. T. (1983). Theoretical and experimental investigation of a microstrip radiator with multiple lumped linear loads. *Electromagnetics*, *3*(3), 371–385. doi:10.1080/02726348308915194

Richards, W. F., Lo, Y. T., & Harrison, D. D. (1981). An improved theory of microstrip antennas with application. *IEEE Transactions on Antennas and Propagation*, *AP-29*(1), 38–46. doi:10.1109/TAP.1981.1142524

Richards, W. F., & Long, S. A. (1986a) Adaptive pattern control of a reactively loaded, dual-mode microstrip antenna. In *Proceedings of the International Telemetering Conference*, pp. 291-296. October 1986, Las Vegas.

Richards, W. F., & Long, S. A. (1986b) Impedance control of microstrip antennas utilizing reactive loading. In *Proceedings of the International Telemetering Conference*, pp. 285-290. October 1986, Las Vegas.

Shen, L. C., Long, S. A., Allerding, M. R., & Walton, M. D. (1977). Resonant frequency of a circular disc, printed-circuit antenna. *IEEE Transactions on Antennas and Propagation*, *AP-25*(4), 595–596. doi:10.1109/TAP.1977.1141643

Stutzman, W. L., & Thiele, G. A. (1998). *Antenna Theory and Design*. New York: John Wiley & Sons.

Uzunoglu, N. K., Alexopoulos, N. G., & Fikioris, J. G. (1979). Radiation properties of microstrip dipoles. *IEEE Transactions on Antennas and Propagation*, *AP-27*(2), 853–858. doi:10.1109/TAP.1979.1142173

Wong, K. L. (1999). *Design of Nonplanar Microstrip Antennas and Transmission Lines*. New York: John Wiley & Sons. doi:10.1002/0471200662

Wong, K. L. (2002). *Compact and Broadband Microstrip Antennas* (pp. 232–273). New York: John Wiley & Sons. doi:10.1002/0471221112.ch7

Yang, S. L. S., Kishk, A. A., & Lee, K. F. (2008). Frequency reconfigurable U-slot microstrip patch antenna. *IEEE Antennas and Wireless Propagation Letters*, *7*, 127–129. doi:10.1109/LAWP.2008.921330

Zurcher, J. F., & Gardiol, F. E. (1995). *Broadband Patch Antennas*. Boston, MA: Artech House.

Chapter 10
Active Integrated Antenna

Kim Huat Yeap
Westlake International School, Malaysia

ABSTRACT

This chapter discusses the active integrated antenna (AIA), which is the integration of an antenna with an active circuitry that acts as a radiating element as well as to perform additional functions simultaneously. The designs of AIA can be generalized into three different classifications, the amplifier type, the oscillator type, and the frequency conversion type. An AIA is classified as the amplifier type when its active device functions as an amplifier. Correspondingly, an AIA is classified as the oscillator type when its active device offers the function of an oscillator while the frequency conversion type of AIA integrates an active device with a passive antenna element for the purpose of frequency translation.

INTRODUCTION

The terminology 'active antenna' simply means that active devices are used in passive antenna to improve the antenna performance. 'Active integrated antenna (AIA)' indicates specifically that on the same substrate, the passive antenna and the active circuitry are integrated together as a single entity (Lin & Itoh, 1994), without the need of a matching circuit or interconnecting cables. This differs from the traditional design, where the antenna and the RF front-end circuits are different components, connected with standard 50-Ω transmission lines or waveguides. The AIA not only functions as a radiating element, it also renders built-in signal and wave processing capabilities, such as amplification and frequency multiplication.

The AIA provides a new paradigm for designing modern microwave systems. It consists of one or more active devices, such as diodes and/or transistors, integrated together with a passive radiating element, such as a microstrip patch, a printed dipole, a bowtie, or a slot antenna (Chang *et al.*, 2002). It offers many advantages, which include compactness, low cost, low profile, minimum power consumption, and multiple functionalities (Flynt *et al.*, 1996). Such integration also reduces the inherent losses and noise figure (NF) of the system as well as realizes the ease of machine manufacturability. Unlike conventional antennas, the addition of active devices makes an AIA system non-reciprocal. Thus, it can only function either as a transmitting or a receiving device, but not both simultaneously (Itoh *et al.*, 2001).

DOI: 10.4018/978-1-7998-2381-0.ch010

The use of AIA can be traced back to as early as the 1920s (Lin & Itoh, 1994), where an electron tube was first integrated into the antenna of a radio broadcast receiver, to eliminate the lossy and bulky interconnection between the active devices and the antenna (Wheeler, 1975). At that time, the application of AIA was limited because the usage of electron tubes was expensive to build and maintain. This status quo remained until the transistor was founded in 1948 (Tranter, 1995) and the concept of AIA was resurrected in the early 1950s. Solid-state solutions soon played a dominant role in AIA research, due to their reliability, superior inherent voltage standing wave ratio (*VSWR*), lower cost, and smaller size.

Since then, several pioneering works in AIA had been reported. Copeland and Robertson (1961) demonstrated a mixer-integrated antenna, which they called an 'antennaverter'. They used a travelling wave antenna together with tunnel diodes, to operate as a travelling wave amplifier, which they called an 'antennafier' (Copeland *et al.*, 1964). Meinke and Landstorfer (1968) integrated a field effect transistor (FET) to the terminals of a dipole to serve as a Very High Frequency (VHF) amplifier, for reception at 700 MHz. Meanwhile, Ramsdale and Maclean (1971) used bipolar junction transistors (BJTs) and dipoles for transmission applications. Lindenmeier (1976) then introduced the general theory behind the optimum bandwidth of signal-to-noise ratio (SNR) for receiver systems integrated with small antennas.

The first modern AIA was developed by Thomas *et al.*, (1985). It was a Gunn diode integrated with a patch antenna, operating at the X-band. Today, the technological maturity in Monolithic Microwave Integrated Circuit (MMIC) has enabled rigorous integration between the antenna and the circuit component. AIA with various functions can now be packed into a mini footprint. Through manipulation of the appropriate configuration, an infinite potential on multiple communication and sensor applications can be realized. Despite its original role serving as a radiating element, the active antenna can now be made injection locked, frequency tunable, power amplified, or mutually coupled. For instance, the AIA had been applied as a large-scale spatial power combiner, in which the radiation of many AIAs combines coherently. Such combiners had been demonstrated at the X-band (Chew & Itoh, 1995) and at microwave frequencies (York & Popovic, 1997). In the works of Cryan *et al.* (1997) as well as Andrew and Hall (2002), AIAs had been used to achieve synchronous operation of a number of integrated antenna oscillating elements. By controlling the phase distribution of each AIA element, the array could be transformed into a beam steering system. Likewise, Lin and Itoh (1994) and Hall *et al.* (2002) explored on simultaneous transmit-receive active arrays with duplex operation. Good transmit-receive isolations were reported in their literatures.

In general, based on the different functions of the active device applied, the AIA can be classified into the amplifier type, the oscillator type, and the frequency conversion type.

AMPLIFIER TYPE

An AIA is classified as the amplifier type when its active device functions as an amplifier. The integration of the amplifier with the antenna increases the antenna gain and bandwidth as well as improves the noise performance of the overall AIA module. When the antenna is connected to the input port of the amplifier, it acts as the source impedance of the active device and the AIA functions as part of a receiver. The goal of the receiving AIA is to obtain a low NF. Correspondingly, when the antenna is connected to the output port of the amplifier, it acts as a load impedance of the active device and the AIA functions as part of a transmitter. The goal of the transmitting AIA is to achieve high linearity and efficiency.

Over the years, much research work had been dedicated to the amplifier type AIA by integrating a power amplifier (PA) with the antenna. Radisic et al. (1997a) integrated a Class B PA with a patch antenna using the probe feed method to suppress the second harmonic. A 55% power-added efficiency (PAE) was obtained. Later, Radisic *et al.* (1997b) integrated a Class F PA with a circular sector patch antenna and obtained a PAE of 63%. The design was based on the roots of Bessel function to optimize the power loss for the second and third harmonics. Weiss and Popovic (1999) eliminated the output matching network and achieved 62% of PAE, by integrating a class E PA with a slot antenna that acted as the harmonic load to the PA. Similarly, Chung *et al.* (2003) developed an AIA using an Aluminium Galium Nitride / Galium Nitride (AlGaN/GaN) High Electron Mobility Transistor (HEMT) PA with harmonics termination. The design yielded a linear gain of 8 dB and 45% of PAE. Soon after, Kim *et al.* (2005) proposed a direct integration method by combining a Class F PA with an antenna that provided optimum impedance at the fundamental frequency, a short circuit at the second harmonic, and an open circuit at the third harmonic. The design obtained a PAE of 67.5%.

Recently, interest in the amplifier type AIA is growing extensively due to its potential in power combining applications (Lin & Itoh, 1994; Garg *et al.*, 2001). The idea of integrating a power-combining patch antenna was first presented by Deal *et al.* (1999). The antenna is designed to serve as an out-of-phase combiner, driven by a pair of push-pull PAs. The PAs were placed directly at the antenna platform. Unlike the conventional push-pull front-end, the balun was removed at the output of the amplifier. This led to lower output losses and a compact transmitter front-end. Furthermore, this design also increased efficiency and reduced heat-sink size. The antenna structure and its mode profile are shown in Figure 1. It had a pair of microstrip feeds placed at both opposite radiating edges. The length of the antenna was about half a wavelength, similar to a single feed patch antenna. Due to the symmetrical and opposite electric fields along the excitation direction, the differential-mode signals from both ports were superimposed and radiated out of the antenna. During common-mode excitation, however, a magnetic wall was formed across the center of the patch. Power flowing from the first feed port to the second feed port was thus negligible. Since the active antenna was designed to operate in differential-mode only, there was no reflected common mode back to the amplifier. With this configuration, the output power was doubled compared to using a single-ended amplifier and a PAE of 55% at 2.5 GHz was achieved.

For the case of a receiving module, Ormiston *et al.* (1998) proposed an amplifier type AIA using a quarter-wavelength, inset-fed patch, directly integrated with a low noise amplifier (LNA). Complex impedance matching for low noise performance was provided by adjusting the distance from the feed point to the gate connection. Meanwhile, the distance from the shorted edge to the source also provided series inductive feedback. This design yielded a NF of 0.5 dB and a gain of 24 dB at 1.33 GHz.

OSCILLATOR TYPE

An AIA is classified as the oscillator type when its active device offers the function of an oscillator while its antenna functions as both a radiator and a load for the active device (Chang *et al.*, 2002). Its design is based on its oscillating structure (Colpitts, Hartley, series or parallel configuration) and the function of its antenna (as a resonator, filter, or negative feedback path). The oscillator type AIA usually adopts the diode for high power applications, but it is affected by low direct-current-to-radio-frequency (DC-to-RF) efficiency. Hence, heat dissipation is a critical concern for such design. On the other hand, the transistor, when used in the oscillator type AIA, offers high DC-to-RF efficiency and the ease of integration with

Figure 1. Power-combining patch antenna proposed by Deal et al. (1999)

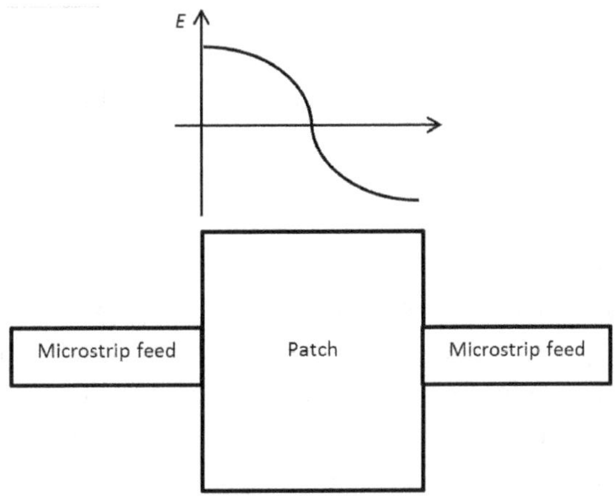

planar circuit structure. However, its operation range is limited by its lower cut-off frequency. Typically, for frequencies less than 3 GHz, BJT is utilized while for higher frequencies, FET is more suitable.

In 1989, Birkeland and Itoh proposed a leaky wave antenna that also acted as an oscillator. It was made of a transmission line with periodic perturbation that exhibited a leaky wave stopband when the physical period was one guided wavelength long. The antenna input impedance was highly reactive and was used as a frequency selective feedback. When connected to an active device in the negative resistance region, it functioned as both a radiating element and a resonator to determine the oscillating frequency. On the other hand, Martinez and Compton (1994) proposed a quasi-optical oscillator type AIA. The antenna acted as part of the feedback loop for the oscillating FET. The radio frequency (RF) output was amplitude-modulated by applying a modulating signal at the drain, which was then radiated through the patch.

The oscillator type AIA tends to suffer from narrow bias tuning range and large output power variations. This can be improved by using varactor diodes connected to the radiating element. Bhartia and Bahl (1982) described the use of varactor diodes in microstrip patches as a mean of changing the resonance frequency while Navarro *et al.* (1992) portrayed an active notch antenna, tuned by a varactor diode, in which the source oscillator was a Gunn diode. Meanwhile, Kitchen (1987) depicted varactor diode tuning applied on the transistor oscillator, in which the devices were located in the gate and source, alongside a microstrip load in the drain.

FREQUENCY CONVERSION TYPE

The frequency conversion type AIA integrates an active device with a passive antenna element for the purpose of frequency translation, either up- or down-convert. It can function either as a modulator in a transmitter or a mixer in a receiver. The active device adopted in the design can be either a diode or a transistor. The main advantages of using a diode are its simplicity in circuit design and less board space consumption. Meanwhile, a transistor exhibits conversion gain (CG) in the design and is compatible with MMIC technology.

In the work of Stephan *et al.* (1983), a slot ring mixer was presented. It integrated a slot ring antenna with two mixer diodes, which were placed physically at 90° apart. The structure received the incoming RF signal in the horizontal polarization while the local oscillator (LO) signal was injected in the vertical polarization. The diode polarities were arranged in such a way that it was aligned with the RF signal polarization and opposite to the LO signal polarization, so as to realize balanced mixing. In conjunction to this, the slot ring antenna acted as a receiving antenna as well as a diplexer to enable correct phasing of the diodes.

Another leading application of the frequency conversion type AIA is the non-contact RF interrogation detection (ID) transponder card, proposed by Pobanz and Itoh (1994). It is an AIA compact transponder base on the subharmonically-pumped quasi-optical mixer. This design constitutes of a planar bowtie antenna mounted with an antiparallel Schottky diode pair and a slot antenna coupled to a rectifying diode. A 10-MHz low-frequency oscillator is modulated by a digital data generator on the back side of the transponder. When illuminated by a 6 GHz signal, the slot antenna activates the transponder from a power-conserving standby mode. Subsequently, the diode pair at the bowtie generates a 12 GHz signal, which mixes with the 10 MHz modulated signal specific to the card and is then re-transmitted from the card. Since the interrogation signal and the response signal are not harmonically related, this design minimizes the occurrence of false detection. Furthermore, simultaneous interrogation of more than one card is possible as the 10-MHz signal can also be tuned to vary the digital code specific to each different card.

It is also noted that the antenna array is popularly adopted in the frequency conversion type AIA. One such example is the low noise active antenna array, proposed by Brauner *et al.* (2003). The design acted as an active down-converter. It consisted of a LNA that was integrated to each antenna element, such that the design was compact and robust. The directional couplers allowed the injection of an identical calibration signal in the front-end into all four branches of the antenna array. Amplified through the LNA, the signal at each branch passed through a bandpass filter (BPF), which provided much inherent attenuation on the image band. A LO buffer was added before the mixer that provided backward isolation to prevent unwanted crosstalk and compensate for power divider loss at each branch. Finally the resultant intermediate frequency (IF) output was amplified by an IF amplifier. This design produced a very low NF of 3.5 dB and a CG of 30 dB. It was thus found useful in a multi-dimensional channel sounding system.

SUMMARY

The AIA is the integration of an antenna with an active circuitry that acts as a radiating element as well as to perform additional functions simultaneously. The designs of AIA can be generalized into three different classifications, the amplifier type, the oscillator type, and the frequency conversion type. In this chapter, each AIA type is elaborated along with design examples.

REFERENCES

Andrews, J. W., & Hall, P. S. (2002). Phase-locked-loop control of active microstrip patch antennas. *IEEE Transactions on Microwave Theory and Techniques*, *50*(1), 201–206. doi:10.1109/22.981266

Bhartia, P., & Bahl, I. J. (1982). Frequency agile microstrip antennas. *Microwave Journal*, *25*, 67–70.

Birkeland, J., & Itoh, T. (1989). FET-based planar circuits for quasi-optical sources and transceivers. *IEEE Transactions on Microwave Theory and Techniques*, *37*(9), 1452–1459. doi:10.1109/22.32230

Brauner, T., Kung, R., Vogt, R., & Barchtold, W. (2003) 5-6 GHz low-noise active antenna array for multi-dimensional channel-sounding. In *Proceedings of the 2003 SBMO/IEEE MTT-S International Microwave and Optoelectronics Conference*, pp. 297-301. September 2003, Foz do Iguacu. IEEE. 10.1109/IMOC.2003.1244874

Chang, K., York, R. A., Hall, P. S., & Itoh, T. (2002). Active integrated antennas. *IEEE Transactions on Microwave Theory and Techniques*, *50*(3), 937–944. doi:10.1109/22.989976

Chew, S., & Itoh, T. (1995) A 2x2 beam-switching active antenna array. In *Proceedings IEEE MTT-S International Microwave Symposium Digest*, pp. 925-928. May 1995, Orlando. IEEE.

Chung, Y., Hang, C. Y., Cai, S., Qian, Y., Wen, C.-P., Wang, K. L., & Itoh, T. (2003). AlGaN/GaN HFET power amplifier integrated with microstrip antenna for RF front-end applications. *IEEE Transactions on Microwave Theory and Techniques*, *51*(2), 653–659. doi:10.1109/TMTT.2002.807685

Copeland, J. R., & Robertson, W. J. (1961). Antennaverters and antennafiers. *Electronics (Basel)*, 68–71.

Copeland, J. R., Robertson, W. J., & Verstraete, R. G. (1964). Antennafier arrays. *IEEE Transactions on Antennas and Propagation*, *AP-2*(2), 227–233. doi:10.1109/TAP.1964.1138196

Cryan, M. J., Hall, P. S., Tsang, S. H., & Sha, J. (1997). Integrated active antenna with full duplex operation. *IEEE Transactions on Microwave Theory and Techniques*, *45*(10), 1742–1748. doi:10.1109/22.641721

Deal, W. R., Radisic, V., Qian, Y., & Itoh, T. (1999). Integrated-antenna push-pull power amplifiers. *IEEE Transactions on Microwave Theory and Techniques*, *47*(8), 1901–1909. doi:10.1109/22.780389

Flynt, R. A., Fan, L., Navarro, J. A., & Chang, K. (1996). Low cost and compact active integrated antenna transceiver for system application. *IEEE Transactions on Microwave Theory and Techniques*, *44*(10), 1642–1649. doi:10.1109/22.538955

Garg, R., Bhartia, P., Bahl, I., & Ittipiboon, A. (2001). *Microstrip Antenna Design Handbook*. Boston, MA: Artech House.

Hall, P. S., Gardner, P., & Ma, G. (2002). Active integrated antennas. *IEICE Transactions on Communications*, *E85-B*(9), 1661–1667.

Itoh, T., Haddad, G., & Harvey, J. (2001). *RF Technologies for Low-Power Wireless Communications*. New York: Wiley. doi:10.1002/0471221643

Kim, H., Yoon, I. J., & Yoon, Y. I. (2005). A novel fully integrated transmitter front-end with high power added efficiency. *IEEE Transactions on Microwave Theory and Techniques*, *53*(4), 3206–3214.

Kitchen, J. (1987). Octave bandwidth varactor-tuned oscillators. *Microwave Journal*, *30*(5), 347–353.

Lin, J., & Itoh, T. (1994). Active integrated antennas. *IEEE Transactions on Microwave Theory and Techniques*, *42*(12), 2186–2194. doi:10.1109/22.339741

Lindenmeier, H. (1976, October). Design of electrically small broadband receiving antennas under consideration of nonlinear distortions in amplifier elements. In *Proceedings IEEE Antennas and Propagation Society International Symposium*, pp. 242-245. Amherst, MA. IEEE. 10.1109/APS.1976.1147544

Martinez, R. D., & Compton, R. C. (1994). A quasi-optical oscillator/modulator for wireless transmission. pp. 839-842. IEEE. 10.1109/MWSYM.1994.335226

Meinke, H., & Landstorfer, F. (1968). Noise and bandwidth limitations with transistorized antennas. In *Proceedings IEEE Antennas and Propagation Society International Symposium*, pp. 245-246. September 1968, Boston, MA. IEEE. 10.1109/APS.1968.1150608

Navarro, J., Shu, Y.-H., & Chang, K. (1992). Broadband electronically tunable planar active radiating elements and spatial power combiners using notch antennas. *IEEE Transactions on Microwave Theory and Techniques*, *40*(3), 323–328. doi:10.1109/22.120105

Ormiston, T. D., Gardner, P., & Hall, P. S. (1998). Compact low noise receiving antenna. *Electronics Letters*, *34*(14), 1367–1368. doi:10.1049/el:19980524

Pobanz, C., & Itoh, T. (1994) A microwave non-contact identification transponder using subharmonic interrogation. In *Proceedings IEEE MTT-S International Microwave Symposium Digest, pp.* 753-756. May 1994, San Diego, CA. IEEE. 10.1109/MWSYM.1994.335246

Radisic, V., Chew, S. T., Qian, Y., & Itoh, T. (1997a). High-efficiency power amplifier integrated with antenna. *IEEE Microwave and Guided Wave Letters*, *7*(2), 39–41. doi:10.1109/75.553052

Radisic, V., Qian, Y., & Itoh, T. (1997b) Class F power amplifier integrated with circular sector microstrip antenna. In *Proceedings IEEE MTT-S International Microwave Symposium Digest, pp.* 687-690. June 1997, Denver. IEEE. 10.1109/MWSYM.1997.602884

Ramsdale, P., & Maclean, T. (1971). Active loop-dipole aerials. *Proceedings of the Institution of Electrical Engineers*, *118*(12), 1698–1710. doi:10.1049/piee.1971.0313

Stephan, K. D., Camilleri, N., & Itoh, T. (1983). A quasi-optical polarization-duplexed balanced mixer for millimeter-wave applications. *IEEE Transactions on Microwave Theory and Techniques*, *31*(2), 164–170. doi:10.1109/TMTT.1983.1131452

Thomas, H., Fudge, D., & Morris, G. (1985). Gunn source integrated with a microstrip patch. *Microwaves & RF*, *24*(2), 87–89.

Tranter, Z. (1995). *Principles of Communications*. New Jersey: John Wiley & Sons.

Weiss, M. D., & Popovic, Z. (1999) A 10 GHz high-efficiency active antenna. In *Proceedings IEEE MTT-S International Microwave Symposium Digest*, pp. 663-666. June 1999, Anaheim, CA. IEEE.

Wheeler, H. A. (1975). Small antennas. *IEEE Transactions on Antennas and Propagation*, *AP-12*(4), 462–469. doi:10.1109/TAP.1975.1141115

York, R. A., & Popovic, Z. B. (1997). *Active and Quasi-Optical Arrays for Solid-State Power Combining*. New York: Wiley.

Section 5
Applications of Cosmic Sources

Chapter 11
Fundamentals of Meteor Burst Communication

Ryoji Wakabayashi
Tokyo Metropolitan College of Industrial Technology, Japan

ABSTRACT

When space dust rushes into the atmosphere, oxygen and nitrogen are ionized by frictional heat. Along the dust flight path, a very long cylindrical plasma tube, 10 meters in diameter and several kilometers long is formed. The long plasma tube is called "meteor burst" and is a good reflector for radio waves in the VHF band. Non-line-of-sight communication performed using this reflector is called "meteor burst communication". In this chapter, the basics of meteor burst communication and its applications are outlined.

INTRODUCTION

Research on the influence of meteor on radio wave propagation has been started from the early 20th century. In 1910, it was predicted and expected that radio waves would be reflected by meteor showers when passing Halley's Comet, and observations using a spark transmitter were carried out between New York and Massachusetts, but unfortunately no useful results were obtained (Pickard, 1931). In 1921, it was discovered that there is a relationship between the passage of meteor showers and the arrival of radio waves using radio waves from European LF band stations, but the results were presented in 1931 with caution. After World War II, research on meteor burst communication (MBC) began to be actively conducted in the United States, Canada, the United Kingdom, and the USSR as VHF band technology progressed. At that time, the secrecy of MBC was regarded as important, and since it was developed as one of military communication, the result was undisclosed. In particular, they are more useful for military communications than HF band communications because they are hardly affected by ionospheric disturbances caused by nuclear explosions in the air. In 1957, the vast majority of data on the nature of the previously undisclosed meteor burst channel was released from IRE (for example, Vincent *et al.*, 1957), causing the primary MBC boom.

DOI: 10.4018/978-1-7998-2381-0.ch011

In 1953, the Radio Physics Laboratory of the Canadian Department of Defense have completed a MBC experiment system "JANET" (Forsythe *et al.*, 1957). The first system was built as a teletype link between Ottawa and Port Arthur to demonstrate the possibility of MBC. It is a duplex channel system using two frequencies. Both stations always transmit probe signals, and data modulation is started when a strong radio wave from the opposite station is detected, and then data transmission is stopped when the received signal is lower than the threshold level of the receiver. The procedure is the basis adopted in recent MBC systems. The throughput of about 10 to 20 bps is obtained using a 5-element Yagi-UDA antenna with transmission power 500 W. At the same time, many studies were conducted for example by the National Bureau of Standard (NBS) (Carpenter *et al.,* 1959) and by the Stanford Research Institute (SRI). After the launch of Sputnik satellite by the USSR in 1957, and the realization of the communication satellite Telster by NASA in 1962, research on satellite communication became active, and the research on MBC had gone down. However, even so, ARQ (Automatic Repeat reQuest) technique and space diversity, frequency diversity, height diversity techniques are used in COMET system operated in Europe by STC (SHAPE Technical Center) of NATO. The average throughput about 150 bps for a day is obtained between stations located in the Hague, The Netherlands and South France (Bartholome *et al.,* 1968).

In the 1970s, with the development of computers, small and inexpensive MBC communication devices became available. The propriety of MBC to the small capacity data acquisition system from a large number of terminals placed in the location where simply can not access the telephone network for example mountain areas, was reviewed and it rushed into the second meteor MBC boom. The second boom was triggered by SNOTEL (Johnson, 1987), a system that collects several weather data including snow amount from over 500 terminals installed in the Rocky Mountains. The SNOTEL was built by the MCC in the United States, and has been operated by the Ministry of Agriculture until now.

METEOR BURST

In outer space, as the comets orbiting the sun approach the sun, dust is released and becomes a meteor shower. Since the comet is made of ice and dust, when it approaches the sun, the ice evaporates from the surface and tiny dust is repelled, and the one that overcomes the attraction of the character flows out. The meteor shower spreads like clouds around the orbit of the mother comet, and the dust spreads further to the entire orbit. Furthermore, as time passes, the dust gets further apart each other due to the influence of perturbations by giant planets such as Jupiter and the collision with scattered meteors. At this stage, it cannot be recognized as a meteor shower, and each meteor will be a scattered meteor that appears to be moving by itself. That is, they become scattered meteors and will be present everywhere in the solar system. The above process is outlined in Figure 1. The distribution density of scattered meteors is not uniform. In the revolution orbit, the density is high around the outer space corresponding to summer in the northern hemisphere, and the lower density around the outer space corresponding to winter. Therefore, the meteor activity will show annual fluctuations.

When it encounters the scattered meteors due to the rotation and revolution of the earth, dust rushes into the atmosphere. As a result, space dust that falls on the earth per day is said to reach about 1 ton in weight and about 1 trillion in number. Space dust that has entered the atmosphere ionizes the surrounding gas at an altitude of about 80 to 120 km by frictional heat, and the ionized gas spreads along the path of the meteor. This height just corresponds to the height of the E layer of the ionosphere. In

Figure 1. The transition from a comet to a scattered meteor through a meteor shower

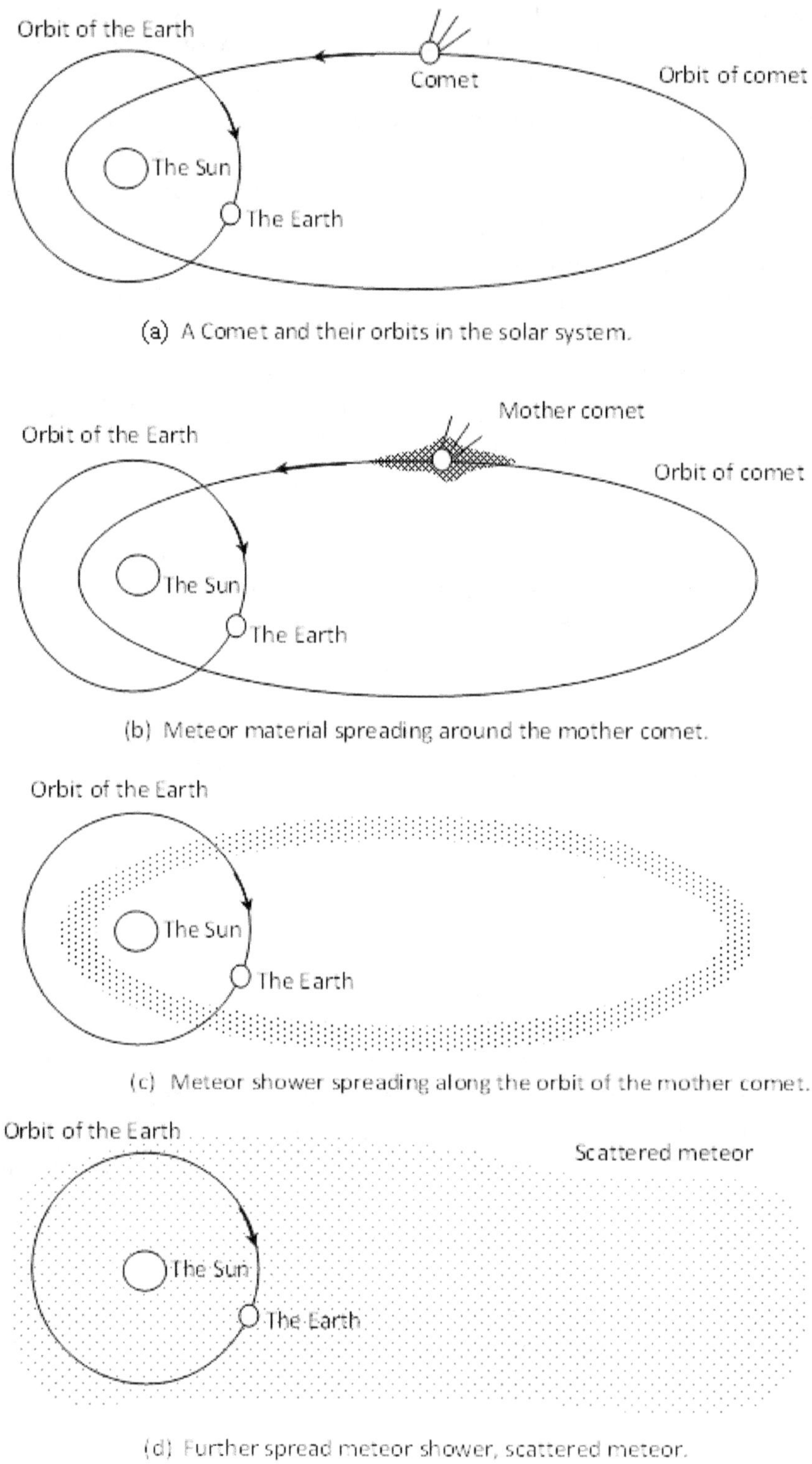

Figure 2. Meteor bust occurring near 80 ~ 120 kilo meters above the ground

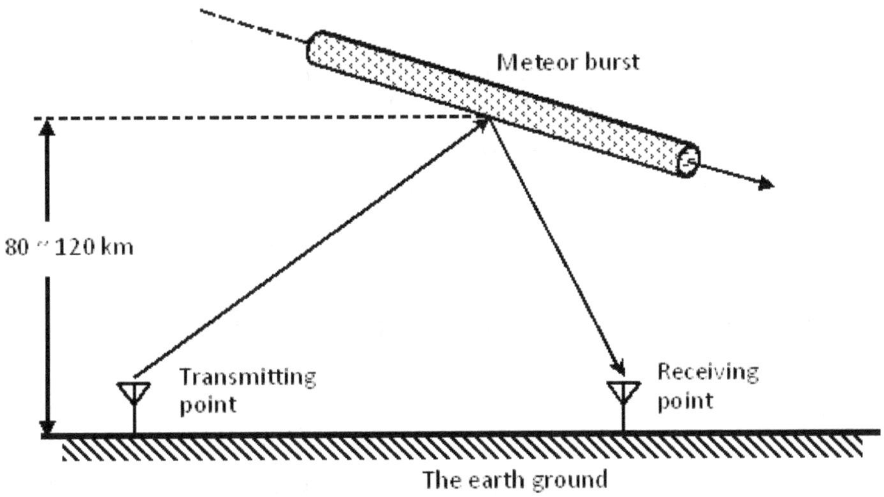

addition to the ionization phenomenon, when the dust is large a light emission phenomenon occurs, and the emitted light is visible as a shooting star or a fireball. The gas diffuses into a cylindrical shape, precisely into a very long rotating body of parabola, and when the gas density is moderately high and its diameter is several meters to several tens of meters, it becomes a good reflective object (or scattering object) of radio waves in the VHF band. The electron density in the tube is proportional to the mass of the original dust. Most meteor bursts evaporate completely before they reach 80 km above ground level and disappear. As shown in Figure 2, Meteor burst communication refers to out-of-sight communication using this meteor burst as a good reflecting (or scattering) object of VHF waves located at the same elevation as the E layer of the ionosphere.

Meteor bursts can be classified into two types, underdense bursts and overdense bursts, depending on the electron linear density. Electron density is the number of free electrons per axial unit length of a meteor burst, and this number independent of meteor diffusion. The boundary between overdense burst and underdense burst is where the electron linear density is about 2×10^{14} electrons/m. If the density is larger than the value, that is classified as overdense burst, and smaller as underdense burst.

METEOR BURST COMMUNICATION

Meteor burst communication is one of the non-line-of-sight communication methods using the plasma tube as an element of the reflector made by the dust of the universe that exists infinitely in nature. Therefore, meteor burst communication can be said to be an economical communication line with no cost for laying the line. The meteor burst communication path is probabilistic in its generation time and duration, but it is very stable while the communication path exists, and is less affected by multipath and Doppler shift. Unlike the ionosphere, the propagation state of meteor burst channels does not change depending on the season or time zone, so there is no need to change the operating frequency or adjust the antenna according to the season or time zone. In the case of communication using ionospheric reflection in the shortwave band, it is necessary to change the operating frequency every season or every hour, and the cost of the transmitter, receiver and the antenna are several times higher. Meteor burst communication can be operated at a single frequency throughout the year, and its operation is also simple.

The upper limit of the communication distance by meteor burst communication can be easily calculated from the height of the reflector (about 100 km) and the curvature of the surface (the radius of the earth is about 6,400 km), and the value is about 2,000 km. Theoretically, although the usable time of each meteor burst becomes shorter as the distance between transmitting and receiving points becomes shorter, there is no limit to the short distance and there is no skip zone, which is one of the features of the meteor burst channel.

INTENSITY OF REFLECTED WAVE FROM METEOR BURST

Underdense Burst

In the beginning of 1950, Eshleman, V. R. et al. gave the following equation for determining the reflected wave $P_R(t)$ from an underdense burst (Eshleman *et al.,* 1954),

$$P_R\left(t\right) = \frac{P_T G_T G_R \lambda^3 q^2 \sin^2 \alpha r_e^{\,2}}{16\pi^2 R_T R_R \left(R_T + R_R\right)\left(1 - \sin^2\phi \cos^2\beta\right)} \times \exp\left(-\frac{8\pi^2 r_0^{\,2}}{\lambda^2 \sec^2\phi}\right) \times \exp\left(-\frac{32\pi^2 Dt}{\lambda^2 \sec^2\phi}\right) \quad (1)$$

where

λ is wave length,
P_T is transmission power,
G_T is the gain of the transmitting antenna,
G_R is the gain of the receiving antenna,
q is the electron linear density,
r_e is the classical radius of the electron ($= 2.8178 \times 10^{-15}$),
R_T is the distance between the transmitting antenna and the meteor burst,
R_R is the distance from the receiving antenna to the meteor burst,
ϕ is the incident angle of the radio wave to the meteor burst (the same reflection angle),
α is the angle between the electric field vector of the incident wave to the meteor burst and the direction vector to the receiving antenna,
β is the angle between the wave propagation plane and the meteor burst,
r_0 is the initial radius of the meteor burst,
D is the diffusion coefficient.

Figure 3 shows above each parameter including the transmitting pointand receiving point and meteor burst.

In the case of an underdense burst, which is an ionized gas tube with a low electron density, radio waves pass almost unaffected by the ionized gas tube, but they excite electrons when passing through the

Figure 3. Incident wave to a meteor burst and reflected wave from it

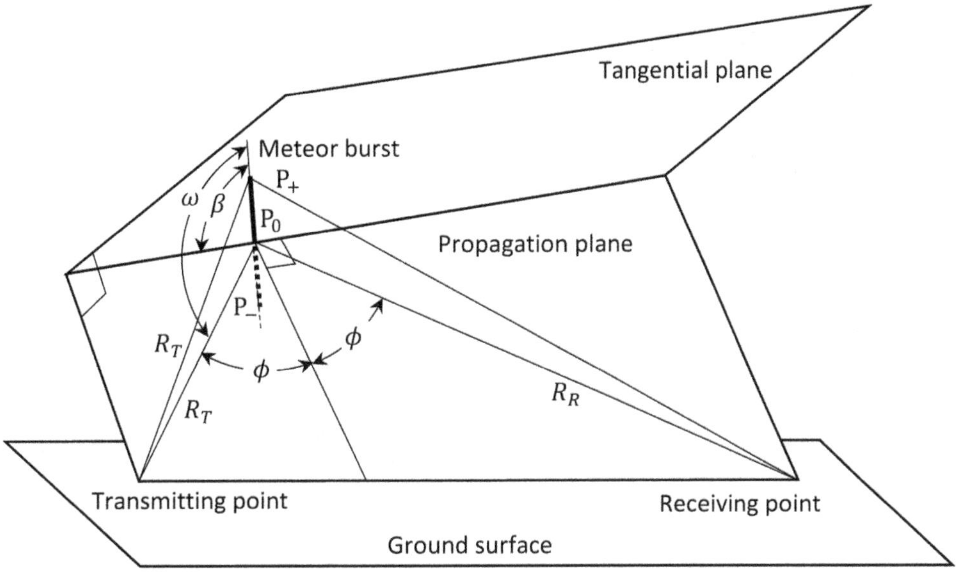

tube. At this time, equation (1) is derived on the assumption that each electron is excited independently and does not affect each other. In addition, the electron density in the tube is assumed to have a uniform distribution in the axial direction and a Gaussian distribution in the radial direction.

In Figure 3, the central point of reflection on the burst is P_0 and two points P_+ and P_- in the meteor burst shows the path difference within $\lambda/2$ from the path length of the radio wave reflected from P_0. It can be seen that the length $2L$ between P_+ and P_- corresponds to the Fresnel zone, and L is given by the following equation.

$$L = \sqrt{\frac{\lambda P_T R_R}{\left(R_T + R_R\right)\left(1 - \sin^2 \phi \cos^2 \beta\right)}} \tag{2}$$

where the factor in the square root of L is found in equation (1).

In equation (1), the first of the two exponential terms is determined by the initial radius of the meteor burst, and the second shows the reflection intensity attenuation due to the radius increase of the burst by diffusion. It is reported in the recent experimental data that the attenuation is almost exponential, and the average value τ_a of the attenuation time constant is expressed by the following equation (Weitzen, 1987).

$$\tau_a = \frac{\lambda^2 \sec^2 \phi}{32\pi^2 D_a} \tag{3}$$

Overdense Burst

On the other hand, the reflection from the overdense burst is best known by the following formula published by Hines and Forsyth (1957).

$$P_R(t) = \frac{P_T G_T G_R \lambda^2 \sin^2\alpha}{32\pi^2 R_T R_R (R_T + R_R)(1 - \sin^2\phi\cos^2\beta)}$$

$$\times\sqrt{\frac{4Dt + r_0^2}{\sec^2\phi} \times \log_e\left(\frac{r_e q \lambda^2 \sec^2\phi}{4\pi^2 Dt}\right)} \tag{4}$$

Each parameter is the same as in the case of under dense burst. This equation is valid when $P_R(t) \geqq 0$. that is, in a period of

$$T_0 = \frac{r_e q \lambda^2 \sec^2\phi}{4\pi^2 D} \tag{5}$$

and takes the maximum value at time T_0/e as follows.

$$P_{R\,max} = \frac{P_T G_T G_R \lambda^3 \sqrt{q} \sin^2\alpha \sqrt{r_e / e}}{32\pi^3 R_T R_R (R_T + R_R)(1 - \sin^2\phi\cos^2\beta)} \tag{6}$$

In the case of an overdense burst, as the electron density in the burst is high, the reflected wave by the burst can be treated as a reflection from a cylindrical metal object. However, in the case of an overdense burst, it is often deformed under the influence of the wind blowing in the long duration, and fading occurs. Then fluctuation as in equation (4) is rarely observed. In the early stage of burst generation where the electron density is sufficiently high, strong reflection is obtained with the increase of the burst diameter by its own diffusion.

Typical power transition patterns of the equations (1) and (4) are shown in Figure 4.

Figure 4. Typical power transition pattern of reflected wave from overdense burst and underdense burst

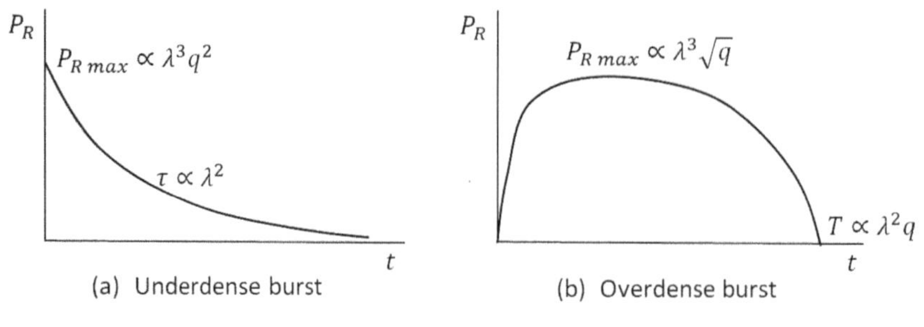

HOT SPOT AND FOOTPRINT OF METEOR BURST COMMUNICATION

Hot Spot

If the transmitting point and the receiving point are fixed, the existence of an airspace region is limited to some extent for the contribution to the meteor burst communication path.

If there is a meteor burst that contributes to communication just above the straight line connecting the transmitting and receiving points, the meteor must enter the atmosphere horizontally, but it is clear that such a meteor does not exist. Therefore, the sky above the midpoint between the transmitting and receiving points is the airspace with the lowest probability. The probability of meteor bursts that contribute to communication increases in the direction away from the straight line, and the probability decreases again as the distance from the line increases. The contour lines obtained by calculating the equations of the paper by Eshleman, V. R. and Manning, L. A. are roughly as shown in Figure 5. The airspace with the highest probability where a meteor burst that satisfies the reflection condition exists is an ellipse whose major axis is a line segment connecting the transmitting and receiving points, and whose minor axis is about twice the altitude at which the airspace is formed. The probability is known to be about 5%. The hatched area in Figure 5 is an area where a particularly high probability is obtained.

Figure 5. Contour lines showing the probability of meteor bursts that satisfy the reflection conditions

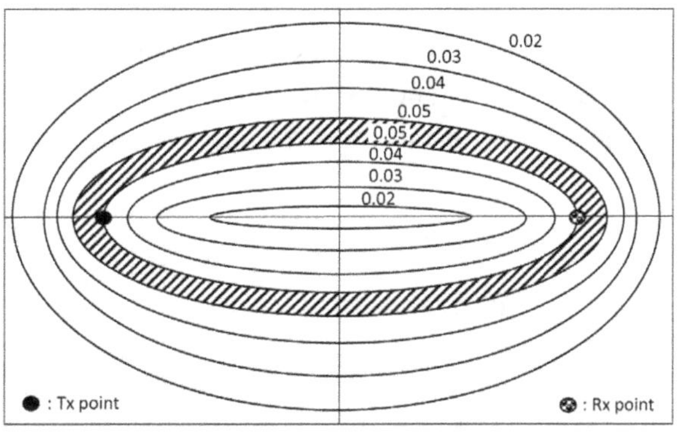

The above is just a discussion about the number of meteor bursts. As can be seen from equations (1) and (4), the duration of a meteor burst is approximately proportional to $\sec^2\phi$, so the duration is maximum above the midpoint of the transmitting and receiving stations. When this is combined with the number of arguments discussed above, the airspace or hot spot that contributes most to the duty ratio of the communication path in meteor bursts is as shown in Figure 6.

Footprint

The range is called a footprint in the meteor burst communication where a radio wave reaches a ground after transmitted from a single transmission station and reflected by a single meteor burst.

Figure 7 illustrates how radio waves incident and scattered by meteor bursts reach the ground. In Figure 7, if A~B is a meteor burst, the Huygens principle indicates that there is a reflected wave in the

Figure 6. Hot Spot of meteor burst communication at the height of 80 ~ 120 km above the ground

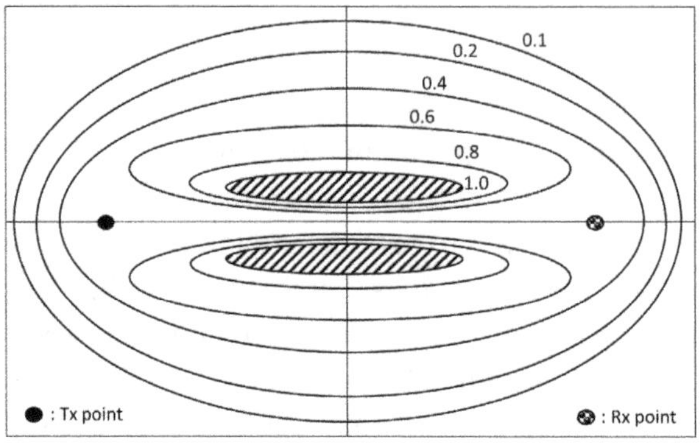

spatial region between the two cones. A footprint is the hatched area as shown in Figure 7 where the region intersects the ground. The shape and size of the footprint varies depending on the position, angle and length of the burst, but it is known that the footprint is often a long and narrow horseshoe shape with a width of about 10 km.

Note that it takes time for the footprint to be generated. For example, when a 50 km long footprint is formed by a meteor flying at a speed of 25 km/s, there is a time difference of 2 seconds between both ends of the footprint. Furthermore, since the meteor burst lasts on average only about 0.25 seconds after the generation, the part of about 6 km long that reflects the radio waves in the burst moves at a speed of 25 km/s. In other words, instead of thinking that a 50 km long meteor burst is stationary, it is necessary to think that a 6 km long burst travels a distance of 50 km. In Figure 8, it should be noted that a small footprint (instantaneous footprint) moves at high speed on the ground and forms a composite (or total) footprint.

Figure 7. Footprint at the ground surface of single meteor burst from single transmitting station

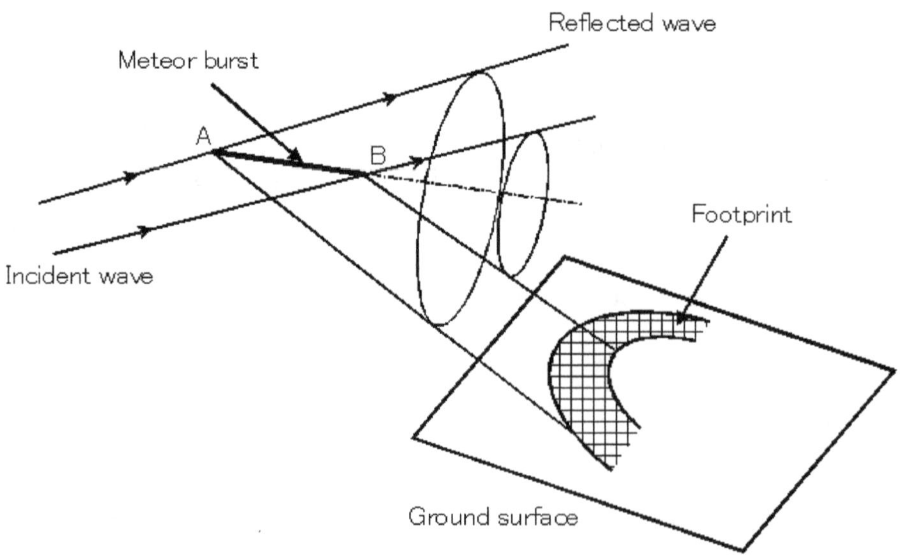

Figure 8. Instantaneous footprint and composite footprint

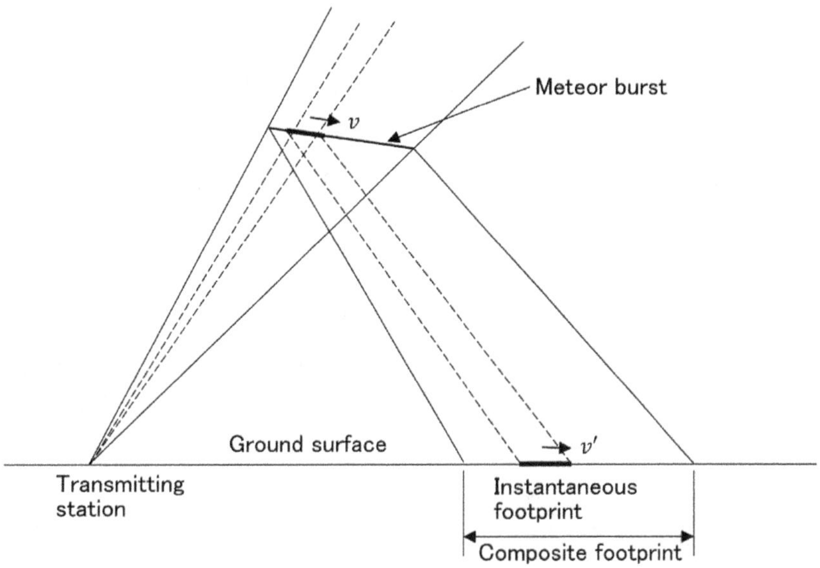

CONCLUSION

This chapter outlines the basics of meteor burst communication. The formula taken up here is based on classical theory, but it is still a basic theory that can be used sufficiently when considering meteor burst communications.

A meteor burst communication system can be operated with a transmitter with a power of about 100 W, a receiver, a 5-elements Yagi-Uda antenna, a solar cell and a secondary battery. A lot of high-level experiments are continuously conducted among amateur radio operators, and high-performance digital communication software is being developed and released. Due to the spread of the Internet lines and mobile lines, the demand for meteor burst communication is not so high because of its lack of immediacy and low speed, but it seems to be necessary as an emergency line at the time of disaster and so on. The author believes that it is desirable to continue to study the effective use and dissemination of meteor burst communication fused with various modern digital mobile communication technology.

REFERENCES

Bartholome, P. J., & Vogt, I. M. (1968). COMET - a new meteor-burst system incorporating ARQ and diversity reception. *IEEE Transactions on Communications*, *16*(2), 268–278. doi:10.1109/TCOM.1968.1089833

Carpenter, R. J., & Ochs, G. R. (1959). The NBS meteor burst communication system, *IRE Trans.*, CS-7, 4, pp. 263-271.

Eshleman, V. R., & Manning, L. A. (1954). Radio communication by scattering from meteoric ionization. *Proceedings of the IRE*, *42*(3), 530–536. doi:10.1109/JRPROC.1954.274813

Forsythe, P. A., Vogan, E. L., Hansen, D. R., & Hines, C. O. (1957). The principles of JANET: A meteor burst communication system. *Proceedings of the IRE*, *45*(12), 1642–1657. doi:10.1109/JRPROC.1957.278296

Hines, C. O., & Forsyth, P. A. (1957). The forward scattering of radio waves from overdense meteor trails. *Canadian Journal of Physics*, *35*(9), 1033–1041. doi:10.1139/p57-115

Johnson, D. E. (1987) Ten years experience with the SNOTEL meteor burst data acquisition system, In *Proceedings of MBC Symposium*, pp. SII.5-SII.20

Pickard, G. W. (1931). A note on the relation of meteor showers and radio reception. *Proceedings of the IRE*, *19*(7), 1166–1170. doi:10.1109/JRPROC.1931.222442

Vincent, W. R., Wolfram, R. T., Sifford, B. M., Jaye, W. E., & Peterson, A. M. (1957). A Meteor-Burst System for Extended Range VHF Communications. *Proceedings of the IRE*, *45*(12), 1693–1700. doi:10.1109/JRPROC.1957.278324

Weitzen, J. A. (1987). A data base approach to analysis of meteor burst data. *Radio Science*, *22*(1), 133–140. doi:10.1029/RS022i001p00133

Chapter 12
Practice of Meteor Burst Communication

Ryoji Wakabayashi
Tokyo Metropolitan College of Industrial Technology, Japan

ABSTRACT

A lot of research and experiments on meteor burst communication described in the previous chapter have been done by research institutions, educational institutions, and amateur radio operators. The contents of research and experiments are roughly classified into two categories. One is to investigate the frequency and duration of meteor bursts themselves, and the other is to study data transmission using meteor bursts. This chapter introduces some experiments on meteor burst communications conducted in Japan and the method of QSO by meteor scatter communications conducted by amateur radio operators.

INTRODUCTION

Currently, the main institutions conducting research on meteor burst communications in Japan are Shizuoka University, Numazu National College of Technology, Tokyo Metropolitan College of Industrial Technology, National Defense Academy, High Tech Research Inc., and Geosports Co., Ltd.

Shizuoka University, Numazu National College of Technology, National Institute of Technology, National Defense Academy have been conducting long-term research with the cooperation of the Institute of Low Temperature Science at Hokkaido University. Through a single-tone transmission experiment, we investigated the propagation status of meteor burst communications in various parts of Japan, and developed the IT-5000S, an original MBC transceiver, jointly with Intertech Co., Ltd., taking advantage of our experience. The initial model had a built-in digital signal processor (DSP) control unit and controlled transmission and reception timing. This first model was actually trans-

DOI: 10.4018/978-1-7998-2381-0.ch012

ported to Antarctica and played an active role in field experiments conducted by graduate students of Shizuoka University and faculty member of Numazu National College. The second machine, IT-5000N, is a transceiver equipped with a single-board computer (SBC) that runs on a Linux OS instead of DSP on the IT-5000S, and it can perform internal processing using software radio technology. Modulation and demodulation can be performed by the built-in SBC, and various experimental data can be acquired and stored. Both IT-5000S and IT-5000N are still in operation, and not only single-tone transmission experiments but also various digital data transmission experiments are being carried out utilizing the functions of software defined radio.

Also, in the joint research titled "Development of the Okhotsk Sea Area Environmental Information Collection System" with the Institute of Low Temperature Science of Hokkaido University, the MCC radio equipment used in SNOTEL (i.e. the US Department of Agriculture system described in the previous chapter) was used, and a meteorological observation data collection system with a master station in Sapporo (Hokkaido University) has been constructed.

On the other hand, High Tech Research Inc. has built a meteorological observation system with a master station in Ibaraki Prefecture, and a data collection system that transmits data from Okinotorishima, the southernmost part of Japan, to the master station in Ibaraki by meteor burst communication. The total distance of this communication channel is 2,150 km。 Since it exceeds the limit of meteor burst communication, a 2-hop line with a relay station is constructed along the way. All the radio equipment owned by Hi Tech Research Inc. has now been transferred to Geosports Co., Ltd., and the radio station business has been taken over by that company.

In addition, amateur radio operators have studied meteor burst communication (which is more commonly referred to as meteor scatter communication by amateur radio operators) for a long time, and some excellent software have been developed and released as free software due to recent improvements in personal computer processing capabilities.

Here, the single-tone transmission experiment system and the Okhotsk sea area environmental information collection system conducted by the author and the other collaborators are introduced, and the observation data obtained by them are outlined. In addition, the current state of meteor scatter communications developed by amateur radio operator is also described.

DATA COLLECTION SYSTEM

The Institute of Low Temperature Science Hokkaido University in Japan has been conducting comprehensive research of various natural phenomena occurring in the cryosphere and cold environments. One of the objects of research is drift ice, and it is indispensable to conduct a thorough survey of the weather in the Pan Okhotsk region to carry out the research. The method of recording in a data logger and collecting data at a later date involves a significant time delay unless the data is collected fairly frequently. The method of transmitting data using a satellite communication line allows data to be collected in real time, but the system is quite expensive and the antenna adjustment is also required to be precise. Therefore, data collection using meteor burst communication lines began to be considered in 2002. This is because the cost of using the line is free and data can be collected several hours after the data acquisition although real-time data transmission is not always possible.

In order to investigate the weather in the Sea of Okhotsk, it is necessary to install sensors around the mouth of the Amur River and Sakhalin. As a preliminary experiment, a sensor was installed on the shore of Lake Saroma, and a meteor burst communication line between there and the Institute of Low Temperature Science located in Sapporo. Lake Saroma is located in the northeastern part of Hokkaido, has an area of approximately 152 square kilometers and a perimeter of 87 km where scallops are cultivated. Local weather conditions are required by aquaculture fishery cooperative of Lake Saroma to determine when young shellfish of scallop should be released. For that reason, fortunately, permission was granted to install some weather sensors and data transmitting system in a corner of the local quasi-national park by providing observation data to the fishery cooperative. The distance from Lake Saroma to the Institute of Low Temperature Science is about 236 km as shown in Figure 1. Propagation through mountain diffraction is thought to be possible because Mt. Daisetsuzan range is located between them, but considering the certainty, meteor burst communication seems to be the best choice.

Figure 1. Position of remote station and master station across Mt. Daisetsuzan Range

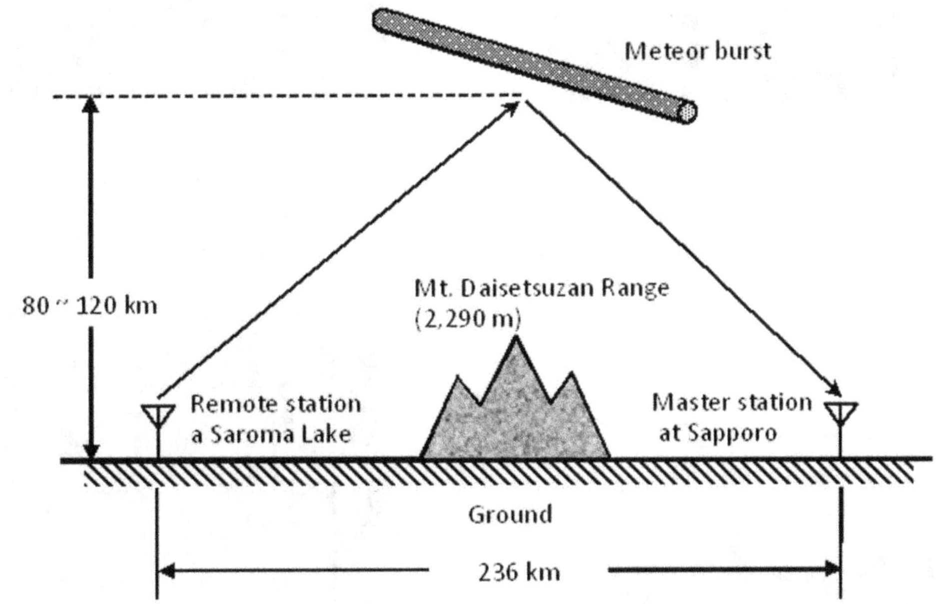

In a data collection system using meteor burst communication, a central station that collects data is called a "master station", and a station that acquires data from a sensor and transmits it to the master station is called a "remote station". Figure 2 shows a photograph of an antenna, solar cells and a set of weather sensor of a remote station installed on the shore of Lake Saroma. The transceiver is a dedicated one for meteor burst communication manufactured by Meteor Communications Corporation (MCC) in Seattle, and is housed in a box under the blue vinyl sheet in Figures 2 (a) and (c).

Figure 2a. Lake Saroma remote station: 5-element Yagi-Uda antenna and solar panel

Figure 2b. Lake Saroma remote station: Various weather sensors

Figure 2c. Lake Saroma remote station: Upper; System overview and Lower; backup scene from data logger to note PC. The data logger is installed in a duralumin box

Figure 2d. Lake Saroma remote station: Transceiver (called RF MODEM) MCC-545A manufactured by MCC, installed in a duralumin box

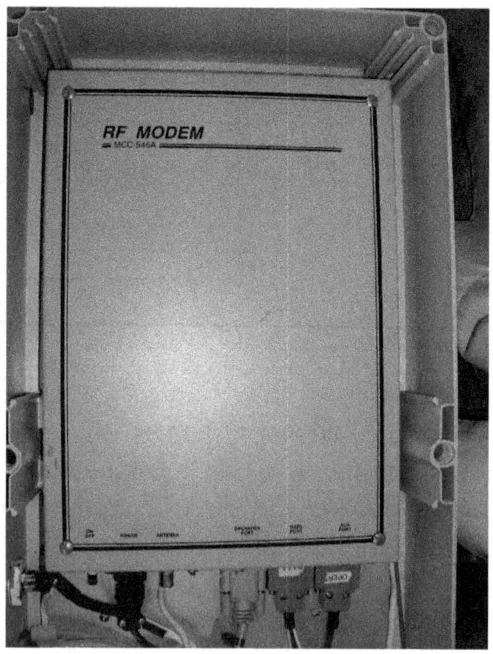

Sensors installed on the lakeside are to measure the temperature, atmospheric pressure, humidity, wind power, wind direction, sunshine degree and battery voltage. These data are measured every hour and stored in a data logger. The data is then transmitted toward the master station when a meteor burst occurs. The communication procedure is shown below.

First, in order to check the occurrence of meteor bursts, probe packets are sent from the master station to the remote station (or stations). This packet is for checking whether a meteor burst that contributes to communication has occurred or not. It is a very short packet of 28 ms. On the remote station side, it is only necessary to know the ID number of the transmitting station and to ascertain that the received packet is a probe packet at this stage. Each probe packet is transmitted with an interval of 45 ms in consideration of processing on the receiving side. At this time, the remote station always checks whether the probe packet has arrived or not. Normally, the probe packet does not reach the remote station, but when a meteor burst occurs, it is reflected by the meteor burst and reaches the remote station.

When a probe packet is received, the remote station immediately transmits a data packet to the master station. This data is transmitted right after the last data is successfully transmitted by the previous meteor burst. The master station checks the arrival of the data packet signal from the remote station at the 45 ms time interval that is described above. Each data packet has a data part of length 40 ms which consist of 10 data words each with 2 bytes length. When the data packet is correctly demodulated at the master station, the master station returns an acknowledgment (Ack) packet to the remote station and waits for reception of the next data packet. When the remote station receives the Ack packet, the point of the last data that was successfully transmitted is updated and the next data is then sent to the master station.

If the meteor burst spreads and disappears and, as a result, the master station fails to receive the next data packet within the 45 ms duration, the master station will shift to the mode in which probe packets are continuously transmitted. Similarly, if the remote station fails to receive an Ack packet within a certain time, the remote station will shift to the standby state again. The time flow of the above procedure is shown in Figure 3.

Figure 3. The communication procedure and timing for a data packet transmission

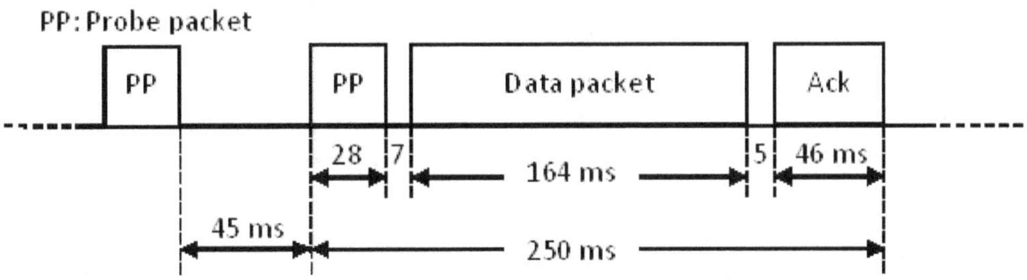

By repeating this procedure, the data accumulated in the data logger is transferred from the remote station to the master station in the order of measurement date and time. However, in consideration of the case where the propagation condition is not good, data that has passed 24 hours from the measured date and time is deleted from the transfer queue. When the meteor burst communication system was first set up, data from the meteorological sensors were acquired several times per hour.

When the meteor burst communication system was initially set up, the frequency of acquiring data from weather sensors was set several times per hour. The frequency was later reduced to once every hour since data communication at this rate is found to be efficient enough. In the case of SNOTEL described in the previous chapter, data collection from each remote station to the master station is once a day. Similarly, it is sufficient for the aquaculture and fishery cooperatives in Lake Saroma to get the data measured at a fixed time of about three times a day such as in the morning, noon and night. However, in order to check the performance of the system, the frequency of acquiring data was set once an hour. Table 1 shows the specifications of the master station and the remote station. The transmission frequency is 43.85 MHz, the modulation method is BPSK, and its data transfer rate is 4 kbps. Both antennas are orientated such that the directivity of them faces each other. Figure 4 shows the experimental system diagram of the data collection system.

Table 1. Specifications of the master station and the remote station

	Master Station	Remote Station
Transmitter power	100 W	100 W
Antenna	5-element Yagi-Uda	5-element Yagi-Uda
Antenna height	12 m	5 m
Polarization of antenna	Horizontal	Horizontal

Figure 4. The experimental system diagram of the data collection system

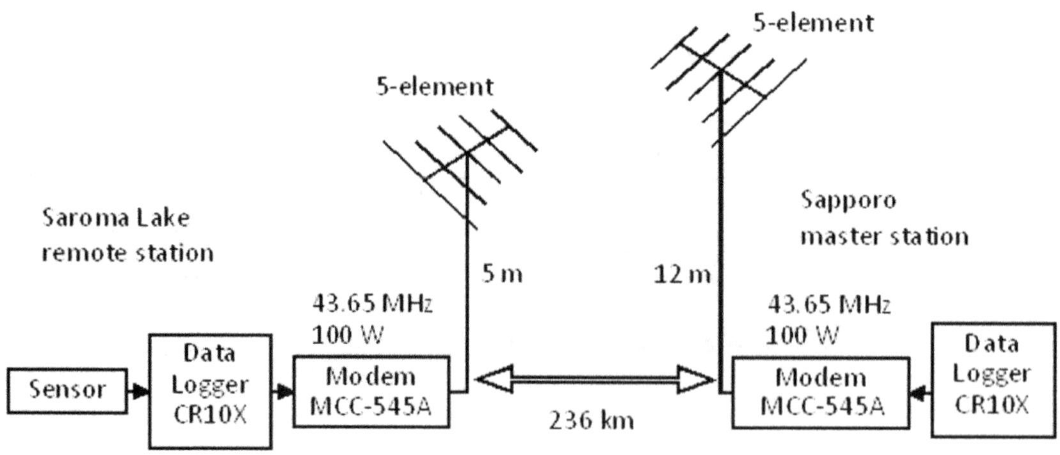

At the beginning of the experiment, the antennas of the master station and the remote station were installed horizontally, but the data recovery rate from the remote station was not satisfactory. Considering that the distance from Lake Saroma to Sapporo is 236 km and the meteor burst occurs at an altitude of about 100 km, it was considered appropriate to set the elevation angle of the antenna to 45 degrees. Therefore, the data acquisition rate was examined by changing the elevation angles of both antennas at 25 degrees, 45 degrees and 60 degrees at appropriate intervals. Figure 5 shows the measurement results from May 15, 2009 to March 12, 2014, and Table 2 shows the settings of the antenna elevation angle for each period. In the graphs depicted in Figure 5, the period in which the data is missing corresponds to the period in which transmission is stopped due to a power failure, antenna failure and so on.

From the figure, it can be seen that a 100% data acquisition rate was obtained over a long period from November 25, 2011 to May 30, 2013 when the elevation angle of both antennas was set to 45 degrees. However, the data acquisition rate has fallen in winter due to fewer scattered meteors. As there is no period in which the 100% data acquisition rate declines continuously outside this period, it is considered that the meteor bursts to contribute data transmission between the two stations exist in the airspace above the midpoint connecting the two points where both stations exist in this experiment.

Table 2. Antenna elevation angle setting and measurement period

Date		Elevation Angle (Degree)	
from	to	MASTER STATION	Remote Station
May 15 2009	Nov. 21, 2011	0	0
Nov. 22, 2011	Nov. 24, 2011	45	0
Nov. 25, 2011	May 30, 2013	45	45
May 31, 2013	June 3, 2013	65	45
June 4, 2013	Nov 5, 2013	65	65
Nov 6, 2013	Nov 9, 2013	25	65
Nov 10, 2013	March 12, 2014	25	25

Figure 5a. Data Acquisition Rate By The Master Station In Sapporo: May 15, 2009 to Dec. 31, 2009

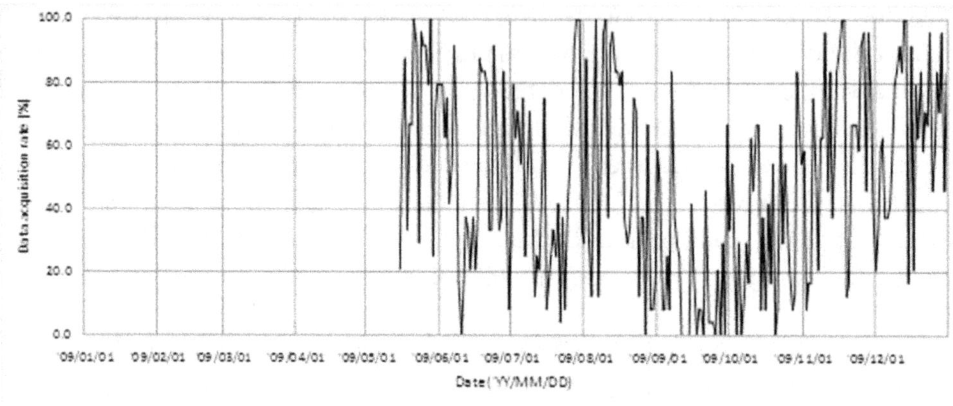

Figure 5b. Data Acquisition Rate By The Master Station In Sapporo: Jan. 1, 2010 to Dec. 31, 2010

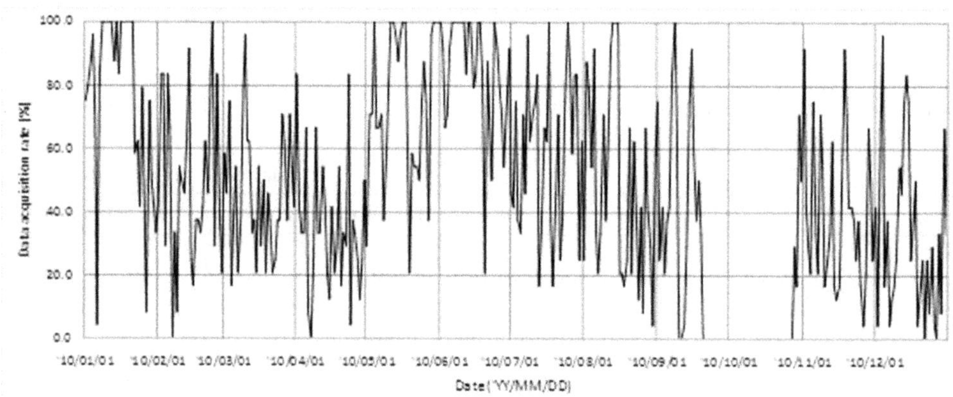

Figure 5c. Data Acquisition Rate By The Master Station In Sapporo: Jan. 1, 2011 to Dec. 31, 2011

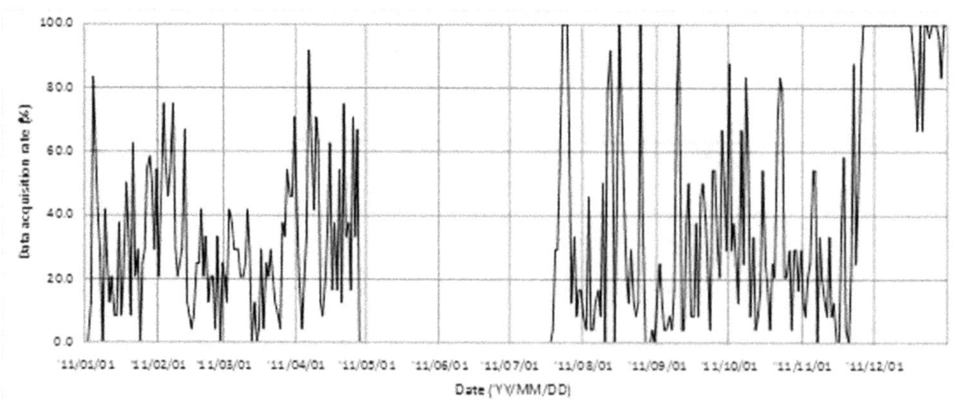

Figure 5d. Data Acquisition Rate By The Master Station In Sapporo: Jan. 1, 2012 to Dec. 31, 2012

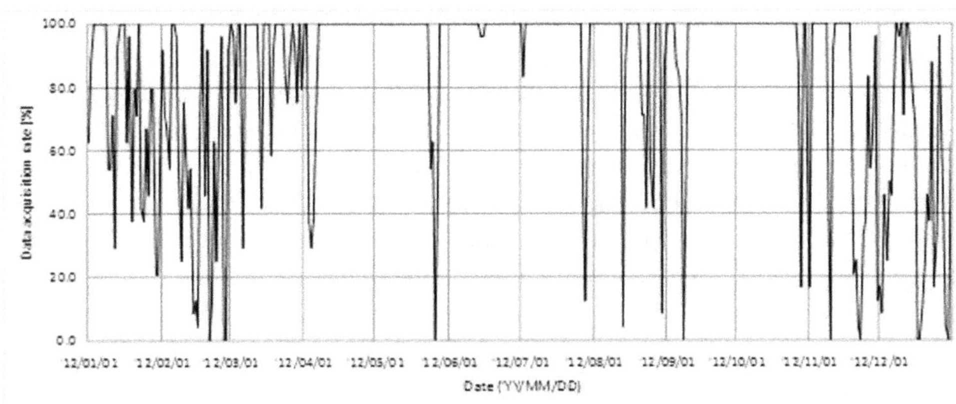

Figure 5e. Data Acquisition Rate By The Master Station In Sapporo: Jan. 1, 2013 to Dec. 31, 2013

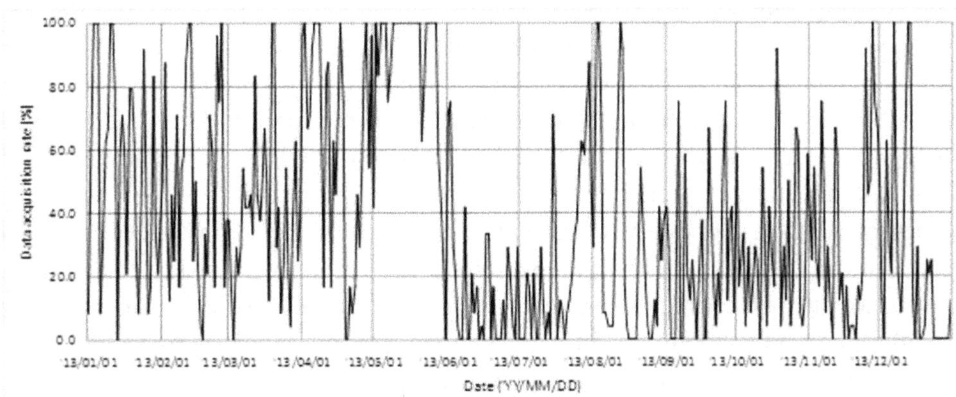

Figure 5f. Data Acquisition Rate By The Master Station In Sapporo: Jan. 1, 2014 to March 12, 2014

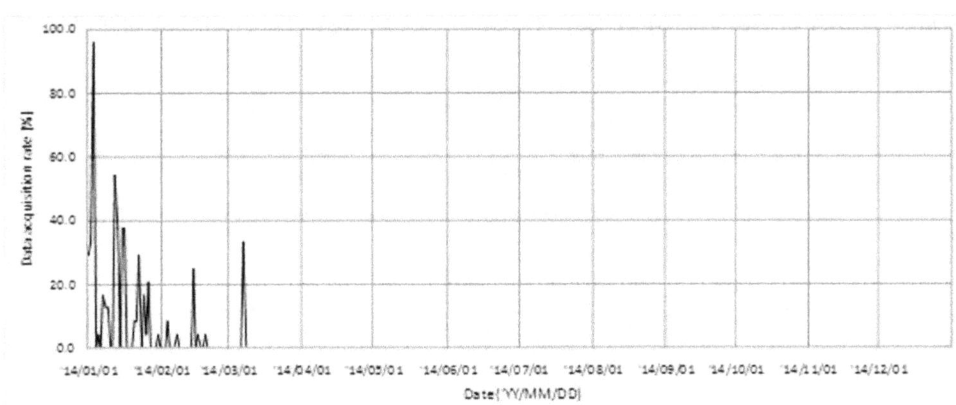

SINGLE TONE TRANSFER EXPERIMENT

The data collection system described in the previous chapter is the actual digital data transmission using meteor burst communication. In this section, the experiments conducted by the author and his research team to examine the frequency and duration of meteor bursts are illustrated. In the experiments, a non-modulated carrier wave is continuously transmitted from the transmitting station at a constant intensity at 48.37145 MHz. The receiving station receives this carrier wave and demodulates it as a 48.3702 kHz USB modulated wave, thereby obtaining an audio signal of 1250 Hz. Thus, since a single tone is used, this experiment is called a tone transfer experiment. A 1250 Hz frequency component is extracted from the received signal using a digital filter. When the intensity exceeds a certain threshold set by the noise level when no signal is received, a meteor burst is determined to have occured. The received signal processing is not performed in real time, but is analyzed by a personal computer (PC) after being recorded.

Figure 6 shows the time chart of the transmitting station and receiving station. Since the transmitting station is operated at 100 W near the maximum output, it will periodically stop for 7 minutes after continuously transmitting the carrier wave for 3 minutes. In consideration of the time lag with the transmitting station, the receiving station starts recording 10 seconds earlier before the 3 minutes transmission (denoted as as #1 in Figure 6) and ends recording 10 seconds later after the transmission. In the receiving station, the output from the receiver is recorded for 1 minute (denoted as #2 in Figure 6) during the interval when the tone signal is not transmitted. The noise level recorded at #2 is used to determine the above-mentioned threshold value. For this reason, the number of detected tone signals, i.e. meteor bursts, is very dependent on the noise environment around the receiving station.

In a nutshell, the receiving station will take 1 minute to record the noise level and allocate another 20 seconds to account for the time lag before and after each 3 minutes carrier wave transmission. Hence, the total time spent to transmit and receive the carrier wave is 5 minutes and 40 seconds. All of the tone signals are detected and the start time, end time, and their duration are calculated by the PC in the receiving station. With the current PC processing capacity, the time given for this analysis has a sufficient margin.

Figure 6. Timing chart of transmission and reception voice recording in tone transfer experiment

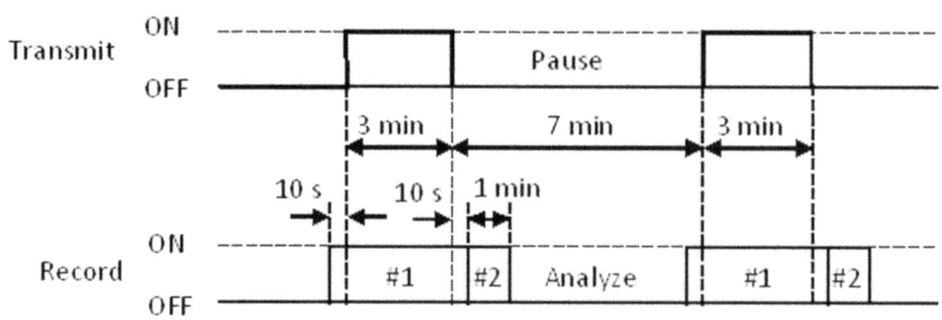

Figure 7. A sample of received signal reflected by meteor burst

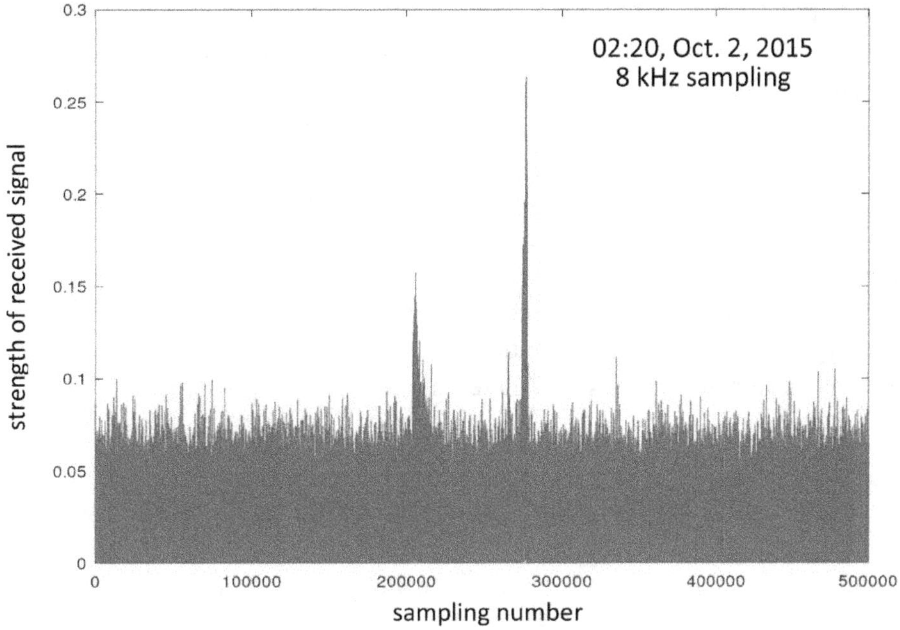

(a) in the case of underdense burst

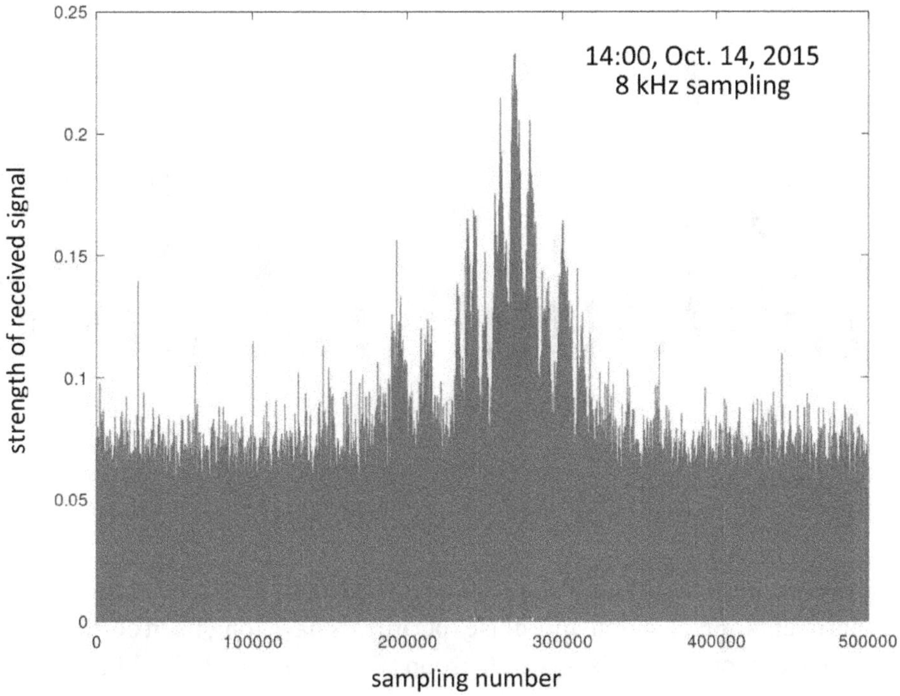

(b) in the case of overdense burst

Figure 7 shows an example of a typical received signal in the overdense and underdense case recorded in October 2015. Since the audio signal is recorded by sampling at 8 kHz, the horizontal axis of the figure is the sample number; the unit of which is 1/8,000 sec. Upon close inspection on Figure 7(a), two underdense bursts can be confirmed, and it can be seen that the second one disappeared in a very short time. By referring to the equation (1) in the previous chapter, it is considered that the second burst is detected with a value of ϕ close to 90 degrees in Figure 3 of the previous chapter. Also, from Figure 7(b), it is confirmed that fading occurs in this overdense bursts. This is presumably caused by fluctuations in the meteor burst itself which could be caused by the airflow above the meteor burst or reflections of other flying objects in the vicinity of the meteor burst.

Figure 8 shows a sample of changes in the number of meteor bursts per day during the month of June 2012. The large fluctuation in the number of occurrences from day to day could be attributed to the difference in the density of scattered meteors.

Figure 8. An example of daily fluctuation of meteor burst occurrence

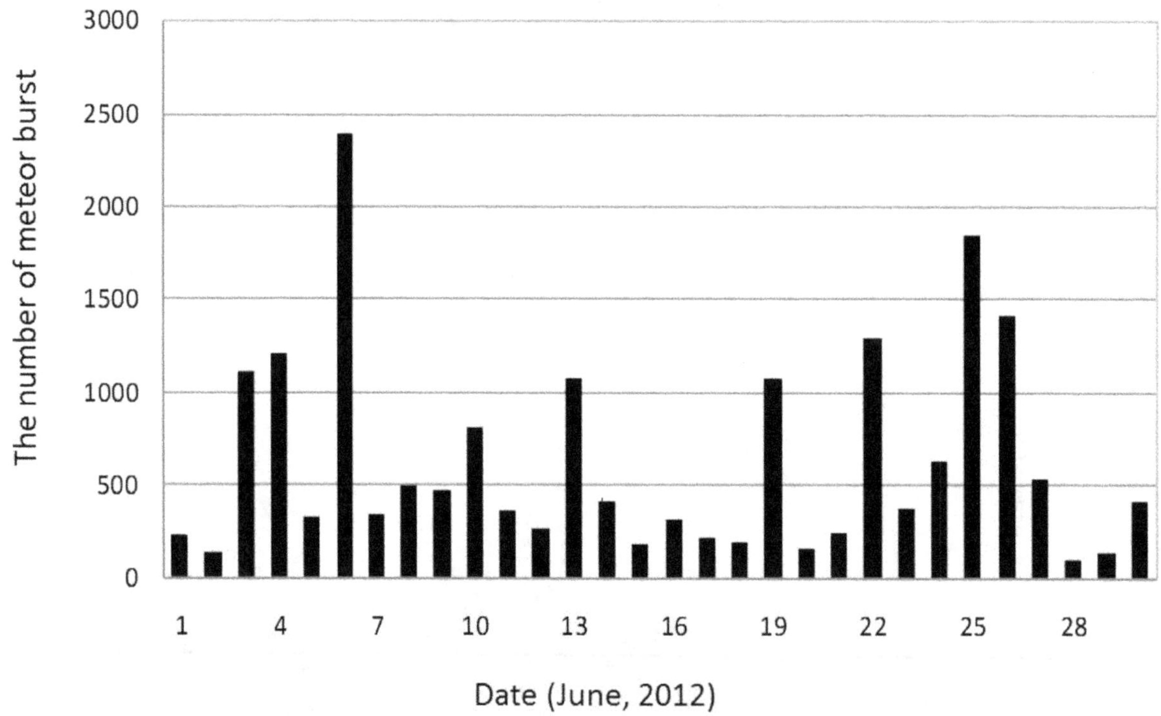

Tone transmission experiments have been conducted at a lot of locations in Japan. Indeed, the tone transmission experiments and data transmission experiments have long been commonly conducted in Antarctica, before the author joined his research group (Fukuda *et al.*, 2003; Mukumoto *et al.*, 2005).

Figure 9a. Two types of transceiver developed by the author's research group: IT-5000S(powered by DSP)

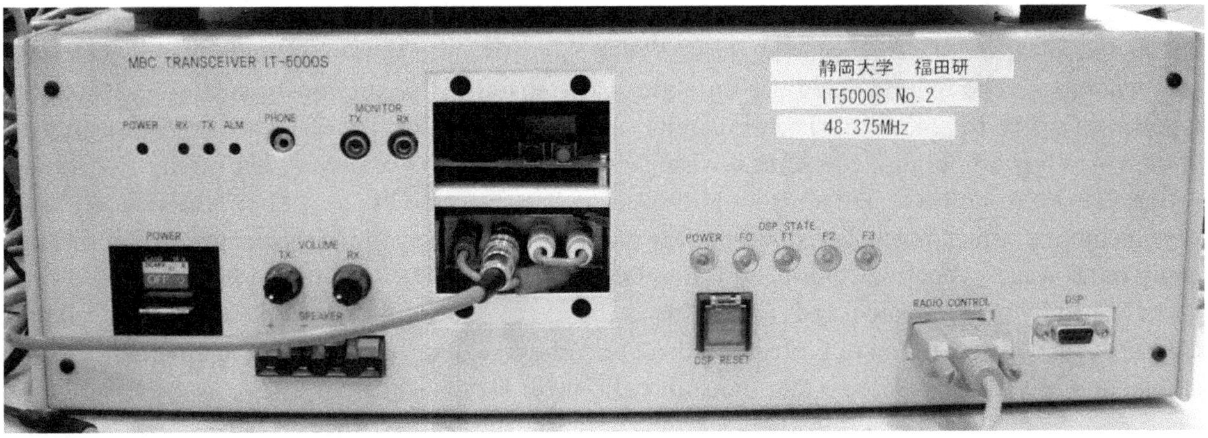

Figure 9b. Two types of transceiver developed by the author's research group: IT-5000N(powered by SBC)

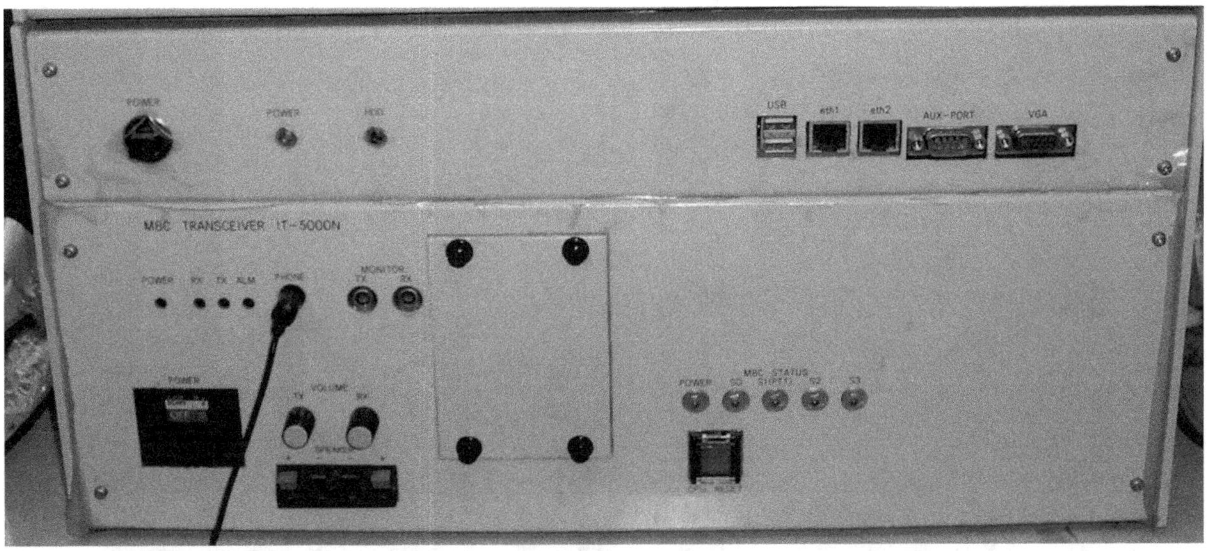

The meteor burst transceiver developed in Japan by both the author's research group and Intertech Co. Ltd., is shown in Figure 9. Figure 9 (a) shows an early model IT-5000S that was also active in Antarctica, with a digital signal processor (DSP) for transmission and reception control. BPSK and QPSK baseband signals can be input from the front panel terminals. Figure 9(b) shows a new model equipped with a single board computer (SBC) so that transmission and reception control, data packets generation, modulation and demodulation can be performed internally. The SBC runs on a Linux operating system, which eliminates the need for external PC control and enables it to operate the transceiver as a software defined radio.

EXPERIMENT BY AMATEUR RADIO STATION

Experiments similar to the tone transmission experiment described in the previous section have been conducted in Japan for a long time by the JA9YDB amateur radio station of Fukui National College of Technology. The unmodulated carrier wave was transmitted continuously at 53.750 MHz with a power of 50 W for 24 hours. The transmitter and the antenna of this beacon carrier transmitting system are shown in Figure 10. After it was taken over by Fukui Prefectural University's corporate office, JH9YYA, however, the amateur radio station is now transmitted at 53.755 MHz with 50 W. This technique, which is employed by amateur radio operators to examine the occurrence of meteors using radio waves, is called Ham-band Radio Observation (HRO). This technique has indeed been widely used by many amateur radio enthusiasts and astronomical observation enthusiasts in Japan. Currently, in addition to Fukui Prefecture, there are other amateur radio stations in Hokkaido, Fukushima, and Miyazaki Prefecture that transmit radio wave for the same purpose.

Figure 10. The beacon carrier transmitting system for HRO of JA9YDB in Fukui National College of Technology. The photographs were taken in Sept. 12, 2019

(a) The beacon carrier transmitting system for HRO in JA9YDB. The IC-7400 made by ICOM Inc. is used for transmitter.

(b) 2-element cross Yagi-Uda antenna set elevation angle as 90 degrees for transmitting the beacon carrier at 53.750 MHz. The Left side is Professor Maegawa and the right side is the author.

Besides observing meteors using radio waves in the 50 MHz band, communication among amateur radio stations is also performed by digital packet transmission using scattering radio wave by meteors in the VHF or UHF band. The WSJT-X developed by K1JT, call sign of amateur radio station, is a well-known software that realizes this meteor scatter communication in the amateur radio frequency band. This software has various modulation modes, but in meteor scatter communications, a mode called MSK144 based on OQPSK modulation is often used (Taylor, 2001; Steven, 2017). In addition, this software has a very useful function that shows the hot spot described in the previous section as an angle when the positions of the local station and the communication partner station are input. At the same time, the distance between the two points, the antenna direction and elevation angle to be set are also shown. These position data are input by the Grid Locator i.e. Maidenhead Locator System. In meteor burst communication, a probe packet was transmitted for 28 ms and a waiting time of 45 ms was provided for reception. However, in the operation of this software, a call signal is continuously transmitted for 10 seconds, and a waiting time of 10 seconds is provided for confirming the response. It is possible to change this 10 seconds to 5 seconds by software setting. After the response is confirmed, the mode shifts to a mode in which communication is terminated in a very short time. For this reason, the clocks of both PC need to be accurately matched with each other. So, it is essential for the PC connected to the radio station to be synchronized by a network time server. In view of such a flow of time, it seems that communication by the WSJT-X assumes overdense burst.

CONCLUSION

In this chapter, a meteorological observation data collection system using meteor burst communication and a single-tone transmission experiment system that investigates the occurrence time and duration of meteor bursts, which are conducted by the author's research group were outlined and some of the experimental data was introduced. At present, the data collection rate of the data collection system has not yet reached 100% when the number of meteor showers is decreasing, especially during the winter season in Japan. Endeavours to improve the collection rate are therefore necessary in the future. The results of the single-tone transmission experiment shown in this chapter were measured in the mid-latitude region. The results may vary since they are dependent on the electromagnetic noise environment around the system, the positional relationship, distance, and direction of the two points of the transmitting and receiving stations. When constructing a meteor burst communication system, it is necessary to estimate in advance how much throughput can be expected from the tone transfer experiment. The transceiver for the meteor burst communication experiment built by the author's research team is also shown in this chapter. Since meteor burst communication is often used in areas where infrastructure is not sufficiently developed, it is necessary to reduce the size and power consumption of the transceiver in the future.

Apart from the radio systems used by the industry and academic institutions, some of the recent progress in meteor scatter communications made by amateur radio enthusiasts was also introduced at the later part of this chapter. Compared to the very primitive software that has been created and used, the WSJT-X is astounded for its superior performance and wide range of functions. The author hopes that collaborative research could be performed together with amateur radio operators in the world.

ACKNOWLEDGMENT

The experiment in this chapter was supported partly by the Grant for Joint Research Program of the Institute of Low Temperature Science, Hokkaido University.

REFERENCES

Fukuda, A., Mukumoto, K., Yoshihiro, Y., Nagasawa, M., Yamagishi, H., Sato, N., ... & Jin, L. J. (2003). Experiments on meteor burst communications in the Antarctic, *Advances in polar upper atmosphere research*, 17, 120-136.

Mukumoto, K., Fukuda, A., Nagasawa, M., Yoshihiro, Y., Nakano, K., Ohichi, S., ... & Yao, M. W. (2005). VHF data transmission experiments using MBC equipment conducted during the period from JARE-43 to JARE-45. National Institute of Polar Research. *Advances in polar upper atmosphere research*, 19, 89-105.

Steven, J. F., & Joseph, H. T. (2017). *The MSK144 Protocol for Meteor-Scatter Communication.* Retrieved from http://www.arrl.org/files/file/QEX_Next_Issue/SeptOct2017/FrankeTaylor.pdf

Tayler, J. (2001). *WSJT: New Software for VHF Meteor-Scatter Communication.* Retrieved from http://physics.princeton.edu/pulsar/K1JT/WSJT_QST_Dec2001.pdf

Section 6
Cosmic Sources

Chapter 13
The Radio Sun and Planets

Hui Chieh Teoh
Universiti Tunku Abdul Rahman, Malaysia

ABSTRACT

The radio emissions from the Sun and planets provides a new window to explore and improve our understanding of these celestial objects and its magnetic activities. Radio emissions allow us to see things that are not seen in other wavelengths. The radio emissions observed from the Sun and planets can generally be divided into thermal and non-thermal. The thermal emission is related to the temperature of the objects, i.e. the blackbody radiation and the non-thermal emission is usually associated with the magnetic fields.

THE RADIO SUN

Our Sun is one of the earliest celestial objects studied by radio astronomers. The radio emissions from the Sun provides a new window to explore and improve our understanding of the solar atmosphere and its magnetic activities. Solar radio astronomy has become an emerging field that combines solar physics and radio astronomy, where large amount of observations data of the radio Sun coupled with plasma physics has been used to understand the complex and dynamics processes of the origin of solar radio emissions.

Short History

The detection of radio waves from extraterrestrial sources such as our Sun were unsuccessful in the earlier years. This was due to the primitive technology of the radio telescope during its early stage of development and also of the absence of knowledge of the existence of a screening ionosphere. Another hold back that the Sun's radio emission was not detected almost a decade after celestial radio waves were detected was due to the missing knowledge that solar activity affects the radio emission of the Sun, and some experiments unfortunately were carried out during the periods of low solar activity.

So it was not until 1942 that the first detection of the solar radio burst was made, by accident. During World War II, on February 26 to 28, 1942, the British army received strong interference that initially was suspected to be jamming signals transmitted by the Germans. But soon it turned out that it was caused

DOI: 10.4018/978-1-7998-2381-0.ch013

by the Sun. James Stanley Hey noted that the radiations were between dawn and sunset, and were not observed at night. He concluded that the disturbance was of solar origin by going through the bearings and elevations of the receiving sets and found that the bearings moved throughout the day and was always within a few degrees of that of the Sun. It is also noted by Hey that the radiation appears to have been associated with the occurrence of a big sunspot group on the Sun at that time.

Earlier in 1940, Grote Reber observing the Sun at a wavelength of 187 cm reported negative results. Again, in a paper published in 1942, he characterized his results as inconclusive. However, in 1944, he finally reported the detection of solar radiation with his new and improved apparatus. Reber suggested that this long-wave radiation could be set up in the corona of the Sun. In 1945, George Clark Southworth published his results on the observation of microwave radiation coming from the Sun during the summer months of 1942 and 1943. The observation was carried out at three widely spaced wavelength in the region between 1 and 10 cm.

In the years following World War II, the development of radar antenna and receiver technology due to the war has helped to flourish the field of solar radio astronomy. The need for continuous and systematic observations of the Sun was recognized due to the great variability of the Sun's radio emission. The two leading countries that have made significant contributions to the field then were England and Australia, where the basic techniques of observations, such as spectrographic measurement, flux measurement and polarization measurement were developed.

Radio Emission from The Sun

The radiation emitted by our Sun is not only limited to visible light. In fact, it emits radiation over a very wide frequency range in the electromagnetic spectrum from ultraviolet to infrared and to radio waves. It is one of the strongest radio sources in the sky. The Sun emits radio waves through two mechanisms: thermal and non-thermal. The emission through thermal mechanism is due to its high temperature and at higher frequencies. The outermost solar atmosphere, i.e. the corona, is too tenuous to be detected by other ground-based observations except for its radio emissions. These emissions are caused by the gyrations and collisions of thermal electrons in the magnetic field that are able to produce sufficient power to be detected on the ground. The non-thermal mechanism emits at lower frequencies and is due to synchrotron radiation when electrons spiral around magnetic field lines. The violent and dynamic events happened in the solar atmosphere (such as flare and corona mass ejection) are characterized by their non-thermal radio emissions.

The radio emission from the Sun is rather complicated. From analyzing the solar radio flux density as a function of wavelength (Figure 1), it can be seen that for wavelength greater than 1 cm (frequency less than 30 GHz), the solar radio emission is divided into two parts, identified as the "quiet Sun" (constant component) and the "active (or disturbed) Sun" (variable component). The quiet Sun is the background solar radio emission due to heat at the time of minimum sunspot activity and is relatively stable. Thus, it is often used for calibration. The active Sun (radio burst) is due to strong magnetic activity such as flare and sunspots, which can exceed the quiet Sun emission by several orders of magnitude. There is a third component known as the slowly varying component, the slow day-to-day variation in radio emission, which is associated mainly with active regions, visible or invisible, on the Sun's disk.

The Quiet Sun

The thermal radio emission of the quiet Sun is mainly free-free emission (thermal bremsstrahlung) from electrons interacting with ions in the presence of a relatively weak magnetic field. The emission is called free-free radiation because electrons scatter off ions without being captured, i.e. the electrons are free both before and after interaction with ions. Free-free emission arises from the atmosphere of the Sun which consists of layers of ionized gas with different electron density and temperature. This emission affects the opacity of the Sun. A certain electron density is related to a certain critical radio frequency below which propagation of radio waves is impossible.

The so-called surface of the Sun, the photosphere, primarily emits light in the visible spectrum. The temperature of the photosphere is about 5800 K and the gas here produces an almost perfect blackbody spectrum because it is in thermal equilibrium with the photosphere. The gas density drops as the distance from the photosphere increases and the temperature reaches a minimum of 4000 K at a height of approximately 400 km (temperature minimum region). Above the surface is the chromosphere, which extends from about 400 km to 2100 km. In this layer, as the height increases, the temperature rises to

Figure 1. Solar flux density across the electromagnetic wavelength showing the solar radio emission of quiet sun and active (disturbed) Sun

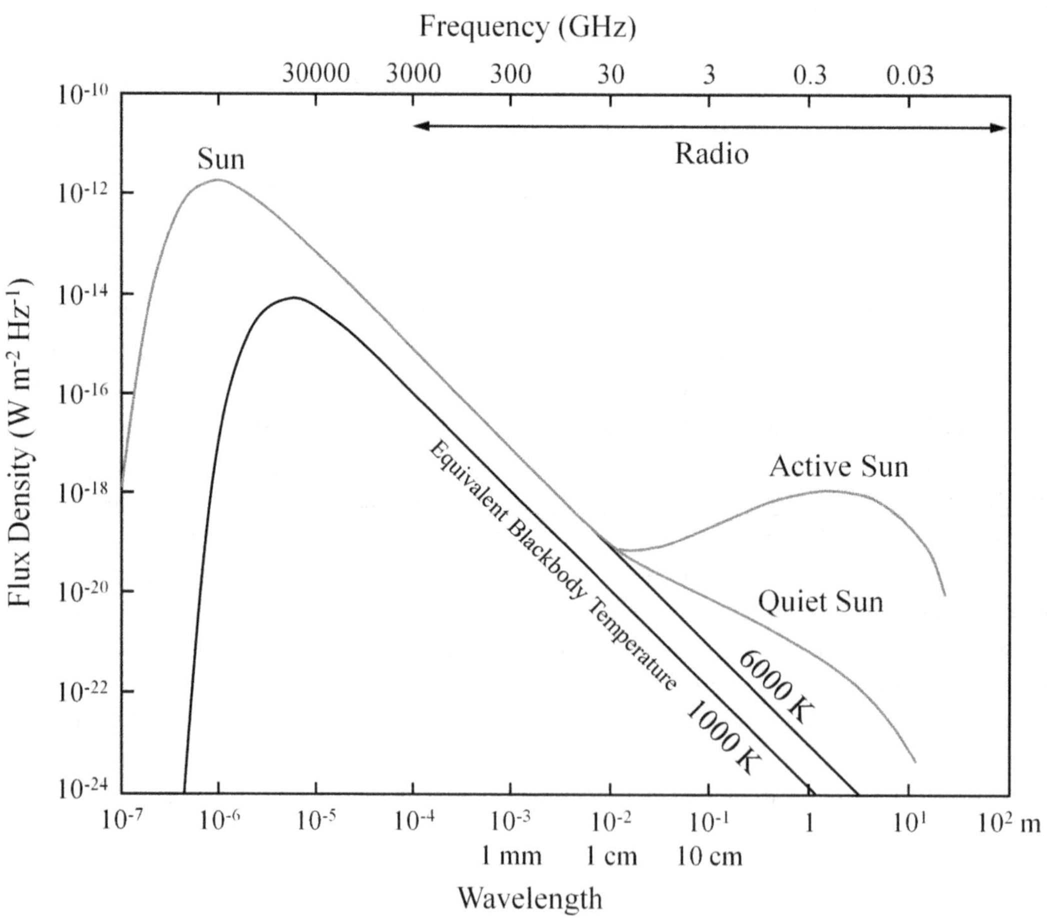

8000 K at the top. The temperature continues to rise as the height increases into the corona, the outermost atmosphere of the Sun, where temperature can reach over a million K. At the same time, the gas density drops significantly and the gas in the chromosphere and corona become very tenuous.

The radio emission from a certain height of the atmosphere can only be observed if the frequency is higher than the plasma frequency. Radiation cannot propagate if its frequency is lower than the plasma frequency. Plasma frequency is a function of electron density and is given by

$$\nu_p = 8.98 \times 10^3 \sqrt{N_e}$$

where ν_p is the plasma frequency in Hz and N_e is the electron density in cm^{-3} (Melrose, 1985). Since the plasma frequency increases with electron density, at lower height (i.e. nearer to the photosphere) where the electron density is higher, only higher frequency emission (shorter wavelength) can be observed. This emission characteristic allows the frequency spectrum to be an indirect measure of height in the solar atmosphere if the density distribution is known. In addition to the refraction effect, the opacity of the Sun will increase as the wavelength of observation increases.

Solar radio radiation originated above the photosphere where emission at millimetric wavelength (higher frequency) is mostly from the chromosphere and metric wavelength (lower frequency) is from the corona. These different wavelengths emitted from different parts of the Sun cause the diameter of the observed Sun to vary. The Sun at millimeter wavelength (less than 100 mm or above 3 GHz) appears similar though slightly greater in size than its optical counterpart since it originated originates closer to the photosphere. At decimeter wavelength (between 100 mm to 3 m), the apparent size of the Sun is larger and the intensity at the edge (corona) is higher than that at the centre (photosphere) giving the effect of limb brightening. At meter wavelength (more than 3 m or below 0.1 GHz), the observed Sun looks much larger compared to shorter wavelength. The lower frequency radio waves originate in the corona gives the observed Sun a larger dimension. The apparent diameter of the Sun can be twice as large than the photosphere at even lower frequencies. The Sun at this wavelength is also brighter in the centre and more variable. The limb brightening effect vanishes as the photosphere disk at the centre is lost behind the bright corona. Figure 2 shows the radio Sun at millimeter wavelength.

The Active Sun

Our Sun has an 11-year activity cycle, also known as solar cycle, where the levels of solar radiation, number of sunspots, coronal mass ejections (CME), solar flares and other solar activities show synchronize fluctuation. The solar cycle has an active ~ 7-year period during which solar activities are high and a quiet ~ 4-year period during which solar activities are low.

Solar activities are all in some ways related to the strong magnetic fields of the Sun and it will always disturb the solar atmosphere. The violent and dynamic events happen in the solar atmosphere such as CME and flare during the active period of the solar cycle are characterized by their non-thermal radio emissions, which are evidence by the observed brightness and the speed with which the intensity changes. A coronal mass ejection (CME) is a large ejection of plasma which can expulse billions of tons of material consisting primarily of protons and electrons and carry embedded magnetic field from the solar corona into the solar wind. CME usually originated from active regions, such as sunspot groupings, on the photosphere and took about one to five days to reach Earth. They frequently, but not always, follow

Figure 2. The radio Sun imaged by the Very Large Array (VLA) radio telescope. (a) The quiet Sun at 4.6 GHz where red and yellow are regions of million-degree gas coincide with sunspots, (b) the Sun at 1.4 GHz with strong emissions from active regions near the equator (Courtesy: NRAO / AUI / NSF)

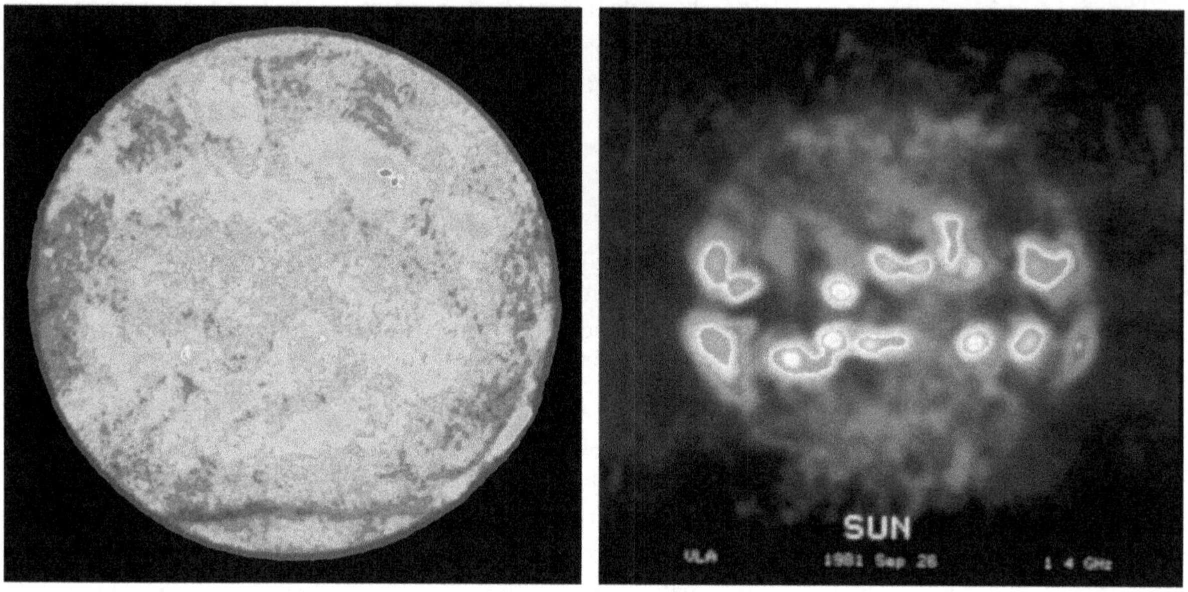

(a) (b)

solar flares. Solar flares are sudden outburst of electromagnetic radiation that can last from minutes to hours. Solar flares, like CME, are usually associated with active regions where the magnetic fields are strong. These magnetic fields can become unstable as they evolve and release their energy that can be observed as sudden flash of increased brightness on the Sun. The electromagnetic radiation travels at the speed of light and reaches Earth at the same time the event is observed, which is approximately eight minutes. Solar flares can be classified into two types: impulsive and eruptive. Impulsive solar flares only develop in the lower layer of the solar atmosphere and have durations of seconds to minutes, with fast increase in the intensity of centimeter and decimeter radio emissions. Eruptive solar flares, on the other hand, have emissions from minutes to hours, and can eject large quantities of matter and energy into interplanetary space.

During the period when the Sun is active, CME and flares (bursts and storms) will occur more often than usual. The Sun can dominate across all frequencies, being several orders of magnitude more than the quiet Sun at lower frequencies during storming and outbursts activities. Optical observations of solar flares have been carried out by astronomers since the 1930s and it was known that large optical flares were accompanied by intense bursts of radio emission over a range of wavelengths originated from the acceleration of electrons in the corona. Although flares emit radiation across the electromagnetic spectrum, X-ray and extreme ultraviolet (EUV) emission are undetectable due the low electron density in the solar corona causes very low levels of emission.

The meter-wave solar radio bursts (at frequencies below a few hundred MHz) are classified into five spectral types as presented in Figure 3 despite the wide variety exhibited by their spectra (Wild et al., 1963).

- **Type I**: This type of radio bursts is a non-flare related phenomenon. The bursts are non-drifting, short duration (~ 1 second) and narrow band. This event usually occurs in large numbers, forming irregular structures and superimposed on a broader band continuum (continuous emission) often referred to as a "noise storm". While the burst might be short, the storm may last for few hours to days. This long duration is a characteristic that distinguish Type I from the other types.
- **Type II**: This type of radio bursts is always seen in conjunction with flares. They start about 5 – 20 minutes after the emergence of a flare and occur after the maximum of the flare. They are strong events with slow drift from high to low frequencies, which implies an upward movement of exciting agent from the lower layer to the upper layer of the corona. Type II bursts can be easily discerned from Type III bursts by their frequency drift rate, in which the drift rate for Type II bursts is typically two order of magnitude slower than the fast-drift Type III bursts. The bursts are generated by magneto-hydrodynamics shock waves that propagate through the corona into interplanetary space. A typical burst of this type shows two-band emission consisting of the fundamental emission band and the second harmonic frequency structure (Figure 4). This may last for minutes.
- **Type III**: This is the most common type of solar radio bursts and it often accompanies the impulsive (flash) phase of large flares. This type of bursts are strong events caused by the emission of electron beams. The bursts have a rapid drift from high to low frequencies (Figure 4) implying that the beam moves through decreasing electron density regions from the lower layer to the upper layer of the corona. Type III and Type II bursts have similar layer of origin. The electron beams moving up the corona along the magnetic field lines stimulate the electron plasma to oscillate with the local plasma frequency and may exhibit harmonics. The harmonic structure of Type III is harder to recognize compared to Type II because the rapid drift of frequency combined with broad bandwidth causes the merging of fundamental emission band and the second harmonic. Type III burst has short duration, lasting only seconds, but they cluster into groups, which can last for minutes.
- **Type IV**: This type of radio bursts is related to large solar flare and are broad band continual (continuous emission). Hilary V. Cane and Donald Vernon Reames (1988) found that generally Type

Figure 3. Schematic diagram of the basic types of meter-wave solar radio bursts (Dąbrowski, et al., 2016)

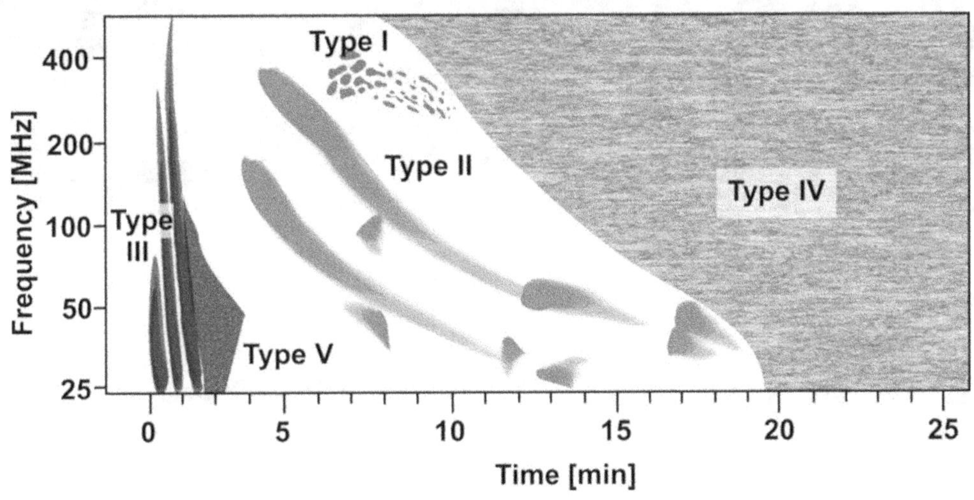

IV bursts are associated with Type II emission. These bursts are relatively rare in the meter-wave spectrum and can be categorized into moving Type IV (mA) and stationary Type IV (mB). The moving type may last from minutes to 2 hours and is characterized by slow drift in frequency. The stationary type can last from few hours up to days and is characterized by long-lived broad band continuum, slow variations with some fine structures.

- **Type V**: This type of radio bursts is rare and can be difficult to distinguish, especially if there are other bursts occurring at the same time. They are broad-band continual (continuous emission) associated with Type III bursts. A Type V burst possibly emerges from the decay phase of the Type III burst ahead of it (Figure 5). They occur at similar heights as Type III, showing similar dispersion of position with frequency. They may last from one to two minutes, with duration increasing as frequency decreases.

Apart from meter-wave solar radio bursts, the study of microwave (centimeter wave) bursts is also important because these bursts take place in the same region of the solar atmosphere as flares. Consequently, these emissions can provide insight on the region of energy release and energetic particles ac-

Figure 4. A group of Type III bursts followed by Type II bursts with fundamental and second harmonic frequency structure. Data obtained from the Green Bank Solar Radio Burst Spectrometer (GBSRBS) (Courtesy: NRAO / AUI / NSF)

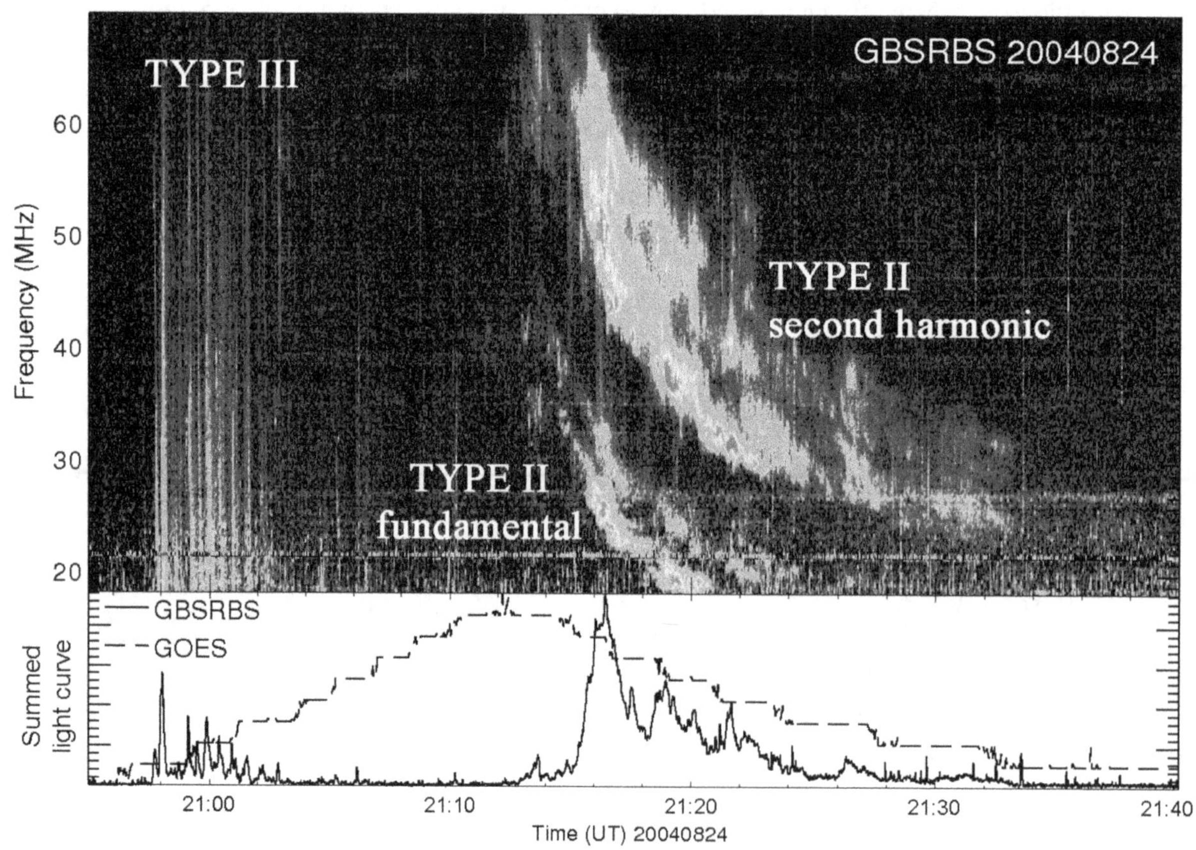

Figure 5. A Type V burst, typically described as extended phase of a Type III burst. Data obtained from the Green Bank Solar Radio Burst Spectrometer (GBSRBS) (Courtesy: NRAO / AUI / NSF)

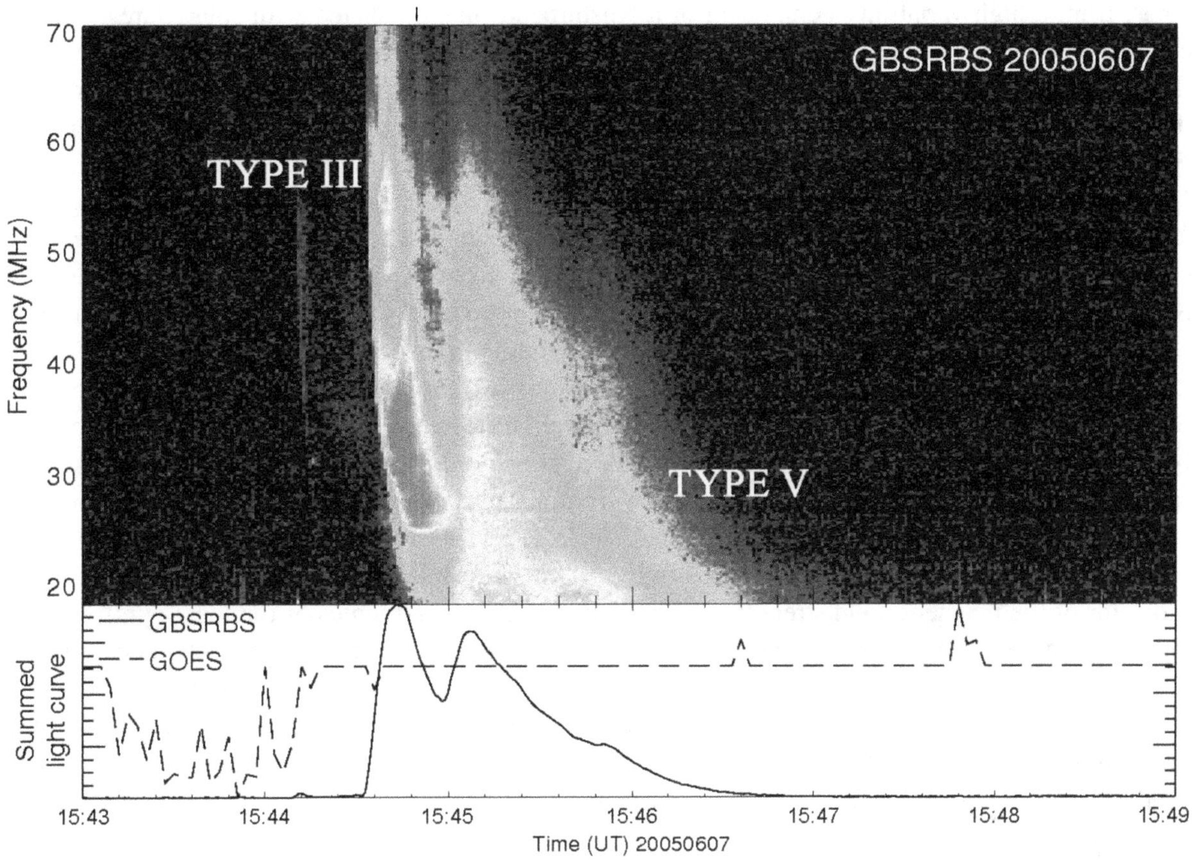

celeration. Microwave bursts are less varied, relatively simpler and smoother, and usually free of details in time and frequency. Wild et al. (1963) classified three main physically significant types of these higher frequency bursts, namely, the gradual burst, the impulsive burst, and the microwave Type IV (IV μ) burst.

The gradual burst, as the name implies, has gradual rise and fall for a duration of a few tens of minutes. It has a slow rise to maximum intensity, on average takes almost half the duration of the burst, and also a slow decline to the pre-burst level. The start of the gradual burst coincides closely with an optical flare onset and its maximum intensity coincides with or after the flare maximum. The impulsive burst, on the other hand, has a sharp rise to a peak intensity and followed by a slower decline to the pre-burst level for a total duration of one to five minutes. The rapid rise on average takes about a quarter of the total duration and also occurs near the start of an optical flare. The meter-wave Type IV (IV m) burst as described above is almost always associated with microwave Type IV burst. Only more intense microwave Type IV burst due to larger flares will develop a strong meter-wave component. Many bursts of this type are of quite low intensity and has duration of five to thirty minutes.

Slowly Varying Component

Apart from a high variability component of a "disturbed Sun" associated with solar flares, which is characterized by strong emissions ranging from seconds to hours as described in the previous section, there exists a lower variability component (slowly varying component) associated with active regions, dominant in the centimeter range with duration of days to months. During the partial solar eclipse in November 1946, the Moon covered a large sunspot group on the solar disk and Arthur Edwin Covington (1947) observed a reduction in the radio emissions from the Sun. From this observation, he suggested a correlation between the solar radio emissions and active regions. The emissions of the slowly varying component have a time dependence of roughly 27 days, which corresponds to the rotational period of the Sun, also indicates that the emission source are local regions on the disk and not from the Sun as a whole.

Figure 6 shows the flux density of the slowly varying component compared to the quiet Sun and active sun. The slowly varying component may have only a slightly higher flux density than the quiet Sun, but it can have much higher brightness temperature. The slowly varying component is characterized by thermal radio emissions. The non-flaring active regions produce emissions through gyro-resonance radiation above sunspots and free-free radiation from hot plasma trapped in active region loops. The gyro-resonance radiation is produced when hot electrons gyrate at a slightly higher electron velocity around the coronal magnetic field where relativistic effect starts to come in. This emission can be a good indicator of the magnetic field strength as it is limited to a few harmonics of the local gyrofrequency. Gyro-resonance radiation is associated with the active regions with bright, compact sunspot groups. However, there are times when radio emissions are detected when no sunspot is visible. These no sunspot-associated emissions are free-free radiation attributed to faculae or plage with low-brightness. In regions with strong magnetic field, the emission is gyro-resonance and in regions that the magnetic field is too weak for gyro-resonance, the free-free radiations dominate.

RADIO EMISSION FROM THE PLANETS

Short History

The radio emissions observed from planets can generally be divided into thermal (approximate blackbody radiation) and non-thermal (cyclotron or synchrotron radiation). In the Solar System, all the planets' emission at visible wavelength is due to the light reflected from the Sun. On the other hand, the radio emission from the planets is mainly due to thermal emission since there is very little reflected "sun-light" at this wavelength. The thermal emission is related to the temperature of the planetary disk – the Planck (blackbody) radiation due to the planets' non-zero temperature. The first thermal radio emission of a planet was detected from Venus in the mid-1956 by Cornell Henry Mayer, Timothy Pendleton Mc-Cullough Jr. and Russell M. Sloanaker using the U.S. Naval Research Laboratory 50-foot reflector at 3.15-cm wavelength. After the successful detection of radio emission from Venus, the group attempted the observations of Mars and Jupiter in the following months when the planets were near opposition in September 1956 and March 1957, respectively. The thermal emissions were successfully measured and the blackbody temperature for these three planets were determined.

Prior to the detection of radio thermal emission from Venus in 1956, the study of planetary radio emissions was started in 1955 with the first discovery of radio emissions from Jupiter by Bernard Flood

Figure 6. The slowly varying component dominates in the centimeter range and may has a higher flux density than the quiet Sun (adapted from Wild et al., 1963)

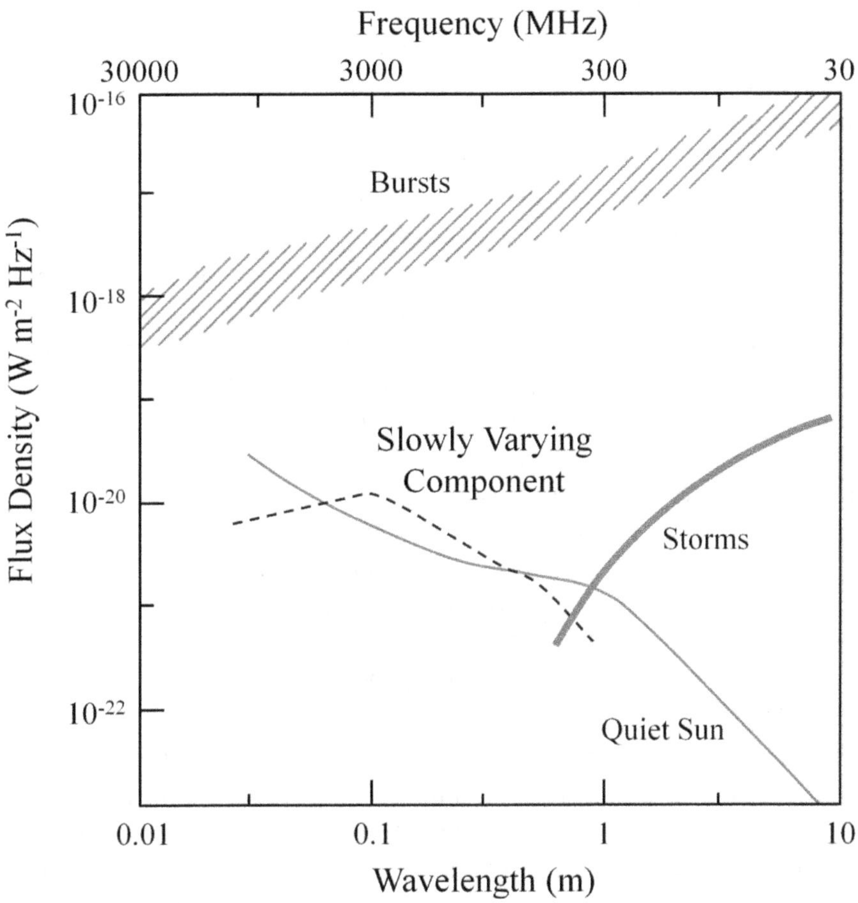

Burke and Kenneth Linn Franklin from the Carnegie Institution of Washington at a frequency of 22 MHz (1.36 decametres). This was an accidental discovery. Franklin later recounted that "Our identification of Jupiter as a radio source is not based directly on reasoning, but more on luck.". At that time, they were observing the Crab Nebula in Taurus using the Mills Cross antenna, which was made up of a pair of 66-dipole linear arrays arranged in a slightly flattened X, and slowly changing the observing direction progressively towards the south. A number of these observations showed brief interference that the scientists were unable to identify the source. Initially, the interferences were thought to be terrestrial, which is very common at these frequencies such as from power lines or car ignitions.

After analyzing the data over many nights, it was found that the source was very unlikely to be terrestrial. The interference appeared to be occurring about four minutes earlier each night, and this is the type of change with time that would be expected from objects in the sky such as stars or galaxies due to our Earth's revolution around the Sun. However, closer inspection of several months of data on the changes in the interference's timing revealed that the source did not quite match up with the movement of stars. This would eliminate the source as star or galaxy or other far away celestial object since these objects will move across the sky at the same rate. It has to be something nearer, such as planets, that will change position with respect to the background stars. During that time, Jupiter happened to be near the Crab Nebula and the mean position of the interference agreed with Jupiter's position in the sky. In the early months of the observation, although Uranus was near Jupiter, it was far outside the range of position indicated by the interference source towards the end of the observation period. Thus, the scientists concluded that the source of radio emission interference must be associated with Jupiter.

After the announcement of this discovery, a radio astronomer from Australia, Charles Alexander Shain, poured through his old records and found that he had actually observed radio emissions from Jupiter five years earlier in 1950, but had attributed it to interference. These pre-discovery data were of great value to understand the Jovian radio emissions and soon the emissions were interpreted as cyclotron emissions. The emission was observed to be circularly polarized and the sense of the polarization was predominantly right-handed. Apart from being thermal radio emitters as expected, the discovery that planets are also non-thermal radio emitter was unexpected. These non-thermal radio emissions indicated the existence of magnetic field around the planet. Subsequent observation by spacecrafts such as Voyagers, Ulysses and Galileo revealed that all planets with magnetosphere will produce intense non-thermal radio emissions, particularly the outer planets.

In 1958, Saturn radio emission was first detected by Frank Donald Drake and Harold Irving Ewen using a 28-foot parabolic reflector at 3.75-cm wavelength. Two years later, Jerald J. Cook, Lloyd G. Cross, Max E. Bair and C. Blake Arnold detected the 3.45-cm wavelength radiation from Saturn using the University of Michigan's Radio Astronomy Observatory 85-foot paraboloidal antenna. The next planet to have its first radio emission detected was Mercury in 1960 – 1961 by William Eager Howard, Alan Hildreth Barrett and Frederick Theodore Haddock using the same radio telescope from the University of Michigan's Radio Astronomy Observatory at 3.45- and 3.75-cm wavelength. The thermal radio emission from Uranus was observed by Kenneth Irwin Kellerman using the CSIRO (Commonwealth Scientific and Industrial Research Organisation) 210-foot radio telescope at 11.3-cm wavelength in 1966. In the same year, together with Ivan Il'ja Karol Pauliny-Tóth, Kellermann again investigated the radio emission from Uranus, but this time using the NRAO (National Radio Astronomy Observatory) 140-foot radio telescope at 1.9-cm wavelength. The observations using the NRAO radio telescope also included the measurement of radio emission from Neptune.

The observations of radio emission from the planets were extended to other wavelengths in the subsequent years and it was interesting to find that Venus is much hotter than expected and this has led to a re-examination of theories of planetary atmosphere. It was also found that Jupiter has two separate non-thermal radio emissions, where the decametric emission (DAM) might be associated with the solid body of Jupiter and thus give the rotational period of the solid body, and the decimetric (DIM) emission suggesting a type of Jovian magnetic field equivalent of the terrestrial van Allen belt. All the new findings on the conditions of the planets are crucial in the design of spacecrafts that were to be sent in the future to explore these worlds at that time. If the conditions of a planet were not known well enough, the first spacecraft sent to the vicinity of the planet might be improperly instrumented, perhaps to the extent that rendered the spacecraft useless.

Radio Emission from Jupiter

The sporadic radio bursts of Jupiter can sometimes be very powerful and surpass the intensity of the Sun (except for strong solar radio bursts), temporary making Jupiter the brightest radio source in the sky. Most of the Jovian radio emissions are polarized, i.e. the emissions travelled with one preferred direction instead of travelling randomly, implying the existence of a magnetic field. The study of Jovian radio emissions also found that the bursts are not emitted in every direction but rather are directional and narrow beams. They depend on which longitude of the planet is facing us, suggesting that the events may be localized, but they did not seem to correspond with any cloud features including the Great Red Spot. However, the emissions match well with the planet's rotation period and stay constant. Combining the observation of localized emission reappearing at constant period and the existence of a magnetic field, it can be inferred that the emissions may be rotating at the same rate as the part of Jupiter that generates the magnetic field.

Some of the directional and narrow radio beams can be associated with sources related only with certain alignments of the magnetic field (Io-independent sources), while others can be attributed to the position of the moon Io in Jupiter's magnetosphere (Io-dependent sources). The intensity and the predictability of Jovian decameter (DAM) emissions are influenced greatly by Io's motion in its orbit. There are some particular positions in Io's orbit where the radio emissions are stronger. Io is the most volcanically active body in the solar system and materials are being ejected into space every second. These materials then become ionized and are trapped within the magnetic field of Jupiter. As Io orbits around Jupiter, Jupiter's magnetic field moves rapidly past it and is disturbed, generating currents that fuel the decametric emissions. The Io-independent sources are called Non-Io-A or Non-Io-B, while Io-dependent sources are called Io-A, Io-B and Io-C, roughly in order of the chances of observing them. Figure 7 shows the occurrence probabilities against Jupiter's Central Meridian Longitude (CML) (top left) and Io-phase (Φ_{Io}) (bottom right). CML is defined by the longitude of Jupiter facing the Earth at a particular time, whereas Io phase is defined by the orbital position of Io. When Io is directly behind Jupiter, the phase is 0°. The phase increases as Io orbits around Jupiter until when it crosses in front of Jupiter as seen from Earth, the phase is 180°. It can be seen in Figure 7 that the occurrence probabilities of the radio emissions vary with the CML and were modulated by the magnetic field's rotation. The Io-independent components can be detected regardless of the Io phase, but the Io-dependent components can only be detected at certain Io phase. The Io-dependent sources have also higher likelihood to being observed than the Io-independent sources.

The Jovian decametric emissions are due to radio wave amplification by the cyclotron maser instability (CMI). They are complex and variable, and have two noticeably different durations, i.e. the short S-bursts which have duration from one to ten milliseconds and the long L-bursts with timescales from seconds to minutes or hours or even up to months. The S-bursts sound like popping of popcorn or crackling of a campfire, whereas L-bursts sound more like ocean waves breaking up against a beach. Some emissions only produce L-bursts while some others might produce a mixture of L-bursts and S-bursts, such as Io-B and Io-C. Apart from the timescale, the frequency drift rate for these two types of bursts are also different. The S-bursts drift from high frequency to lower frequency due to the electrons that are spiraling out from Jupiter's magnetic field progressively encounter regions of weaker magnetic field; and the burst frequency reduces to reflect the reducing strength of the magnetic field. For L-bursts, the frequency can either drift to higher or lower frequency. The bursts contain modulation lanes due to diffraction effects and interplanetary and ionospheric scintillations.

Apart from having decametric emission, Jupiter is also a source of steady decimetric (DIM) emission. The continuous decimetric emission is a non-thermal synchrotron emission caused by extremely energetic electrons moving within the radiation belts at nearly the speed of light. Figure 8 shows the details of Jupiter's radiation belts, indicating that the region around Jupiter is one of the harshest radiation environments to be found in the solar system. The radiations were mapped by NASA's Cassini spacecraft at a frequency of 13.8 GHz (wavelength of 2.2 cm). The image of Jupiter is superimposed to show the relative size and orientation of the planet compared to the belts and the three views show that the belt position changes as Jupiter rotates. The belts appear to be tilted in relation to the cloud bands (equator) because they are a result of Jupiter's magnetic field, which is tilted roughly 10° from the rotational axis. These non-thermal radio emissions from the radiation belts can only be distinguished from the thermal emission from the Jovian atmosphere by Cassini due to its proximity; the emission from the belts would be swamped out by the thermal emission if observation was done using Earth-based telescopes.

Jupiter has three different systems of longitude (rotation periods), namely System I, II and III. System I refers to the rotational speed at the equator while System II refers to the rotational speed at higher latitudes, both of which used for measuring features in the cloud layers. The variation in these rotational periods at different latitudes arises because Jupiter is not a solid object. System III was established when radio bursts were detected on Jupiter and was found that the emission sources have a different rotational speed with either System I or II. System III is the rotational period of the Jovian magnetic field and is used to describe the motion of the magnetosphere. System III is also thought to be the rotational period of the planet's core (or the solid body of Jupiter) since there is where the magnetic field is generated. The current most accurate estimate on the rate of rotation of Jupiter is based on decimetric radio observations.

Radio Emission from Saturn

Attempts to detect decametric radio emissions from Saturn after the discovery of similar emission from Jupiter produced inconclusive results initially. In 1975, Larry W. Brown reported the observation of non-thermal radio emissions from Saturn by the Interplanetary Monitoring Platform-6 (IMP-6) spacecraft during the period 1971 to 1972. The detection of this non-thermal radio emission, with a peak flux density at 1 MHz, is the first direct evidence that Saturn has an internally generated magnetic field.

Saturn and its magnetosphere were studied in situ by a total of four spacecrafts that had visited it, namely Pioneer 11, Voyager 1 and 2, and Cassini. Pioneer 11 reached the point of closest approach to Saturn on September 1979, Voyager 1 in November 1980, Voyager 2 in August 1981 and Cassini in July 2004. The first three missions were flyby whereas Cassini is an orbiter. The Cassini Orbiter has a Radio and Plasma Wave Science (RPWS) instrument onboard, a highly capable instrument designed to measure the magnetic and electric fields of radio emissions and plasma waves from 1 Hz to 12 kHz for magnetic fields and from 1 Hz to 16 MHz for electric fields. RPWS is used to conduct in-depth studies of Saturn's radio emissions. Prior to Cassini, the understanding of Saturn's radio emissions is largely based on Voyagers and Ulysses spacecrafts' observations. The Voyagers were first to establish that Saturn is an intense radio emitter. Although Ulysses was far away from the planet (the spacecraft primary mission is to study the Sun), it has made important contribution to the understanding of Saturn's radio emission through its unified radio and plasma wave instrument.

There are three primary types of radio emission from Saturn, i.e. Saturn kilometric radiation (SKR), low-frequency narrowband emission and Saturn electrostatic discharge (SED). Saturn kilometric radiation (SKR), as the name implies, has wavelength of a few kilometers, with frequency ranging from 10 kHz to 1.3 MHz. It has a broad peak between 100 and 400 kHz. SKR is generated via the cyclotron maser instability, an auroral radio emission which are closely related to the auroras near the poles of the planet, and is similar to the auroral kilometric radiation (AKR) at Earth and Jupiter. SKR is second in flux density only to Jupiter, as can be seen in Figure 9 based on Voyager measurements, but its spectral range is very similar to the other planets with magnetosphere.

Figure 7. Distribution of emissions in Jupiter's Central Meridian Longitude (CML) – Io phase (Φ_{Io}) and their occurrence probabilities (adapted from Marques et al., 2017). The contours show the probability of a source undergoing a radio storm

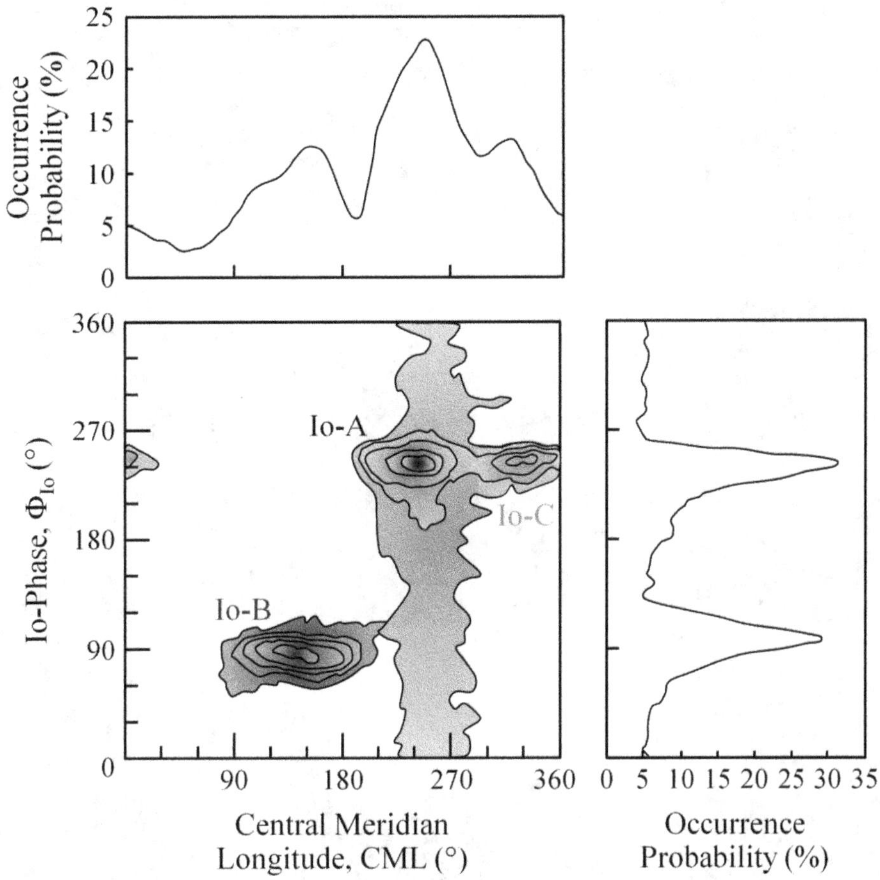

The intensity of SKR is modulated by the rotation of Saturn, just as with the other giant planets, and the modulation is commonly used to determine the rotational period of the core of the fluid giant planets that have no visible surface. The modulation period of SKR has been adopted as the rotational period of Saturn since the magnetic field generated deep inside the planet is controlling the charged particles responsible for this radio emission. However, this modulation period is not constant, as measured by Ulysses and later confirmed by Cassini. In the decades between Voyagers and Cassini observations, there was a substantial shift in the period. In fact, the modulation period varies as short as a timescale of 20 – 30 days, with a longer term variation. This fluctuation was shown to be correlated with solar wind speed at Saturn. One possible reason may be that the perfectly alignment of Saturn's magnetic dipole along its rotational axis fails to induce sufficient rotational wobble to control the modulation and the magnetospheric plasma slipped relative to the planet. In sharp contrast, Jupiter, with its magnetic field tilted roughly 10° from its rotational axis, has a constant rotational period for more than five decades.

Cassini, using its RPWS instrument, began to detect SKR in April 2002 when it was on its journey to the planet. The instrument has provided high resolution observations of these emissions, showing an

Figure 8. The radiation belts of Jupiter mapped by NASA's Cassini spacecraft at a frequency of 13.8 GHz during its flyby of the planet in January 2011. The strength of the emission is coded by color, with red and yellow being the most intense (Credit: NASA / JPL)

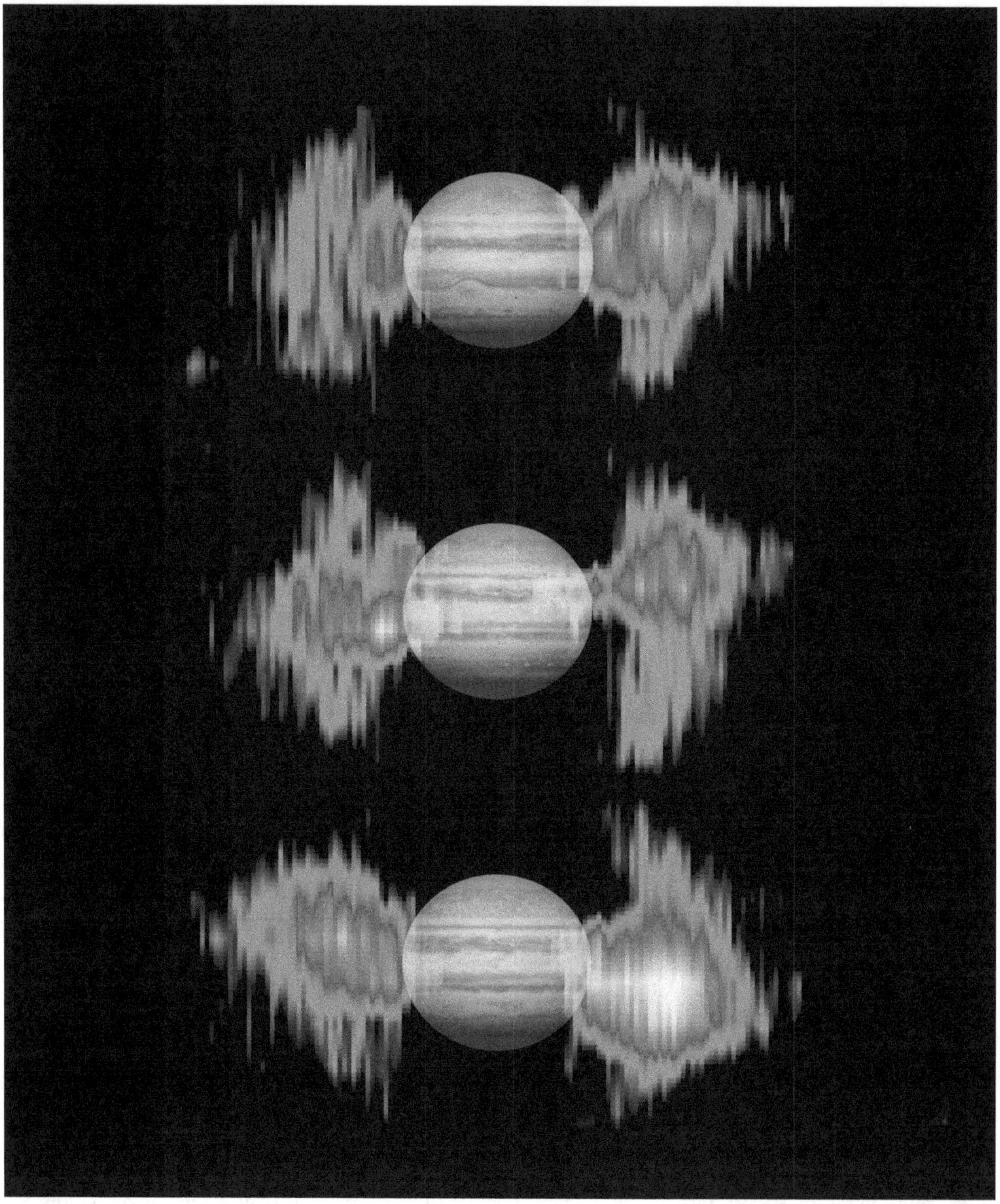

Figure 9. Spectra of auroral radio emissions of five magnetized planets. The average emissions are scaled to a distance of 1 AU from the source (adapted from Zarka, 1998). KOM – kilometric radiation, HOM – hectometric radiation, DAM – decametric radiation, SKR – Saturn kilometric radiation, AKR – auroral kilometric radiation, TKR - terrestrial kilometric radiation, UKR – Uranus kilometric radiation, NKR – Neptune kilometric radiation.

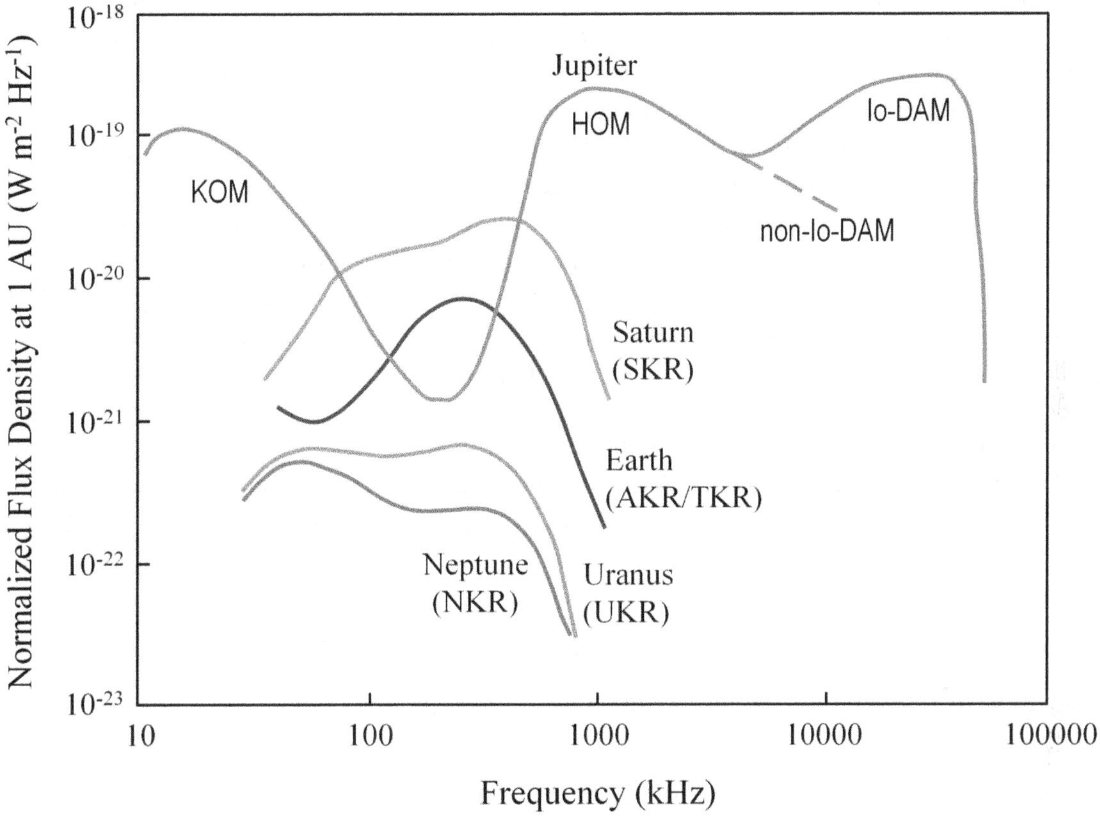

Figure 10. Saturn kilometric radiation (SKR) recorded by Cassini's Radio and Plasma Wave Science (RPWS) instrument (Credit: NASA / JPL / University of Iowa)

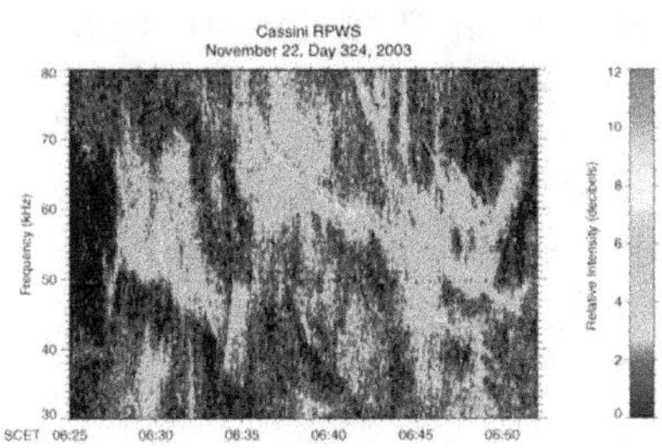

amazing array of variations in frequency and time. Figure 10 shows an example of the complex radio spectrum recorded by Cassini RPWS instrument. The radio spectrum was converted into an audio signal by shifting the frequencies of the radio emissions downward by a factor of 44 to the audio frequency range. The rising and falling tones indicated that there are numerous small radio sources moving along magnetic field lines threading the auroral region.

Apart from the auroral kilometric radiation generated via cyclotron maser instability, radio emissions from planetary magnetospheres also consists of low-frequency narrowband emissions which are generally believed to be generated via mode conversion from electrostatic upper hybrid waves in the inner region of the magnetosphere. The low-frequency narrowband emission at Saturn was first reported after Voyager 1's encounter with the planet, in the frequency range of 3 to 30 kHz. Cassini also has detected numerous narrowband emissions, at frequency of 5 Hz and between 20 and 30 kHz, after its arrival at Saturn. Typically, the emissions follow an intensification of SKR and last for several days.

The third type of radio emission from Saturn is the Saturn electrostatic discharge (SED) that originated from lightning in the atmosphere of Saturn. These impulsive radio bursts, between 100 kHz and 40 MHz, were detected by both the Voyagers during the flybys of the planet. Previously, the evidence that SEDs is of lightning origin in Saturn's atmosphere has been indirect. Most of the SED bursts occurred near the equator and have a repetitive period which corresponded to the rotational period of clouds. This phenomenon suggested that the bursts came from a storm system that was being convected around the planet. Cassini later confirmed that SEDs were of lightning origin by comparing the SEDs detected with the RPWS instrument and images of convective-looking clouds on Saturn obtained with the Imaging Science Subsystem. The comparison with prominent cloud features has resulted in visual evidence that SEDs and lightning storm clouds in Saturn's atmosphere are related. The first direct observation of lightning flashes in Cassini night side images was taken in 2009 during Saturn's equinox.

REFERENCES

Bhattacharya, A., & Raha, B. (2013). Detection of Radio Emissions from Jupiter. *Oriental Journal of Physics*, *5*, 1–12.

Brown, L. W. (1975). Saturn Radio Emission Near 1 MHz. *The Astrophysical Journal*, *198*, L89–L92. doi:10.1086/181819

Burke, B. F., & Franklin, K. L. (1955). Observations of a variable radio source associated with the planet Jupiter. *Journal of Geophysical Research*, *60*(2), 213–217. doi:10.1029/JZ060i002p00213

Burke, B. F., Graham-Smith, F., & Wilkinson, P. N. (2019). *An Introduction to Radio Astronomy*. Cambridge University Press. doi:10.1017/9781316987506

Cane, H. V., & Reames, D. V. (1988). Some statistics of solar radio bursts of spectral types II and IV. *The Astrophysical Journal*, *325*, 901–904. doi:10.1086/166061

Cook, J. J., Cross, L. G., Bair, M. E., & Arnold, C. B. (1960). Radio Detection of the Planet Saturn. *Nature*, *188*(4748), 393–394. doi:10.1038/188393a0

Covington, A. E. (1947). Micro-Wave Solar Noise Observations During the Partial Eclipse of November 23, 1946. *Nature*, *159*(4038), 405–406. doi:10.1038/159405a0

Dąbrowski, B. P., Krankowski, A., Błaszkiewicz, L., & Rothkaehl, H. (2016). Prospects for Solar and Space Weather Research with Polish Part of the LOFAR Telescope. *Acta Geophysica*, *64*(3), 825–840. doi:10.1515/acgeo-2016-0028

Drake, F. D. (1961). Radio emission from the planets. *Physics Today*, *14*(4), 30–34. doi:10.1063/1.3057498

Drake, F. D., & Ewen, H. I. (1958). A Broad-Band Microwave Source Comparison Radiometer for Advanced Research in Radio Astronomy. *Proceedings of the IRE*, *46*(1), 53–60. doi:10.1109/JRPROC.1958.286710

Dyudina, U. A., Ingersoll, A. P., Ewald, S. P., Porco, C. C., Fischer, G., & Yair, Y. (2013). Saturn's visible lightning, its radio emissions, and the structure of the 2009–2011 lightning storms. *Icarus*, *226*(1), 1020–1037. doi:10.1016/j.icarus.2013.07.013

Gary, D. E., & Hurford, G. J. (1994). Coronal temperature, density, and magnetic field maps of a solar active region using the Owens Valley Solar Array. *The Astrophysical Journal*, *420*, 903–912. doi:10.1086/173614

Gurnett, D. A. (2005). Radio and Plasma Wave Observations at Saturn from Cassini's Approach and First Orbit. *Science*, *307*(5713), 1255–1259. doi:10.1126cience.1105356 PMID:15604362

Gurnett, D. A., Kurth, W., Kirchner, D., Hospodarsky, G., Averkamp, T., Zarka, P., ... Pedersen, A. (2004). The Cassini Radio and Plasma Wave Investigation. *Space Science Reviews*, *114*(1-4), 395–463. doi:10.100711214-004-1434-0

Hey, J. S. (1946). Solar Radiations in the 4–6 Metre Radio Wave-Length Band. *Nature*, *157*(3976), 47–48. doi:10.1038/157047b0

Hildebrandt, J., Krüger, A., Chertok, I. M., Fomichev, V. V., & Gorgutsa, R. V. (1998). Solar Microwave Bursts from Electron Populations with a 'Broken' Energy Spectrum. *Solar Physics*, *181*(2), 337–349. doi:10.1023/A:1005007321040

Howard, W. E. III, Barrett, A. H., & Haddock, F. T. (1961). The Measurement of Microwave Radiation from the Planet Mercury. *The Astronomical Journal*, *66*, 287. doi:10.1086/108417

Howard, W. E. III, Barrett, A. H., & Haddock, F. T. (1962). Measurement of Microwave Radiation from the Planet Mercury. *The Astrophysical Journal*, *136*, 995–1004. doi:10.1086/147451

Kellermann, K. I. (1970). Thermal Radio Emission From the Major Planets. *Radio Science*, *5*(2), 487–493. doi:10.1029/RS005i002p00487

Kontar, E. P., Yu, S., Kuznetsov, A. A., Emslie, A. G., Alcock, B., Jeffrey, N. L. S., ... Subramanian, P. (2017). Imaging spectroscopy of solar radio burst fine structures. *Nature Communications*, *8*(1), 1–9. doi:10.103841467-017-01307-8 PMID:29142220

Kraus, J. D. (1986). *Radio Astronomy*. Ohio: Cygnus-Quasar Books.

Krüger, A. (1979). *Introduction to Solar Radio Astronomy and Radio Physics*. The Netherlands: Springer. doi:10.1007/978-94-009-9402-7

Kundu, M. R. (1965). *Solar radio astronomy*. New York: Interscience Publishers, John Wiley & Sons.

Kundu, M. R., & Vlahos, L. (1982). Solar microwave bursts - A review. *Space Science Reviews*, *32*(4), 405–462. doi:10.1007/BF00177449

Kurth, W. S., & Zarka, P. (2001). Saturn Radio Waves. In *"Planetary Radio Emissions V, Proceedings of the 5th International Workshop"* pp. 247–259. Vienna, Austria: Austrian Academy of Sciences Press.

Lee, J. (2007). Radio Emissions from Solar Active Regions. *Space Science Reviews*, *133*(1-4), 73–102. doi:10.100711214-007-9206-2

Marques, M. S., Zarka, P., Echer, E., Ryabov, V. B., Alves, M. V., Denis, L., & Coffre, A. (2017). Statistical analysis of 26 yr of observations of decametric radio emissions from Jupiter. *Astronomy and Astrophysics*, *604*, 1–18. doi:10.1051/0004-6361/201630025

Mayer, C. H., McCullough, T. P., & Sloanaker, R. M. (1958). Observations of Venus at 3.15-CM Wave Length. *The Astrophysical Journal*, *127*, 1–10. doi:10.1086/146433

Mayer, C. H., McCullough, T. P., & Sloanaker, R. M. (1958). Observation of Mar and Jupiter at a Wave Length of 3.15 cm. *The Astrophysical Journal*, *127*, 11–16. doi:10.1086/146434

Melrose, D. B. (1975). Plasma emission due to isotropic fast electrons, and types I, II, and V solar radio bursts. *Solar Physics*, *43*(1), 211–236. doi:10.1007/BF00155154

Melrose, D. B. (1985). Plasma emission mechanisms. In *Solar Radiophysics: Studies of Emission from the Sun at Metre Wavelengths* (pp. 177–210). Cambridge, New York: Cambridge University Press.

NASA, Jet Propulsion Laboratory. (2002). PIA03478: Inner Radiation Belts of Jupiter. Retrieved from https://photojournal.jpl.nasa.gov/catalog/PIA03478

NASA, Jet Propulsion Laboratory, University of Iowa. (2005). PIA07966: Eerie Sounds of Saturn's Radio Emissions. Retrieved from https://photojournal.jpl.nasa.gov/catalog/PIA07966

National Radio Astronomy Observatory. (n.d.). Retrieved from https://public.nrao.edu/

National Radio Astronomy Observatory. (n.d.). Green Bank Solar Radio Burst Spectrometer. Retrieved from http://www.astro.umd.edu/~white/gb/

Nelson, G. J., & Melrose, D. B. (1985). Type II bursts. In *Solar Radiophysics: Studies of Emission from the Sun at Metre Wavelengths* (pp. 333–359). Cambridge: Cambridge University Press.

Newkirk, G. Jr. (1961). The solar corona in active regions and the thermal origin of the slowly varying component of solar radio radiation. *The Astrophysical Journal*, *133*, 983–1012. doi:10.1086/147104

Reber, G. (1944). Cosmic static. *The Astrophysical Journal*, *100*, 279–287. doi:10.1086/144668

Roberts, J. A. (1963). Radio emission from the planets. *Planetary and Space Science*, *11*(3), 221–259. doi:10.1016/0032-0633(63)90026-6

Shibasaki, K., Alissandrakis, C. E., & Pohjolainen, S. (2011). Radio Emission of the Quiet Sun and Active Regions (Invited Review). *Solar Physics*, *273*(2), 309–337. doi:10.100711207-011-9788-4

Southworth, G. C. (1945). Microwave radiation from the sun. *Journal of the Franklin Institute*, *239*(4), 285–297. doi:10.1016/0016-0032(45)90163-3

Wang, Z., Gurnett, D. A., Fischer, G., Ye, S.-Y., Kurth, W. S., Mitchell, D. G., ... Russell, C. T. (2010). Cassini observations of narrowband radio emissions in Saturn's magnetosphere. *Journal of Geophysical Research: Space Physics*, *115*, 1–12.

Warmuth, A., & Mann, G. (2005). Lecture Notes in Physics: Vol. 656. *The application of radio diagnostics to the study of the solar drivers of space weather* (pp. 51–70). doi:10.1007/978-3-540-31534-6_3

Warwick, J. W. (1964). Radio Emission from Jupiter. *Annual Review of Astronomy and Astrophysics*, *2*(1), 1–22. doi:10.1146/annurev.aa.02.090164.000245

Warwick, J. W., Pearce, J. B., Evans, D. R., Carr, T. D., Schauble, J. J., Alexander, J. K., ... Barrow, C. H. (1981). Planetary radio astronomy observations from Voyager 1 near Saturn. *Science*, *212*(4491), 239–243. doi:10.1126cience.212.4491.239 PMID:17783837

Weiss, L. A. A., & Stewart, R. T. (1965). Solar radio bursts of spectral type V. *Australian Journal of Physics*, *18*(2), 143–166. doi:10.1071/PH650143

White, S. M. (2007). Solar Radio Bursts and Space Weather. *Asian Journal of Physics*, *16*, 189–207.

Wild, J. P., Smerd, S. F., & Weiss, A. A. (1963). Solar Bursts. *Annual Review of Astronomy and Astrophysics*, *1*(1), 291–366. doi:10.1146/annurev.aa.01.090163.001451

Zarka, P. (1998). Auroral radio emissions at the outer planets: Observations and theories. *Journal of Geophysical Research*, *103*(E9), 20159–20194. doi:10.1029/98JE01323

Zarka, P., & Kurth, W. S. (2005). Radio wave emissions from the outer planets before Cassini. *Space Science Reviews*, *116*(1-2), 371–397. doi:10.100711214-005-1962-2

Zarka, P., Lamy, L., Cecconi, B., Prangé, R., & Rucker, H. O. (2007). Modulation of Saturn's radio clock by solar wind speed. *Nature*, *450*(7167), 265–267. doi:10.1038/nature06237 PMID:17994092

Zheleznyakov, V. V. (1970). International Series of Monographs in Natural Philosophy: Vol. 25. *Radio Emission of the Sun and Planets*. Oxford, UK: Pergamon Press.

Chapter 14
Active Galactic Nuclei and Pulsars

Hui Chieh Teoh
Universiti Tunku Abdul Rahman, Malaysia

ABSTRACT

With the discovery of radio emission of extraterrestrial origin, it had opened a new and broader window in the electromagnetic spectrum to observe the sky. Two of the common sources of radio emissions are active galactic nuclei (AGN) and pulsars. AGN are very luminous at radio wavelength and are powered by the accretion disk surrounding supermassive black holes at the center. AGN are useful as an alternative standard ruler to determine cosmological parameters. Pulsars, on the other hand, is a type of neutron star that is highly magnetized and rotates at enormous speed. This cosmic lighthouse produces a very precise pulsation period that can be used as cosmic clock.

THE DISCOVERY OF THE RADIO SKY

It was 1931. Karl Guthe Jansky was wondering what was the source that was causing a mysterious steady hiss in his antenna. He had ruled out all possible terrestrial sources. So, what could it be?

Few years back in 1928, Jansky joined the Bell Telephone Laboratories in Holmdel, New Jersey. At that time, Bell Labs wanted to use short wavelength radio waves for telephone calls across the Atlantic. While the technology for short wavelength radio waves transmission was quite well established, engineers also required to know sources of noise that might interfere with this transatlantic radio telephone service. Thus, Jansky was tasked to find all the possible sources of static that might interfere with the radio voice transmissions. This young engineer then built an antenna for the job. His equipment consisted of an antenna array that was rotatable, a short-wave measuring set and an automatic intensity recorder, and was tuned to a wavelength of 14.6 meters (frequency of 20.5 MHz). The turntable antenna was rotated about a vertical axis and was highly directive in the horizontal plane. It therefore allowed him to determine the directions of any radio signals. The rotating antenna was jokingly known as "Jansky's merry-go-round".

After several months of observations, Jansky found static from thunderstorms, both local and distant, with strength fluctuating greatly with time. On top of that, his records also showed the presence of a

DOI: 10.4018/978-1-7998-2381-0.ch014

faint persistent hiss of unknown origin and was unable to associate it with any sources of terrestrial origin. This unknown source of static rose and fell once a day, and for some weeks the signal seemed to be the strongest when the antenna was pointed toward the Sun, which logically led Jansky to think that it might be the radiation from the Sun. As Jansky continued to follow the signal for several months, he found that the peak intensity of the radio waves did not quite match up with the movements of the Sun. Instead of repeating every 24 hours as it would for a solar day, the source repeated every 23 hours and 56 minutes, which corresponded with a sidereal day, i.e. the Earth rotational period with respect to the stars. This observation eliminated the source from inside the solar system; it had to be from objects beyond our Sun that is stationary with respect to the stars. In the paper published in 1933, Jansky narrowed down the coordinates of the source to a right ascension of 18 hours and a declination of -10 degrees, in the direction of the constellation Sagittarius. He was confident that the value of the right ascension was accurate, but the measurements of declination had a huge error of ± 30 degrees. At that time, he was unsure of the source due to the huge error in declination, but offered two possible regions fixed in space to be considered – one was the center of our Milky Way galaxy, and the other was the point in space towards which our solar system was moving with respect to the other stars.

Two years later, Jansky published another paper in the Proceedings of Institute of Radio Engineers after further analysis of the data on the mysterious hiss-like static. This time he concluded that the source of this interstellar interference "is located in the stars themselves or in the interstellar matter distributed throughout the Milky Way". Jansky raised a question that if the stars were the source, similar radiation but with stronger intensity would naturally be expected from the Sun since the Sun is also a star and it is located so much closer to us. However, he had detected none of it. Unknown to him, the absence of solar radiation detection was due to the fact that the observations were done during the periods of low solar activity. Jansky eventually identified that the radio emission was coming from the center of our Milky Way galaxy, since the signal was the strongest when the antenna was pointed toward that direction. He wanted to continue the investigation of this discovery on the galactic radio waves, but unfortunately, he was reassigned to another project by Bell Labs, so no further advancement was made by him in this field.

Jansky's work on this new approach to astronomy had fascinated a young ham radio operator named Grote Reber. In 1937, Reber built the world's first radio telescope in his backyard and in the subsequent years, he did the first surveys of radio waves from the sky. He published his data in 1944 as contour maps showing the brightness of the sky at 160 MHz. His map revealed several minor maxima in the constellations Cygnus, Cassiopeia, Canis Major and Puppis. The radio source in Cygnus, known as Cygnus A (Cyg A), would prove to be important later in the studies of radio galaxies.

With the discovery of radio emission of extraterrestrial origin, it had opened a new and broader window in the electromagnetic spectrum to observe the sky. We can now not only see the visible light universe with our eyes, but also "listen" to the radio sky with our ears. Radio waves can penetrate through dust allowing us to peer into regions that are obscured by dust at optical light. Soon, radio astronomy would reveal a dynamic and violent universe not seen with optical astronomy. Radio astronomy has slowly emerged to become a new field in astronomy in the early 1950s, with several groups of radio engineers from World War II turning their attention to the study of radio astronomy after the end of the war.

Jansky's pioneering efforts in radio astronomy were not gone unnoticed. In his honor, the fundamental unit of flux density (the strength of radiation) used by radio astronomers, the jansky (Jy), is named after him.

ACTIVE GALACTIC NUCLEI

The normal galaxies that we see in the optical and near infrared portion of the electromagnetic spectrum is mainly dominated by stars. The radiation is thermal since the emission from the atmosphere of stars is basically in thermodynamical equilibrium. There are, however, galaxies that have emission across the full range of the electromagnetic spectrum, from radio waves to gamma rays. These are known as active galaxies. The broad range of emission from active galaxy mainly originates from the very small region in the center, called the active galactic nucleus (AGN). The name "active galactic nucleus" implies that there is an energetic process happening in the nucleus, or the central region, of a galaxy which cannot be associated directly with stars. The signature of AGN is the presence of a supermassive accreting black hole in the center and the broad bandwidth of AGN spectra suggests that the radiation is non-thermal.

AGNs that we have discovered so far have some commonalities and some apparent differences. They can basically be divided into the following classes: Seyfert galaxies, quasi-stellar objects (quasars), radio galaxies, blazars, low-ionization nuclear emission-line regions (LINERs) and ultra-luminous infrared galaxies (ULIFGs).

Seyfert Galaxies

The first hint of active galactic nuclei was observed by Edward Arthur Fath at Lick Observatory in 1908. Using the 36-inch Crossley reflector telescope, he took the optical spectra of some spiral nebulae (which is now known to be galaxies) and globular star clusters. Although most of the objects he observed showed absorption-line spectrum which could be explained as the combined light produced by stars, he found six strong emission lines from the galaxy NGC1068 (M77). In 1917, Vesto Melvin Slipher at Lowell Observatory obtained a better resolution spectrum of NGC1068 and commented that the emission lines were typical to the ones seen in gaseous nebula. Edwin Powell Hubble also had recorded the emission-line spectrum of NGC1068 and two other galaxies in his 1926 paper.

Only near to two decades later in 1943, the systematic analysis by Carl Keenan Seyfert drew attention to this new class of galaxies. He reported that a very small fraction of galaxies had very luminous nuclei that showed many high-excitation emission lines. The high excitation energy required suggested that it has to be excited by photons that are far more energetic than photons produced by young stars that ionized HII regions. These newly found properties, i.e. broad emission lines from a luminous, stellar-like appearance (extremely high surface brightness) nucleus and covering a wide range of ionization states, defined the new class of objects known as Seyfert galaxies. Seyfert galaxies are typically radio-quiet, its radio emission is weak, and are usually identified by their optical spectra. They are mostly spiral galaxies.

Seyferts can be further classified into two sub-classes, Seyfert 1 and Seyfert 2. Seyfert 1 can be distinguished from Seyfert 2 through the emission lines in their spectra. Seyfert 1 has both broad and narrow emission lines while Seyfert 2 has only narrow emission lines. Apart from visible light, Seyfert 1 also shines brightly in X-ray and ultraviolet parts of the spectrum.

Quasars

Most of the strongest radio emissions at centimeter and meter wavelengths that we detected from the cosmos besides the Sun, come from objects beyond our galaxy. By the time radio astronomy started to gain popularity in the 1950s, radio telescope was able to provide angular resolution that was high enough to identify the strongest radio sources with individual optical objects. Astronomers then were trying to pair the radio sky with the visible universe. In 1949, John Gatenby Bolton, Gordon John Stanley and Owen Bruce Slee successfully achieved the first optical identification of three discrete radio sources. The radio sources Taurus A, Virgo A and Centaurus A were identified to be associated with the visible light sources of M1 (the Crab Nebula), M87 (an elliptical galaxy) and NGC5128 (also an elliptical galaxy), respectively. The partnership of radio and optical astronomy was underway.

Among the radio galaxies detected in the 1950s, there were some with properties that defied explanation at that time. These objects are now known as quasars, which stands for "quasi-stellar radio sources" or "quasi-stellar objects" (QSOs). The term "quasi-stellar" means "star-like", because the signals came from a point-like source, like a star. When quasars were initially discovered with radio telescopes, no optical counterparts could be located. They were strong radio sources of unknown physical origin. It was not until the 1960s that the sources were located optically, and their spectral lines were found to be highly redshifted, indicating that they are very far away from us. It was also found that the luminosity of some quasars was changing very rapidly, providing an upper limit on their size. It was baffling to find that quasars are not much bigger than our Solar System but the amount of energy that they release can be as much as hundreds to a thousand times more than a galaxy with billions of stars. What object could have such enormous power density?

The first quasars ever identified were 3C 273 and 3C 48. The prefix "3C" stands for the Third Cambridge Catalogue of Radio Sources, followed by the entry number. This catalog was first published in 1959 for survey at 159 MHz and was revised with survey at 178 MHz three years later in 1962. Although at that time, hundreds of radio sources were determined and cataloged, not all radio galaxies had an optical match. Thus, the quest to search for optical identities of these objects was taken up by Allan Rex Sandage and Maarten Schmidt from Mount Wilson and Palomar Observatories. They worked together with Thomas Arnold Matthews from the Owens Valley Radio Observatory to obtain accurate positions of the radio sources so that they can be compared with photographs of the same region.

In 1963, Sandage and Matthews published a paper in the Astrophysical Journal identifying the optical match for radio source 3C 48. The direct photograph of the region showed the existence of a star-like object at the position of the radio source, with very faint wisps of nebulosity. The star-like object has quite peculiar color indices and photometry of 3C 48 showed that it is a variable "star". Another strange feature of this object was that its optical spectra has several very broad emission lines, some at unfamiliar wavelength. About the same time, Schmidt had associated an optical counterpart for another radio source, 3C 273. The position of 3C 273 was very accurately determined by Cyril Hazard, M. Brian Mackey and A. John Shimmins the year before during a lunar occultation using the 210-foot Parkes telescope, thus enabled Schmidt to correlate a 13th magnitude "star" and a faint wisp or jet extended away from the star to 3C 273 (Figure 1). Schmidt took an optical spectrum of the star using the 200-inch Hale Telescope on Mount Palomar and again found broad emission lines that were unidentifiable, though they were different from those of 3C 48.

Nevertheless, close examination by Schmidt found that four of the emission bands showed similar pattern with the Balmer series of hydrogen, in terms of their relative strengths and spacing. They might just be ordinary spectral lines of hydrogen. He found that if these lines were considerably redshifted, with a value of $z = 0.158$, they would agree with the expected wavelength of the Balmer series (Figure 2). Schmidt extended his explanation to a line in the ultraviolet part of the spectrum and based on the mentioned redshift, he found that they can be identified to be the Mg II lines. Knowing that redshift could be used to explain the strange spectrum, Jesse Leonard Greenstein and Matthew applied the same method to the spectrum of 3C 48 and found that the Mg II lines had been redshifted by 36.7%. At last, the mystery of the unidentified emission lines was solved. In the next two years, two more quasar redshifts were published by Schmidt and Matthews in 1964, and five more in 1965 by Schmidt. The spectra for these quasars had been found to be redshifted to an extreme degree, with the highest redshift of 2.012 was determined for 3C 9.

Schmidt, in his 1963 Nature paper, provided two suggestions to the objects that would have such a high redshift. The first type of object is a star with a large gravitational redshift and a very small radius

Figure 1. Visible light image of quasar 3C 273 showing its star-like appearance and a jet extended away from the quasar (Credit: ESA/Hubble & NASA)

Figure 2. Visible spectrum of quasar 3C 273 showing the redshifted emission lines in the Balmer series of hydrogen

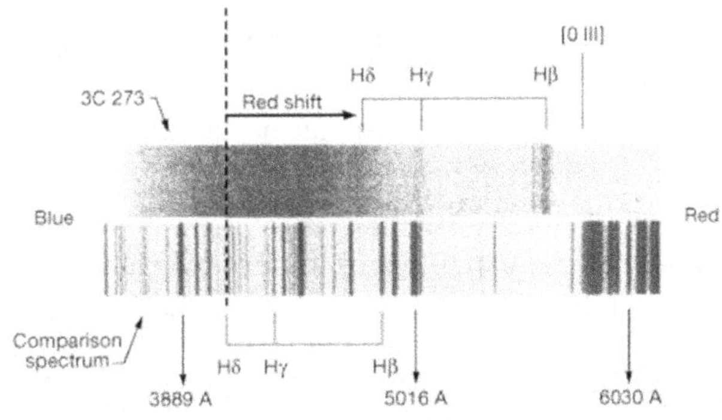

in the order to 10 km. However, he cautioned that it would be extremely difficult to account for the observed lines. The second type is of extragalactic origin – the nuclear region of a galaxy with a cosmological redshift of 0.158. If this was correct, 3C 273, despite appeared very faint, would be far more luminous than any galaxy known at that time due to its large distance (a redshift of 0.158 would place 3C 273 several billions light-year away). Although Schmidt's explanation on high cosmological redshift could fit into the data very well, it was still heavily debated. The major concern was that if these objects were very distant, then ridiculously huge amount of energy had to be emitted by these objects to fit the observations. On top of that, due to the short timescales' variability seen in some of these objects, they had to be comparatively small in size. At that time, no commonly accepted mechanism could account for this. The currently accepted explanation, i.e. the mechanism is related to the supermassive black hole in the center of a galaxy, was not easily accepted at that time because in the 1960s, black hole was still something that was theoretical and too exotic.

It took years of observation and study to gradually demonstrate that the redshift is due to the expansion of the universe, and to realize that these tiny, distant point-like glimmers are actually extremely bright nuclei of active galaxies and can outshine their host galaxies many folds. Quasar are some of the brightest and most energetic objects in the universe apart from gamma-ray bursts and supernovae. The nuclei contain a supermassive black hole with mass ranging from millions to billions of times the mass of the Sun, surrounded by an accretion disk of gas and dust. Although black hole itself does not emit any light, the accretion disk that surrounded the black hole can shine luminously. The tremendous amount of heat and light released from the accretion disk that powered the quasar are caused by friction of matter swirling rapidly around the monster black hole before it gobbles them up. An actively eating black hole is the only credible explanation how quasar can be so powerful and yet so compact, packing such an enormous amount of energy into a space as small as our Solar System. The energy released by a quasar can be observed across the electromagnetic spectrum from gamma rays to radio waves. Quasars may also emit powerful jets of radiation and particles from their central regions. Not all matter swirling around the central black hole will fall into it; a small portion of particles will get accelerated to velocities close to the speed of light and spewn out in narrow bipolar jets along the rotational axis. The radio emission observed is due to synchrotron process, generated by charged particles spiraling around magnetic field lines at relativistic speed. For galaxies with supermassive black holes a billion times more massive than the Sun, the jets can grow powerful enough to expel gas out of the galaxy and turn off star formation.

The large redshift of quasars makes them a very useful tool to answer some of the cosmological questions. Since quasars are found to be very distant, in the order of billions of light-years away, studying them are crucial to our understanding of the early universe billions of years ago. This is because light travels at a finite speed. Thus, the farther away we look, the farther back in time we can see. As of 2017, the farthest quasar identified to date is ULAS J1342+0928. It was found to have a redshift of 7.54, which is translated into a distance of more than 13 billion light-years away. It appeared a mere 690 million years after the Big Bang. Such a young quasar can provide valuable information on how galaxies evolve over time. The supermassive black hole in this quasar is estimated to be 800 million solar masses, and is the farthest black hole ever found to date.

In early 2019, a group of astronomers discovered the brightest quasar ever seen using the Hubble Space Telescope. This quasar, catalogued as J043947.08+163415.7, was formed when the universe was only about a billion years old. Despite being the brightest quasar ever detected, it can only be spotted with the help of strong gravitational lensing. The light from the quasar is bent by a galaxy located between us and the quasar, causing the quasar to appear three times as large and fifty times as bright compared to it would be without the lensing effect. This quasar plays an important role for us to understand the evolution of distant quasars and how the central supermassive black holes affect star formation.

There are two sub-classes of quasars identified by astronomers – radio-loud quasar (RLQ) and radio-quiet quasar (RQQ). RLQs, as the name implies, has strong radio emission primarily generated in the AGN and powerful jets. RQQs have weak radio emission and lack of powerful jets. The bulk of the radio emission is suggested to be generated by star formation. The majority of quasars are actually radio quiet.

Radio Galaxies, Blazars, LINERs and ULIFGs

Radio galaxies are a class of AGN that are very luminous at radio wavelength. Some of the strongest radio sources in the sky can be associated with quasars, but most of them are typically identified with giant elliptical galaxies. The synchrotron radio emission in these radio galaxies typically comes from the bipolar jets – the radio lobes. The structure of the lobes is determined by the interaction between the jets and the surrounding gas. Some jets can be many times larger in extent than the galaxy itself. High energy particles, confining to a well collimated jet, are beam into extragalactic medium at speeds approaching the speed of light, ending in giant radio lobes far away from the galaxy (Figure 3). Frequently, there is also a weak radio source at the compact core.

In 1974, Bernard Lewis Fanaroff and Julia M. Riley first noticed that the relative positions of the brightness regions (hot spots) in the jets of radio galaxies were correlated with their radio luminosity. Radio galaxies with stronger hot spots tend to be significantly more luminous at radio frequencies. They divided the radio galaxies into two classes – Fanaroff-Riley Class I (FRI) sources with low luminosity and Fanaroff-Riley Class II (FRII) sources with high luminosity. FRI sources typically have bright jets in the center. The jets, or radio lobes, are brightest at the center and slowly fade away as the distance from the center increases (Figure 4). The jets fade into the lobes without any bright regions (or hot spots) at the end of the jets. The absence of bright regions suggests that the jets have lesser power and less energy is involved in the collision between the jet particles and extragalactic medium. FRI sources usually exhibit two-sided lobes, whereas FRII sources usually only have one. FRII sources display the opposite luminosity changes from FRI sources. FRII sources have faint jets, but show increasing bright-

Figure 3. Color composite image of radio galaxy Centaurus A with the jets and lobes extending far beyond the optical extent of the galaxy. The optical light shows the stars and dust lane, submillimeter wavelength is shown in orange and x-ray is shown in blue (Credit: ESO/WFI (Optical); MPIfR/ESO/APEX/A.Weiss et al. (Submillimetre); NASA/CXC/CfA/R.Kraft et al. (X-ray))

ness toward the end of the jets, and were brightest at the edges (Figure 5). FRII sources seem to be more efficient in transporting energy to the end of the lobes.

Analogous to the Seyfert galaxies, the optical spectra of radio galaxies can be classified into two sub-classes. The broad-line radio galaxy (BLRG) is the radio-loud equivalent of Seyfert 1, and the narrow-line radio galaxy (NLRG) is the radio-loud equivalent of Seyfert 2, with similar optical emission-line spectrum. Two obvious differences between Seyferts and radio galaxies are that Seyferts are relatively quiet at radio emission, and they are mainly spiral galaxies. On the other hand, none of the strong radio galaxies are spirals. BLRG has bright, stellar-like core surrounded by faint, hazy envelopes, whereas NLRG is giant or supergiant elliptical galaxies. Cygnus A is an example of NLRG.

Blazars are highly variable and have strong linear polarization at visible wavelengths. There are also two sub-classes of blazars – BL Lacertae (BL Lac) and optically violent variables (OVV) quasar. Both also are strong radio emitters. BL Lac is named after the most well-known object in this sub-class which is located in the constellation of Lacerta. Due to its irregular changes in brightness, BL Lac was originally classified as a variable star. BL Lacs spectrum is featureless and is almost devoid of emission lines. The observations of a few faint spectral lines show that BL Lacs have high redshift, implying that they are located at cosmological distance. Majority of them are located in elliptical galaxies. BL Lacs can also

Figure 4. Fanaroff-Riley Class I (FRI) radio galaxy 3C 31 obtained from the Very Large Array (VLA) radio telescopes. The jets were brightest near the center and faded into the end of the radio lobes without any hot spots (Credit: NRAO / AUI)

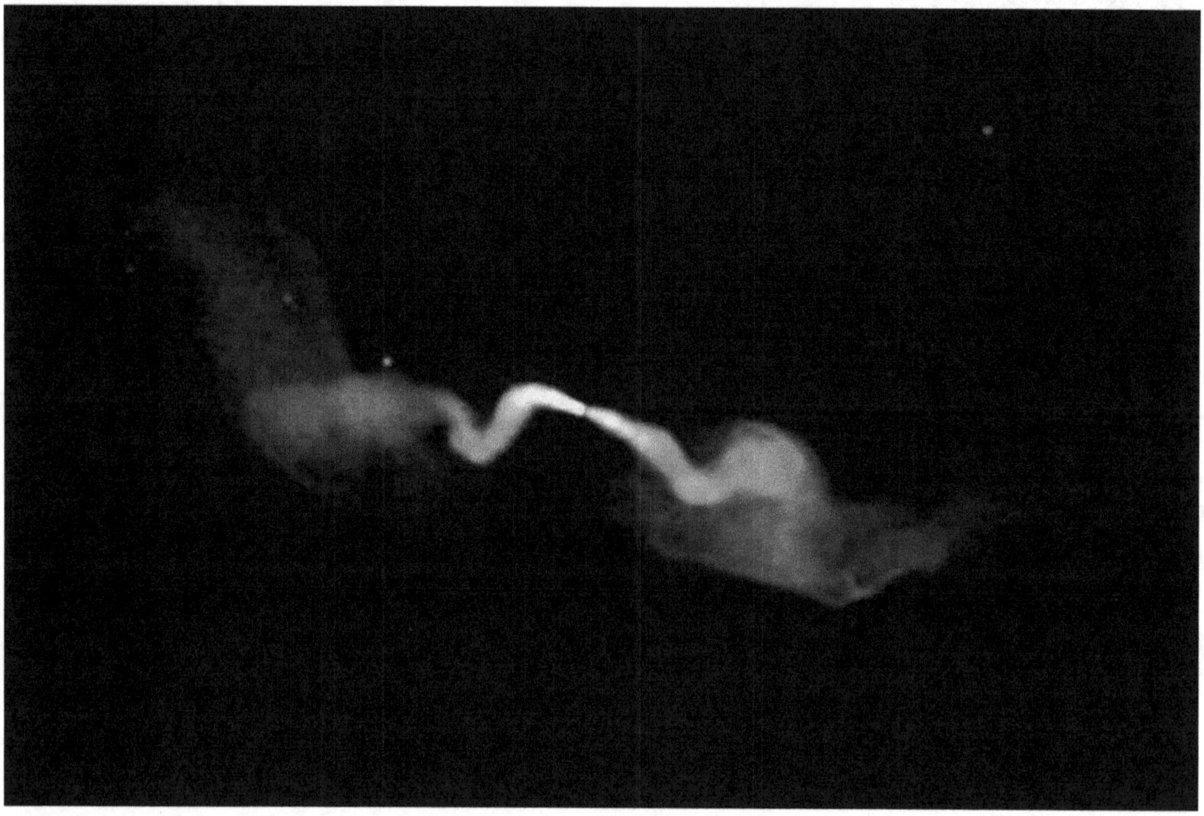

Figure 5. Fanaroff-Riley Class II (FRII) radio galaxy Cygnus A (3C 405) obtained from the Very Large Array (VLA) radio telescopes. The jets show increasing brightness as they moves away from the center and exhibit hot spots at the end of the radio lobes (Credit: NRAO / AUI)

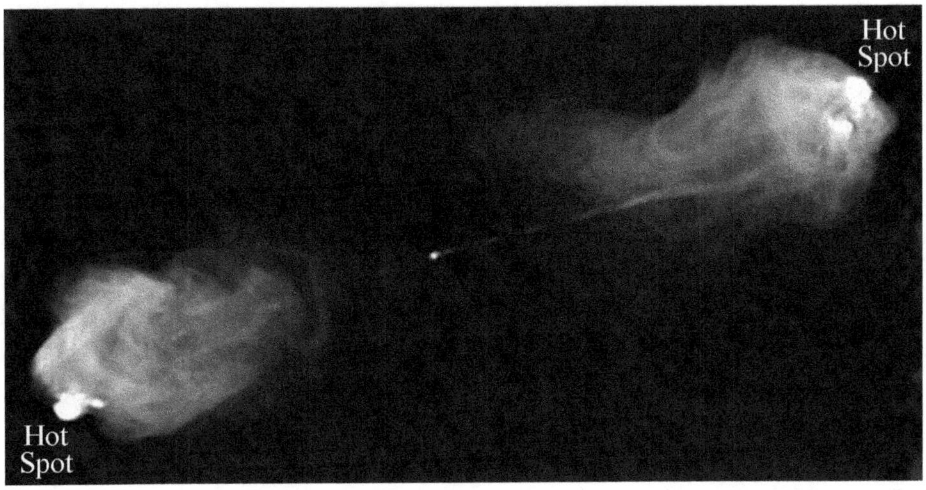

be identified by their strongly polarized power-law continua. OVVs are similar to BL Lacs, except that they have broad and narrow emission lines in their spectra, and they are much brighter than BL Lacs.

The least luminous, and by far the most common class of AGN, are the LINERs (low-ionization nuclear emission-line regions). LINERs can be found in many of the nearby spiral galaxies. Spectroscopically, they are similar to Seyfert 2, but has fairly strong low-ionization emission lines. In contrast, emission lines of strongly ionized ions are either weak or absent. Whether LINERs are AGNs is sometimes questionable, because the low-ionization lines are also detectable in starburst galaxies and HII regions. LINER-like spectra can be produced by various mechanisms that are not associated with black hole. Thus, whether the emission is indeed a signature of an AGN or is powered by star-forming activity remains unclear. If LINERs are truly AGNs, it is also unclear that whether they are a distinct class by themselves, or they are simply very low-luminosity Seyfert galaxies.

Lastly, the ultra-luminous infrared galaxies (ULIFGs) are possibly dust-enshrouded quasars. It is suggested that the dust adsorbs and reradiate the light from the quasar nucleus at infrared wavelengths. Alternatively, it may be starburst galaxies. Infrared galaxies are galaxies that emit more radiation in the infrared than at all the other wavelengths combined.

A Unified Model of AGNs

Despite the fact that AGNs can be classified into many classes with apparent differences, such as the presence or absence of broad emission lines and the strength of radio emission, they also have many properties that are common to all, for example the existence of a bright compact core and a wide continuum. Thus, a question naturally arises as to whether these different classes of AGNs are fundamentally similar objects that only differ in appearance, or whether they are genuinely different fundamentally. The quest to answer this question is on-going, and the pursuit for a unified model of AGNs is on.

The unified model serves as a framework to organize the various components of AGNs and their observations into one place. The model should not only be able to explain all the observations made previously, but should also be able to predict the outcomes of new observational tests. It is now common to all AGNs that a supermassive black hole resides in the center of the galaxies, surrounded by an accretion disk which powers the AGNs by converting gravitational potential energy into synchrotron radiation. The broad lines are emitted from a region nearer to the black hole, and the narrow lines are from much more extended regions. The strong infrared emission suggests the presence of obscuring matter, which is concentrated towards the circumnuclear torus, i.e. the plane in which the accretion occurs (Figure 6). The current understanding is that the different classes of AGNs are due to the different viewing orientations, or the inclination angle of the objects toward us (Figure 7). For examples, if we look directly into the jet, we will see a blazar; if we view the torus at an angle, we can detect both the broad lines and narrow lines, and in this case, we will see either Seyfert 1, quasar or BLRG. However, if the object is orientated in such a way that we are looking from a direction near to the plane of the accretion disk, the view to the center of the AGN is obscured by a large torus of gas and dust preventing us from seeing the broad line region. In this case, we will only see the narrow lines and a Seyfert 2 or NLRG spectrum results.

Although there are still many details that require further studies, the current unified model has been quite successful in describing many of the general features of AGNs.

PULSARS

The Discovery of Pulsars

In the mid-1960s, S. Jocelyn Bell Burnell started her PhD research in University of Cambridge under the supervision of Antony Hewish to study interplanetary scintillation (IPS). They were constructing a large radio telescope at that time, and in July 1967, the radio telescope was brought into operation at the Mullard Radio Astronomy Observatory at a frequency of 81.5 MHz. The purpose of this telescope was to observe IPS to investigate the angular structure of compact radio sources such as quasars. Compact sources will scintillate more compared to extended sources – a common example of this phenomenon is why stars twinkle and planets do not. Stars appear as pinpoints of light in the sky while planets appear as extended disks. The apparent sizes of planets usually are larger than the pocket of air therefore diffractions are canceled out and the effects of scintillation are negligible.

The charts produced from the telescope were initially inspected visually since they were not familiar with the instruments and it is easier for a human to recognize patterns compared to a computer. About two months into the observations, on 28 November 1967, Bell-Burnell found some “scruff” on the record (Figure 8), which neither look like scintillating source nor man-made noise. She further realized that this scruff had been recorded before from the same location in the sky, at right ascension (RA) of $19^h\ 19^m$ in the constellation of Vulpecula. Systematic investigations into this mysterious signal began in November 1967 using a faster chart recorder as it transited. When the signal gotten on the fast recording, Bell Burnell could see that it was a series of pulses and equally spaced (Figure 8). She reported the finding to her supervisor. Hewish, upon going through the recordings, found that the signal kept accurately to sidereal time. Thus, they concluded that it could not be Earth-bound. But the pulsation rate of about 1.337 seconds with each lasted about 0.3 second, as reported in an article published in Nature

Figure 6. The gas and dust disk fueling the black hole in the core of giant elliptical galaxy NGC 4261 (Credit: NRAO / Caltech / Walter Jaffe / Leiden Observatory, Holland Ford / JHU / STScI, and NASA)

in the early 1968, was far too fast for anything as large as a star. The extremely regularity of the pulses ruled out localized features in the atmosphere of star. Rather, it suggested that the pulsation originated from the entire star.

At first, they were considering man-made sources in space such as transmission from deep space probes, planetary radar or reflection of radar from the Moon. However, these explanations could not be accepted since the lack of parallax implied that the source was not within the Solar System. Observation from another telescope by Paul F. Scott and Robin A. Collins also revealed the same signal pulsations. This eliminated that the source was due to instrument noise. John D. H. Pilkington, upon measuring the dispersion of the signal, showed that the source was far outside the Solar System, but was within our galaxy.

At this point, the idea of picking up alien signals crossed their minds – "so were these pulsations man-made, but made by man from another civilization?". Bell-Burnell further said that "we did not really believe that we had picked up signals from another civilization, but obviously the idea had crossed our minds and we had no proof that it was an entirely natural radio emission. It is an interesting problem - if one thinks one may have detected life elsewhere in the universe how does one announce the results responsibly? Who does one tell first?". The source was playfully referred to as Little Green Man (LGM-1).

Figure 7. Unification of active galactic nuclei (AGN) classes. The same objects would appear as different classes of AGNs from different viewing orientations (adapted from Urry & Padovani, 1995)

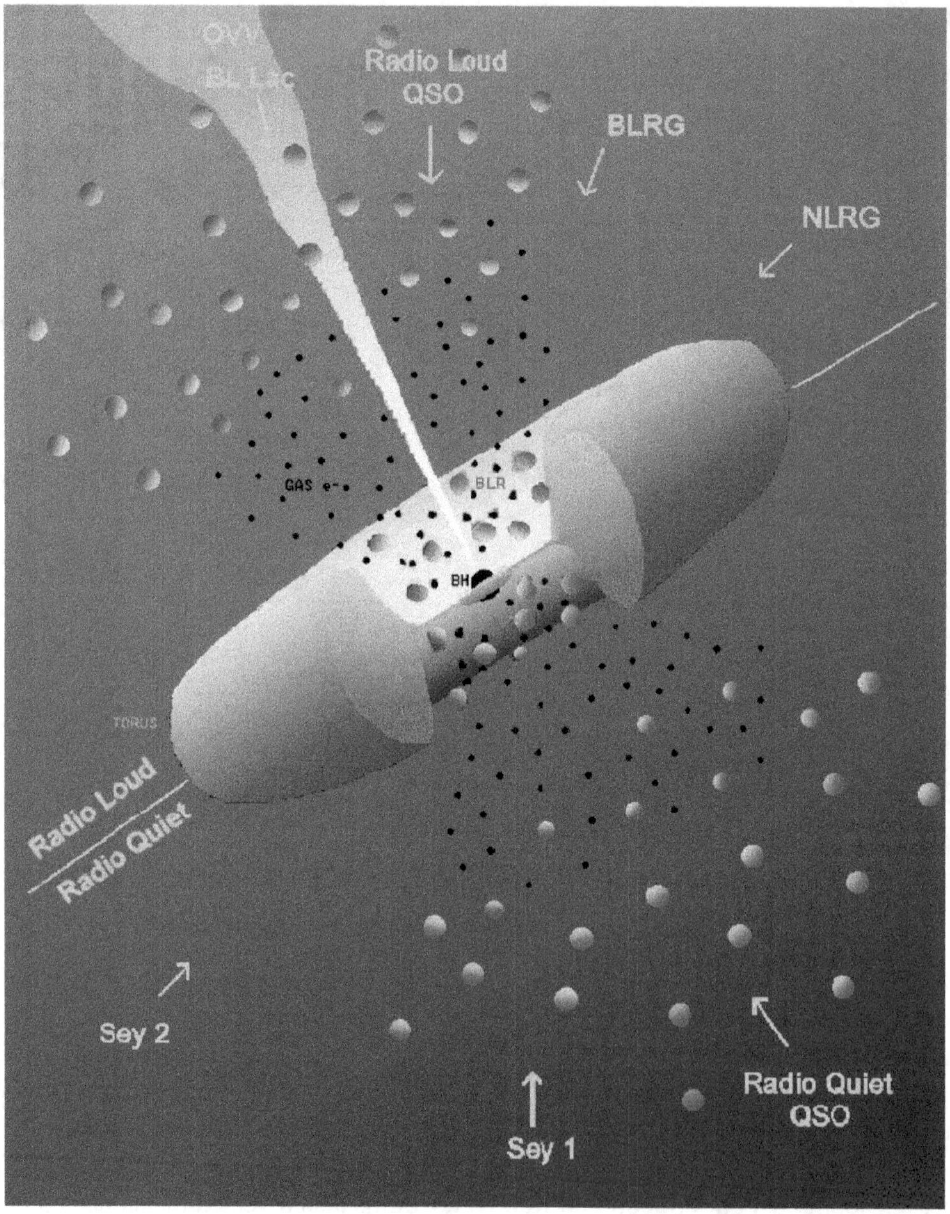

Since nothing could be done much on that peculiar signal, Bell-Burnell continued to analyze the chart recording of another part of the sky. Then again, she suspected she saw some scruff again in the signal from Cassiopeia A and had it reconfirmed by checking through previous recording from the same part of the sky. On that night itself, knowing that this source would transit in the early hours of the morning, she went to the observatory to try to record it. Between the struggles to get the telescope to work on a cold weather, she succeeded to capture the scruff, showing itself with a pulsation rate of 1.2 seconds. This finding of the second pulsating source doubted the fact that aliens were trying to communicate with us.

Figure 8. The "scruff" recorded on the chart which was later identified as pulsar CP1919 (top). The fast recording (bottom) showed that the "scruff" was actually a series of periodic pulses (Credit: Jocelyn Bell Burnell and Antony Hewish)

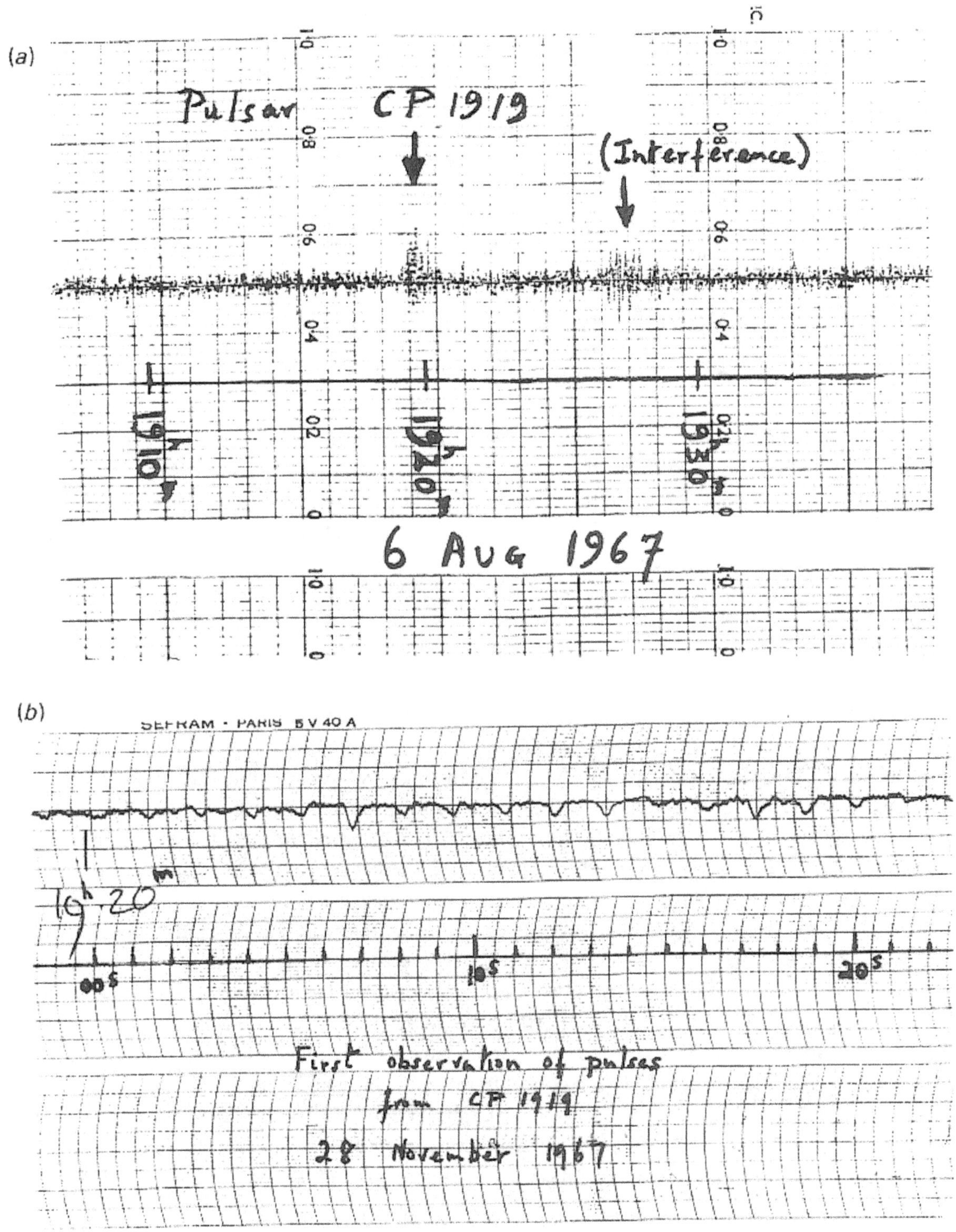

Further recordings and analysis of the charts came up with two other pulsating sources with similar properties which suggested that they might be relatively common. In a seminar given by Hewish in Cambridge on this discovery before the publication of their finding in the journal Nature, Fred Hoyle suggested at the end of the seminar that these objects must be supernova remnants rather than white dwarfs as proposed by the discoverers.

We now call these pulsating radio sources pulsars, and have established that they are rapidly rotating neutron stars with strong magnetic field that emit beams of radiation from their magnetic poles. The first pulsar discovered by Bell-Burnell and Hewish was initially referred to as CP 1919 (where CP stands for Cambridge Pulsar and 1919 stand for RA of 19^h 19^m). Subsequently, it was renamed to PSR B1919+21, or also known as PSR J1921+2153, where PSR stands for Pulsating Source of Radio, B and J stands for coordinates for 1950.0 and 2000.0 epoch, respectively, followed by the coordinates in right ascension (RA) and degrees of declination (Dec.).

In 1974, the Nobel Prize in Physics was awarded one-half to Antony Hewish for his "pioneering research in radio astrophysics and his decisive role in the discovery of pulsars".

Historical Understanding of Pulsars

The existence of neutron stars was first predicted by Wilhelm Heinrich Walter Baade and Fritz Zwicky in 1934. They suggested that supernova process, i.e. the death of very massive stars, represented the transition of an ordinary star into a neutron star. Neutron stars are very small and dense, and are primarily made up of neutrons. Stars like our Sun are not massive enough to form neutron stars, they will only become white dwarf after their fuel have been exhausted and died. Lodewijk Woltjer later predicted in 1964 that if the contraction of a star is spherical and symmetric, and the conservation of magnetic flux is assumed, a neutron star's magnetic field intensity would increase by a factor of 10^{10}.

Franco Pacini then suggested that the magnetic energy could be released by a rotating neutron star and the energy could be pumped from the neutron star into a supernova remnant. In his 1967 Nature article, Pacini pointed out that the energy released by a neutron star would provide an outward momentum to the supernova remnant and accelerate its expansion. He gave Crab Nebula in the constellation of Taurus as an observational evidence, and stated that this process was possible due to the huge amount of energy that could be stored as rotational and magnetic forms in a neutron star. Initially it was thought that neutron stars would be undetectable from Earth. They were thought to be too faint to be observed due to their small size. Besides, their temperature is so high that they will radiate most of their energy in the X-ray or gamma-ray portions of the spectrum, which at that time, the technologies to observe these wavelengths were not available. However, after the discovery of the first pulsar by Bell-Burnell and Hewish, Thomas Gold put forward a strong case in 1968 that neutron stars were responsible for the pulsating radio sources, mentioning that only massive objects, and not plasma physics, would have such short and accurate observed periodicities. It is the high rotational speeds and strong magnetic fields that led to the emission in the pattern of a rotating beacon. He also mentioned that only in the presence of a very strong gravitational field would such concentration of energy be imagined.

In the same year, the discovery of a pulsar in the Crab Nebula (PSR B0531+21) strengthens the model of rotating neutron stars as the source for pulsars. Other proposed models for the emission source of pulsar, such as binary stars or pulsating stars, were unable to cater for the pulsar's short pulse period of 33 milliseconds. The Crab pulsar (Figure 9) was further confirmed by subsequent studies in both radio and optical frequencies. This pulsar is the first pulsar and one of the very few pulsars seen optically.

Although pulsars were initially detected as radio sources, currently we have also been able to observe them in the optical, X-ray and gamma-ray parts of the electromagnetic spectrum.

In 1974, Russell Alan Hulse and Joseph Hooton Taylor, Jr. found the first binary pulsar, the Hulse-Taylor binary (PSR B1913+16), using the 305-meter Arecibo radio telescope. This was a very significant finding, so significant that Hulse and Taylor were awarded the Nobel Prize in Physics 1993 "for the discovery of a new type of pulsar, a discovery that has opened up new possibilities for the study of

Figure 9. Color composite image of the Crab Nebula, which contains the Crab pulsar in the center (red star). The optical light is shown in red and x-ray is shown in blue (Credit: NASA / CXC / HST / ASU / J. Hester et al.)

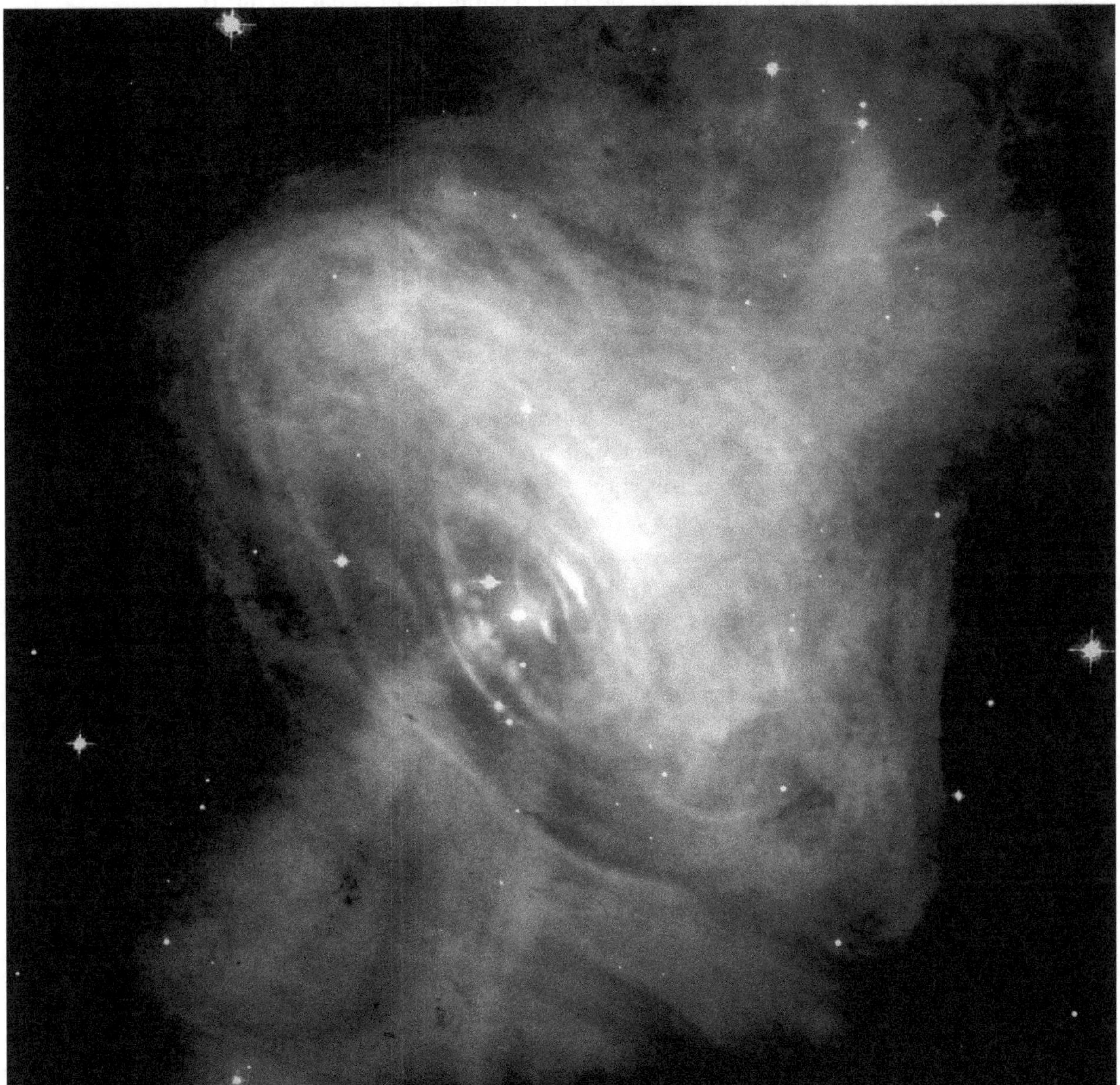

gravitation". The strong gravitational field caused by massive neutron stars allows the theories of gravity and Einstein's theory of relativity to be tested with great accuracy. According to Einstein's general theory of relativity, the orbit of a binary system is predicted to decay due to orbital energy loss in the form of gravitational waves. This system has been investigated for many decades and the observations so far had agreed with prediction. The discovery of a binary pulsar system also allows the studies of the physics of compact objects and the astrophysics of close binary systems.

Amazingly, the first planets found outside the Solar System (exoplanets) were not orbiting a normal star. Instead, in 1992, Aleksander Wolszczan and Dale A. Frail found two exoplanets orbiting a pulsar. Wolszczan and Frail used the Arecibo telescope to make precise timing measurements of the pulses of PSR 1257+12 and showed that instead of having a stellar companion, the pulsar was orbited by two or more planets. In 2016, AR Scorpii was the first pulsar that was identified to be a white dwarf instead of a neutron star. It is a binary system that is made up of a white dwarf and a red dwarf.

Characteristics of Neutron Stars and Pulsars

When a massive star runs out of fuel at its core, it will end its life as a supernova. During this extremely violent explosion, the core will collapse under its own gravity in a fraction of a second and all matter is compressed into a superdense state to become a neutron star, or compressed even further to become a black hole – an object with infinite density – if the star was massive enough. The resulted neutron star from a supernova explosion has a diameter of only 10 to 30 kilometers, but has a mass as much as the Sun. Since the angular momentum of an object must be conserved, a collapsing star must spin faster as it pulls its matter closer to its axis of rotation. The greatly reduced radius of a neutron star compared to its size before it collapses causes its spin to increase tremendously, in some cases it can be as fast as few hundred times per second. The collapse of a star into a size of a city has also squeezed the magnetic field into a smaller volume, with the field increasing inversely as the square of the radius. The highly conducting plasma trapped and concentrated the magnetic flux causing the strength of the magnetic field of a neutron star to increase, in some cases up to a billion folds. Charged particles moving along the magnetic field lines would cause beams of radiation to be emitted from the magnetic poles and the radiation is powered by the loss of rotational energy and magnetic field decay.

As more pulsars were found in the 1960s, the nature of these objects was highly debated. They are definitely not normal stars, as they are too big to pulse that fast. Even white dwarfs were ruled out due to its size. If it was due to a surface feature on a star, it is also impossible for a star to rotate that fast to produce the pulses – the star would fly apart with such rapid rotational rate. Fortunately, the duration of each pulse of a pulsar can help to constrain the upper size limit of the object producing the pulse. It is known that an object is unable to change its brightness appreciably in a duration shorter than the time needed for light to cross its diameter. Thus, knowing that the duration of a pulsar's pulse is no longer than 0.001 second, it gives an object's diameter not bigger than 300 km, which narrowed down the candidate to a neutron star. However, a neutron star is too small to pulsate slowly enough, thus this eliminated the object as a pulsating neutron star. Due to the strong gravitational force of a neutron star, it can spin up to a thousand times without flying apart, thus pulsars are more probable to be a rapidly rotating neutron star rather than a pulsating neutron star.

We are now certain that pulsars are rapidly rotating neutron stars that emit their radiation from their magnetic dipole which is inclined at an angle to their rotational axis (Figure 10). The misalignment between the magnetic dipole and the rotational axis causes the beams of radiation to sweep around the sky at the rotation frequency when the neutron star rotates. The pulses that we observe are when the neutron stars point their beams at us as they rotate, causing a periodic pulse of radiation much like how we observe the regular light pulses from a lighthouse. We will not be able to observe the pulses if the beams do not sweep over Earth. Pulsars have been found with periods ranging from few milliseconds to eight seconds, with a typical period of one second. Although the pulsation period of pulsar is incredibly regular – it can be as accurate as an atomic clock – the period was found to be slowing down ever so slightly due to the loss of energy though its radiation.

Pulsars with periods in the order of milliseconds are known as millisecond pulsars. It is understood that newborn pulsars will spin much rapidly, and as they age, their spin will slow down. However, some old pulsars were found to be spinning very rapidly, including PSR J1748−2446ad, the fastest spinning pulsar to date with a spin rate of 716 times per second. Initially, it was a puzzle as to how an old neutron star could spin so fast. The current understanding of this type of pulsars that have short period and fairly weak magnetic field strength is that they reside in a binary system. The millisecond pulsars gain their rotational energy through mass transfer from their companion star. The gravitational potential energy

Figure 10. Model of a pulsar showing the inclination of the magnetic dipole to the rotational axis (green line). The curve lines indicate the magnetic field lines and the blue cones indicate the emission beams. As the pulsar rotates, the emission beams sweep around the sky at the rotation frequency (Credit: Mysid)

of accreted matter powers the pulsars. As matter falls towards the pulsar from its companion, analogous with water hitting a mill wheel, it spins the pulsar up faster and faster until the pulsar's period becomes milliseconds. Slowly and steadily, the pulsar will continue to pull matter from its companion until it eventually consumes and evaporates its partner.

Since neutron stars are the leftover from supernova explosions, we would naturally expect to find pulsars in supernova remnants. However, only a small handful of pulsars can be associated with supernova remnants. There are a few reasons for this. Firstly, neutron stars can outlive supernova remnants – supernova remnants will disperse and be gone in few tens of thousands of years while pulsar can stay detected for 10 million years or so. Secondly, supernova explosions will likely be asymmetrical, causing a kick to the newly formed neutron star, and possibly ejecting it from the system with a high velocity. It is also possible that some supernova remnants do contain pulsars, but their radiation beams are not directed toward us, causing them to be undetectable by us.

Another characteristic of radio pulsar is the measurement of its pulse dispersion, which can be used to determine its distance. Each radio pulse, like visible light, is made up of radio waves of multiple frequencies. As the pulses travel through the interstellar medium, the electrons that were encountered in the medium will vibrate due to the time-varying electric field of the radio waves and this process will slow down the propagation of radio waves. It is observed that the lower frequencies radio waves will travel slower than the higher frequencies, i.e. greater retardation at lower frequencies. This is known as pulse dispersion and the amount of dispersion depends on the number of free electrons in the interstellar medium. The more distant pulsars have a greater pulse dispersion, because there is bigger space and thus more electrons between us and the pulsars. So, the time delays between the higher and lower frequencies can be used to provide distance estimates for pulsars.

As a pulsar ages and slows down, it is likely that the beam of radiation may turn off after a few million years. The beams may become weaker and weaker until the radio pulses are too faint to be detected, or the magnetic field may decay until it is no longer strong enough to sustain the pulse mechanism. We are not quite sure of what will be a pulsar's final fate. Clearly, much remains to be studied about this cosmic lighthouse.

REFERENCES

Baade, W., & Zwicky, F. (1934). Remarks on Super-Novae and Cosmic Rays. *Physical Review*, *46*(1), 76–77. doi:10.1103/PhysRev.46.76.2

Bañados, E., Venemans, B. P., Mazzucchelli, C., Farina, E. P., Walter, F., Wang, F., ... Winters, J. M. (2018). An 800-million-solar-mass black hole in a significantly neutral Universe at a redshift of 7.5. *Nature*, *553*(7689), 473–476. doi:10.1038/nature25180 PMID:29211709

Bell Burnell, S. J. (1979, Jan). Little Green Men, White Dwarfs, or Pulsars? *Cosmic Search*, *1*(1), 16. Retrieved from http://www.bigear.org/vol1no1/burnell

Bolton, J. G., Stanley, G. J., & Slee, O. B. (1949). Positions of Three Discrete Sources of Galactic Radio-Frequency Radiation. *Nature*, *164*(4159), 101–102. doi:10.1038/164101b0

Buckley, D. A. H., Meintjes, P. J., Potter, S. B., Marsh, T. R., & Gänsicke, B. T. (2017). Polarimetric evidence of a white dwarf pulsar in the binary system AR Scorpii. *Nature Astronomy*, *1*(2), 1–30. doi:10.103841550-016-0029

Cocke, W. J., Disney, M., & Taylor, D. J. (1969). Discovery of Optical Signals from Pulsar NP 0532. *Nature*, *221*(5180), 525–527. doi:10.1038/221525a0

European Southern Observatory (ESO). (n.d.). Centaurus A. Retrieved from https://www.eso.org/public/images/eso0903a/

European Space Agency (ESA). (n.d.). Best image of bright quasar 3C 273. Retrieved from https://www.spacetelescope.org/images/potw1346a/

Fan, X., Wang, F., Yang, J., Keeton, C. R., Yue, M., Zabludoff, A., ... Wu, X.-B. (2019). The Discovery of a Gravitationally Lensed Quasar at z = 6.51. *The Astrophysical Journal*, *870*(2), L11. doi:10.3847/2041-8213/aaeffe

Fanaroff, B. L., & Riley, J. M. (1974). The morphology of extragalactic radio sources of high and low luminosity. *Monthly Notices of the Royal Astronomical Society*, *167*(1), 31–36. doi:10.1093/mnras/167.1.31P

Fath, E. A. (1909). The Spectra of Some Spiral Nebulae and Globular Star Clusters. *Lick Observatory Bulletins*, *149*, 71–77. doi:10.5479/ADS/bib/1909LicOB.5.71F

Gold, T. (1968). Rotating Neutron Stars as the Origin of the Pulsating Radio Sources. *Nature*, *218*(5143), 731–732. doi:10.1038/218731a0

Greenstein, J. L., & Matthews, T. A. (1963). Red-Shift of the Unusual Radio Source: 3C 48. *Nature*, *197*(4872), 1041–1042. doi:10.1038/1971041a0

Hazard, C., Mackey, M. B., & Shimmins, A. J. (1963). Investigations of the Radio Source 3C 273 By The Method of Lunar Occultations. *Nature*, *197*(4872), 1037–1039. doi:10.1038/1971037a0

Hessels, J. W. T., Ransom, S. M., Stairs, I. H., Freire, P. C., Kaspi, V. M., & Camilo, F. (2006). A Radio Pulsar Spinning at 716 Hz. *Science*, *311*(5769), 1901–1904. doi:10.1126cience.1123430 PMID:16410486

Hewish, A., Bell, S. J., Pilkington, J. D. H., Scott, P. F., & Collins, R. A. (1968). Observation of a Rapidly Pulsating Radio Source. *Nature*, *217*(5130), 709–713. doi:10.1038/217709a0

Hubble, E. P. (1926). Extragalactic nebulae. *The Astrophysical Journal*, *64*, 321–369. doi:10.1086/143018

Hulse, R. A., & Taylor, J. H. (1975). Discovery of a pulsar in a binary system. *The Astrophysical Journal*, *195*, L51–L53. doi:10.1086/181708

Jansky, K. G. (1932). Directional Studies of Atmospherics at High Frequencies. *Proceedings of the Institute of Radio Engineers*, *20*(12), 1920–1932. doi:10.1109/JRPROC.1932.227477

Jansky, K. G. (1933). Electrical Disturbances Apparently of Extraterrestrial Origin. *Proceedings of the Institute of Radio Engineers*, *21*(10), 1387–1398.

Jansky, K. G. (1933). Radio Waves from Outside the Solar System. *Nature*, *132*(3323), 66–66. doi:10.1038/132066a0

Jansky, K. G. (1935). A Note on the Source of Interstellar Interference. *Proceedings of the Institute of Radio Engineers*, *23*(10), 1158–1163.

Matthews, T. A., & Sandage, A. R. (1963). Optical Identification of 3C 48, 3C 196, and 3C 286 with Stellar Objects. *The Astrophysical Journal*, *138*, 30–56. doi:10.1086/147615

Matveenko, L. I. (1968). Position of a Source of Small Angular Size in the Crab Nebula. *Soviet Astronomy*, *12*, 552–553.

NASA, Space Telescope Science Institute. (STScI), HubbleSite. (n.d.). Combined x-ray and optical images of the crab nebula. Retrieved from https://hubblesite.org/contents/media/images/2002/24/1248-Image.html

NASA, Space Telescope Science Institute. (STScI), HubbleSite (n.d.). Dust Disk Fuels Black Hole in Giant Elliptical Galaxy NGC 4261. Retrieved from https://hubblesite.org/image/82

Nather, R. E., Warner, B., & Macfarlane, M. (1969). Optical Pulsations in the Crab Nebula Pulsar. *Nature*, *221*(5180), 527–529. doi:10.1038/221527a0

National Radio Astronomy Observatory. (n.d.). Retrieved from https://public.nrao.edu/

Netzer, H. (2013). *The Physics and Evolution of Active Galactic Nuclei* (1st ed.). UK: Cambridge University Press. doi:10.1017/CBO9781139109291

Oke, J. B. (1963). Absolute Energy Distribution in the Optical Spectrum of 3C 273. *Nature*, *197*(4872), 1040–1041. doi:10.1038/1971040b0

Pacini, F. (1967). Energy Emission from a Neutron Star. *Nature*, *216*(5115), 567–568. doi:10.1038/216567a0

Reifenstein, E. C. III, Staelin, D. H., & Brundage, W. D. (1969). Crab Nebula Pulsar NPO527. *Physical Review Letters*, *22*(7), 311. doi:10.1103/PhysRevLett.22.311

Schmidt, M. (1963). 3C 273: A Star-Like Object with Large Red-Shift. *Nature*, *197*(4872), 1040. doi:10.1038/1971040a0

Schmidt, M. (1965). Large Redshifts of Five Quasi-Stellar Sources. *The Astrophysical Journal*, *141*, 1295–1300. doi:10.1086/148217

Schmidt, M., & Matthews, T. A. (1964). Redshift of the Quasi-Stellar Radio Sources 3C 47 and 3C 147. *The Astrophysical Journal*, *139*, 781–785. doi:10.1086/147815

Schneider, P. (2015). *Extragalactic Astronomy and Cosmology: An Introduction* (2nd ed.). Springer-Verlag Berlin and Heidelberg GmbH & Co. KG.

Shields, G. A. (1999). A Brief History of Active Galactic Nuclei. *Publications of the Astronomical Society of the Pacific*, *111*(760), 661–678. doi:10.1086/316378

Slipher, V. M. (1917). The spectrum and velocity of the nebula N.G.C. 1068 (M 77). *Lowell Observatory Bulletin*, *3*, 59–62.

Staelin, D. H., & Reifenstein, E. C. III. (1968). Pulsating radio sources near the Crab Nebula. *Science*, *162*(3861), 1481–1483. doi:10.1126cience.162.3861.1481 PMID:17739779

Urry, C. M., & Padovani, P. (1995). Unified Schemes for Radio-Loud Active Galactic Nuclei. *Publications of the Astronomical Society of the Pacific*, *107*, 803–845. doi:10.1086/133630

Weisberg, J. M., Nice, D. J., & Taylor, J. H. (2010). Timing Measurements of the Relativistic Binary Pulsar PSR B1913+16. *The Astrophysical Journal*, *722*(2), 1030–1034. doi:10.1088/0004-637X/722/2/1030

Wolszczan, A., & Frail, D. A. (1992). A planetary system around the millisecond pulsar PSR1257 + 12. *Nature*, *355*(6356), 145–147. doi:10.1038/355145a0

Woltjer, L. (1964). X-rays and Type I Supernovae. *The Astrophysical Journal*, *140*, 1309–1313. doi:10.1086/148028

Wu, X., Wang, F., Fan, X., Yi, W., Zuo, W., Bian, F., ... Beletsky, Y. (2015). An ultraluminous quasar with a twelve-billion-solar-mass black hole at redshift 6.30. *Nature*, *518*(7540), 512–515. doi:10.1038/nature14241 PMID:25719667

Yusef-Zadeh, F. (2003). The Origin of the Galactic Center Nonthermal Radio Filaments: Young Stellar Clusters. *The Astrophysical Journal*, *598*(1), 325–333. doi:10.1086/378715

Chapter 15
The Cosmic Microwave Background

Hui Chieh Teoh
Universiti Tunku Abdul Rahman, Malaysia

ABSTRACT

The cosmic microwave background (CMB) holds many secrets of the origin and the evolution of our universe. This ancient radiation was created shortly after the Big Bang, when the expanding universe cooled and became transparent, sending an afterglow of light in all directions. It is a pattern frozen in place that dates back to 375,000 years after the birth of the universe. Numerous experiments and space missions have made increasingly higher resolution maps of the CMB radiation, with the aims to learn more about the conditions of our early universe and the origin of stars, galaxies, and the large-scale cosmic structures that populate our universe today.

Our universe began 13.8 billion years ago and has evolved from a single point of extreme heat and infinite density to the rich and complex universe of stars and galaxies that we see today. The early history of our universe was imprinted on the cosmic microwave background, or CMB in short, which is believed to be the relic radiation of our infant universe. Initially known as the primeval fireball, CMB consists of a wealth of information of the universe when it was only 375,000 years old, before any stars or galaxies ever existed. It is a crucial source of information to understand the evolution of our universe. Thus, by studying the properties of this radiation, we can learn about the conditions of our early universe and the origin of stars, galaxies and the large-scale cosmic structures that populate our universe today.

THE DISCOVERY OF THE CMB

The detection of electromagnetic radiation that fills all space, known as the cosmic microwave background (CMB), was again by chance, similar as the accidental detections of radio emissions from our Milky Way galaxy in 1931 by Karl Gurthe Jansky and from our Sun in 1942 by James Stanley Hey. In 1964, Arno Allan Penzias and Robert Woodrow Wilson were working on a 20-foot horn reflector antenna owned by

DOI: 10.4018/978-1-7998-2381-0.ch015

Bell Telephone Laboratories at Crawford Hill, Holmdel, New Jersey. They refitted the antenna and built a Dicke radio receiver for radio astronomy use. After accounted for atmospheric absorption and ohmic losses, their measurements showed persistent noise that could not be accounted for regardless of which direction the antenna was pointing. This excess noise had a blackbody temperature of approximately 3.5 K at 4080 MHz, and Penzias and Wilson were trying very hard to get rid of that noise. They inspected every component of the equipment and initially thought that pigeons that frequented the antenna might be the cause. They even cleaned up pigeon's droppings in the antenna but the noise still remained.

In early 1965, a deeply frustrated Penzias happened to have a conversation with Bernard Burke from MIT. Penzias told him about the antenna excess noise that would not go away, and their fruitless efforts in figuring out the source of the noise. Burke told Penzias that he heard of the work by Robert Henry Dicke and his team members Phillip James Edwin Peebles, David Todd Wilkinson and Peter Roll at Princeton University about the search for a background signal from the early phase of the universe. Peebles is a theoretical physicist, and upon hearing Dicke's suggestion, he investigated and predicted the existence of the primeval radiation that permeates all of space. In order to test Peebles' prediction, Wilkinson and Roll, whom are both experts in radio technology, built an antenna on the roof of the Physics building in Princeton, trying to detect signatures of relic radiation from an early stage of the universe. After hearing of this research, Penzias made a phone call to Dicke and told him about the antenna excess noise. After hanging up, Dicke told his team: "Boys, we've been scooped".

The two teams then met at Crawford Hill. The Princeton team inspected the equipment (ironically, uses the radiometer techniques first devised by Dicke), went through the measurements and confirmed that the excess noise was indeed the background radiation from the early phase of the universe. These resulted in a pair of articles published in the Astrophysical Journal in May 1965. In one of the articles, Penzias and Wilson detailing the measurement of an excess antenna temperature of 3.5 ± 1.0 K that was unaccounted for and pointed to the other article preceded theirs by Dicke and his team for a possible explanation of the source. In the article titled "Cosmic black-body radiation", Dicke et al. discussed that the presence of thermal radiation remaining from the extremely hot early phase of the universe ("the fireball") was to be expected and the possibility of detecting it. They described their experiment at a wavelength of 3-cm although it had not yet produced any results, and also interpreted the findings by Penzias and Wilson at a wavelength of 7.3-cm as confirming the prediction of the "primeval fire-ball" theory. A radiation with a blackbody temperature of 3.5 K was interpreted as the remnant of the extremely hot early universe which then cooled as a result of the universe expansion, while preserving its thermal characteristic.

In 1978, Penzias and Wilson received the Nobel Prize in Physics for "their discovery of cosmic microwave background radiation".

THE EXPANSION OF THE UNIVERSE AND THE PREDICTION OF CMB

The fact that galaxies are moving away from us was first observed in 1917 when Vesto Melvin Slipher noticed that the light of the galaxies (or known as nebulae at that time) that he observed was red shifted, indicative of recession motion of the galaxies. The idea that the universe might be expanding was first demonstrated by Alexander Alexandrovich Friedmann in 1922 after working with Albert Einstein's theory of general relativity. However, Friedmann's work was not well received, and Einstein himself insisted on a static universe. Earlier in 1917, in his obstinacy to have an unchanging universe, Einstein introduced a parameter – the cosmological constant – into his equations to yield a static model of the universe. This parameter only has effect on the largest scales; its effect is negligible on the scales of galaxies and smaller, so it will not conflict with the observations and measurements obtained on those scales. A few years later in 1927, Georges Henri Joseph Édouard Lemaître independently rediscovered the solution of an expanding universe when working on Einstein's equations.

In 1929, Edwin Powell Hubble published his observational results on the relationship between distance and radial velocity of tens of galaxies (extra-galactic nebulae). He found that these two parameters have a linear relationship, i.e. the more distance galaxies are moving away from us at higher velocities, implying that our universe is expanding, as predicted by Einstein's original theory of general relativity. This relationship is widely known as Hubble's law, even though Lemaître was the first to present this relationship two years earlier that the redshift of galaxies can be explained by the expansion of the universe, and he also estimated the value of what is now known as the Hubble constant. The Hubble constant describes the rate of expansion of the universe at different distances from a particular point in space. It is a ratio between the velocity of a galaxy and its proper distance from us. The solid observational evidence of an expanding universe by Hubble means that the addition of the cosmological constant into the equations of general relativity earlier was not necessary, and Einstein called this his greatest blunder (while at that time the cosmological constant is not necessary, today this parameter plays an important role when we discovered that the expansion of the universe is accelerating in 1998). Unfortunately, Friedmann passed away in 1925, he never had the chance to know that his solution of an expanding universe is correct.

Lemaître also put forth an idea of a primeval atom. If we moved the time backward for an expanding universe, there will be a time in the past that all matter comes together, dense and compacted in a primeval atom. This is the origin of the universe, which by decaying gave birth to all matter in the universe. This idea was later known as the Big Bang model and is one of the two more popular cosmological models at that time. The other alternative model is the Steady State model proposed by Hermann Bondi, Thomas Gold and Fred Hoyle in 1948. The pioneering work on the Big Bang model was carried out by Ralph Asher Alpher, Robert Herman and George Gamow. Alpher, together with his PhD supervisor Gamow, studied on a topic now known as the Big Bang nucleosynthesis. It tackled the problem of the formation of chemical elements in the moments following the Big Bang. Alpher, working with Herman, later realized that the early moment of the universe would have a large radiation density due to the formation of light nuclei, and could not have disappeared without a trace. They predicted the intensity of this radiation, and found that it would have a blackbody temperature of about 5 K at the present epoch, which would peak at microwave frequencies. This is the first prediction of the existence of CMB radiation. Between 1948 and 1950, Alpher, Herman and Gamow published a number of different papers regarding this ancient radiation but their works went almost unnoticed. In one of the papers in 1950, Gamow mentioned a temperature of 3 K, although he did not give much explanations for it.

The discovery of CMB radiation by Penzias and Wilson in 1964-65 was the landmark evidence that favored the Big Bang model, showing that the universe did have a denser and hotter early phase, with all matter and radiation coupled together in a single "substance".

STANDARD MODEL OF COSMOLOGY

The current standard model of cosmology is the Lambda Cold Dark Matter (Lambda-CDM or ΛCDM) model. In this model, the universe is homogeneous and isotropic on very large scales and the tiny density fluctuations that arose just right after the birth of the universe resulted in the formation of cosmic structures that we see today. In the standard model, the universe contains three major components. The first is dark energy, associated with the cosmological constant denoted by Lambda (Λ). Dark energy is an unknown form of energy postulated to occupy all space, acting as an anti-gravity force that causes the expansion of the universe to accelerate. The cosmological constant was first introduced by Einstein into his general relativity's equation, using it to balance gravity so as to obtain a static universe. The second component is cold dark matter (CDM). Dark matter is a mysterious unseen element of the cosmos that interacts very weakly with ordinary baryonic matter. Although we might not yet know what this exotic matter actually is, its presence can be felt by its gravitational effects on matter. In the cold dark matter paradigm, structures were formed bottom-up, i.e. small structures were formed through gravitational collapse first and then merged to build larger and more massive structures, in oppose to hot dark matter paradigm where structures were formed top-down through the fragmentation of larger structures into smaller structures. The third component is ordinary matter, or baryonic matter, which refers to all elements that are made of normal atomic matter.

The leading explanation now on how the universe started is the Big Bang model. According to this model, the universe started with a singularity, where all matter was condensed into a single point of infinite density and extreme heat. This point then began expanding, and has expanded ever since to the universe that we know of today. Figure 1 shows the timeline of the universe from the moment of its birth until now, which is about 13.8 billion years later. In the first tiny fraction of a second when the universe was born, it underwent an extremely rapid and accelerated expansion in a very brief period of time, known as inflation, during which the seeds for cosmic structure formation were embedded. The idea of cosmic inflation was first developed in 1979 by Alan Harvey Guth, and in 1981, he formally proposed it. Inflation was proposed to solve two problems: the horizon problem and the flatness problem. The horizon problem arises because the universe is incredibly homogeneous and isotropic on very large scales, in spite the fact that any widely separated regions in the universe were disconnected. These regions are moving away from each other faster than the speed of light and should not have equilibrated because they have not yet had time to communicate with each other to produce a uniform universe. The flatness problem arises because the geometry of the universe that we see today is very flat – too flat that the initial conditions of the universe must be fine tuned to an incredible precision without explanation.

The infant universe was filled with photons and hydrogen, but hydrogen is in its ionized form of free electrons and protons (plasma state) due to the extremely hot early universe. The universe was so hot and dense that photons were constantly interacting with free electrons thus remained trapped in the dense soup of matter. Photons could only propagate a very short distance before they encountered an electron and were scattered off. Thus, at this moment, the universe was opaque to all radiations. Eventually, after about 375,000 years, as the universe continued to expand, it cooled sufficiently for electrons and protons to combine and form neutral hydrogen atom, transitioning from the state of ionized plasma to neutral gas. This epoch is called recombination. During this epoch, another phenomenon took place, i.e. the decoupling of photon and matter (photon decoupling). Since photons interact very weakly with neutral hydrogen, they could now travel freely without being scattered. The universe became transparent at this point, and is called the surface of last scattering since this was the last time photons were scattered off by electrons. Although we are unable to peer back to the moment of our universe's birth due to the opacity of the initial universe, we can observe the remnant radiation that permeates all space, known as the cosmic microwave background (CMB), from this surface of last scattering. We can see the afterglow of the Big Bang, which has a black body temperature of 3000 K at this point. However, due to cosmic expansion, the wavelength of this afterglow has been stretched tremendously. CMB radiation that we observe today is very cold, just 2.725 degrees above absolute zero, so it shines mostly in the microwave portion of the electromagnetic spectrum.

Right after recombination and decoupling, the universe was still filled with the glow of the hot gas. However, as the universe continued to expand and cool, the glow faded. The clouds of hydrogen gases were still collapsing very slowly under gravity to form stars and galaxies, so there were no new sources of visible light, leaving the universe in darkness. This epoch is known as the dark age. It is a period between the time when CMB radiation was emitted and the time when the first stars were born. When the first generation of stars began to shine, they produced sufficient ultraviolet radiation that ionized the surrounding gas and the universe entered a period of reionization, marking the end of the dark age. Early large-scale structures gradually emerged and drawn to the dark matter filaments, i.e. the skeleton of the universe. The dark matter filaments served as a scaffolding for gas to accumulate and stars, galaxies and galaxy clusters to be built. This marks the beginning of the era of stars and galaxies in which we live today.

ANISOTROPIES IN THE CMB

All the cosmic structures that exist in the universe, such as stars, galaxies, galaxy clusters etc., grew from the action of gravity. Matter experience the attraction of other matter; thus, an initial small clump of matter will start to grow bigger and bigger, in turn attracting even more matter, making them cluster more and more effectively as time passes. Although the early universe was always considered to be uniform, a universe with an evenly spread matter will not have matter clustering because the force of gravity will be the same everywhere. Hence, in order to kick-start the aggregation process, an initial clump of mass or "seed" is necessary, i.e. there should be small inhomogeneities in the distribution of matter, to alter the perfect equilibrium. Due to gravitational instability, the gravity around this initial inhomogeneity will be stronger, resulting in aggregation of matter around this point in space that can grow into stars and galaxies.

These small inhomogeneities in the distribution of matter or the fluctuations of density in the early universe were predicted to leave an imprint as small anisotropies in the intensity of the CMB radiation by both Peebles and his student Jer Yu, and Rashid Alievich Sunyaev and Yakov Borisovich Zeldovich in 1970 independently. In his 1982 paper, Peebles discussed the effect on the large-scale anisotropy of CMB radiation and the large-scale fluctuations in the mass distribution in the presence of very massive, weakly interacting particles (dark matter). Peebles' work on CMB and his works among other on cold dark matter and the large-scale structure of the universe earned him the Nobel Prize in Physics in 2019 "for his theoretical discoveries in physical cosmology".

In 1967, the works by Rainer Kurt Sachs and Arthur Michael Wolfe, Martin John Rees and Dennis William Siahou Sciama, and Joseph Silk demonstrated that the CMB temperature fluctuations produced by the density perturbations or primordial cosmic seeds at the time of recombination had to be in the order of one percent of the average to allow galaxies to form by gravitational instability. Even that, the available technologies at that time were not advanced enough to detect such small temperature fluctuations, not to mention now that we know this value is far too optimistic. The temperature fluctuation is so weak that it is actually in the order of a hundred thousand times smaller than the average temperature of about 2.7 K. The failure to observe any traces in the CMB temperature fluctuations put the hypothesis of gravitational instability that built galaxies and the large-scale structure of the universe at risk.

Figure 1. The evolution of the universe. The size of the universe is represented by the vertical extent of the grid. The afterglow light pattern imprinted with the conditions of the early universe, known as the cosmic microwave background (CMB) radiation, has been traveling unobstructed since it was emitted 375,000 years after the birth of the universe (Credit: NASA / WMAP Science Team)

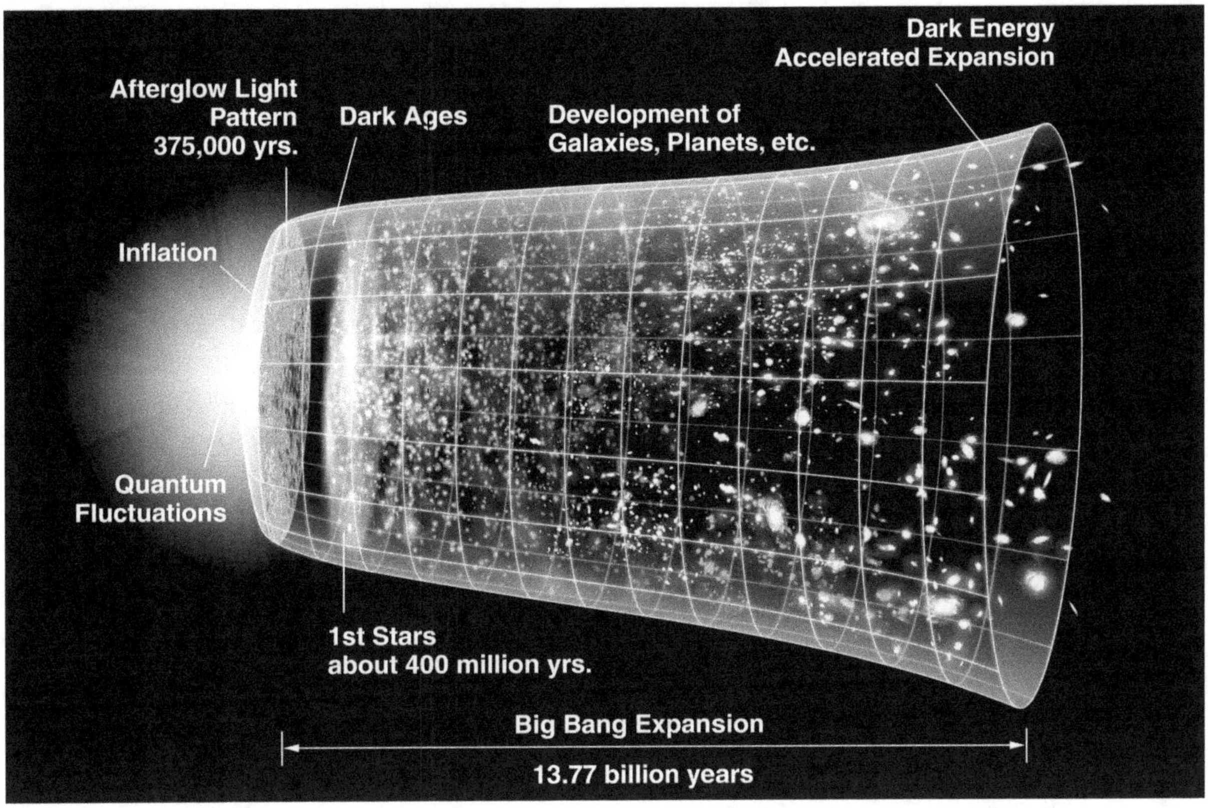

The faintness of this signal is the reason why CMB anisotropies escaped detection for decades. Since the signal is so weak, the equipment built to detect these small imperfections not only have to be very sensitive, but has to be able to discard all possible noises, such as from the Earth itself, its atmosphere, foreground cosmic sources like our own Milky Way Galaxy and other galaxies, and the inherent internal noise of the detectors. One of the ways to reduce the noise is to put the detectors in space. Since the observatory will be away from Earth, it can minimize the noise due to Earth's thermal emission and remove the contamination due to Earth's atmosphere. Earth's atmosphere also blocks most of the electromagnetic radiation from reaching the surface. Thus, by placing an observatory in space, it will allow a wider window to observe the microwave spectrum. Lastly, the sky can be observed in all the directions for longer times from the vantage point in space.

In 1981, the construction of a space-based observatory, the COsmic Background Explorer (COBE) spacecraft, was carried out at NASA's Goddard Space Flight Center. COBE was successfully launched in November 1989, and its objective was to measure the diffuse infrared and microwave radiation from the early universe between 1 μm and 1 cm over the entire sky. COBE carried three instruments, i.e. a Far Infrared Absolute Spectrophotometer (FIRAS), a Differential Microwave Radiometer (DMR), and a Diffuse Infrared Background Experiment (DIRBE). The principal investigators for these three instruments were John Cromwell Mather, George Fitzgerald Smoot and Mike G. Hauser, respectively. The objective of the FIRAS instrument was to measure precisely the CMB spectrum and to observe the dust and line emission from our galaxy. The measurements from FIRAS showed a perfect fit of the CMB spectrum and the theoretical curve for a blackbody with a temperature of 2.725 ± 0.002 K (Figure 2). This finding matches the predictions of the hot Big Bang model very well, and shows that all the radiant energy of the universe was released within the first year after the Big Bang. All theories that attempt to explain the origin of large-scale structure seen in the universe today must now conform to the constraints imposed by these measurements.

The DMR experiment was designed to search for primeval temperature fluctuations between different parts of the sky in the CMB radiation and to map the cosmic radiation precisely. DMR spent four

Figure 2. Cosmic Microwave Background (CMB) monopole spectrum from the Far Infrared Absolute Spectrophotometer (FIRAS) on the NASA Cosmic Background Explorer (COBE). The solid curve indicates the intensity of a blackbody at a temperature of 2.725 K as predicted by the hot Big Bang model. FIRAS measurements show that the CMB spectrum matches the prediction to an extraordinary degree (Credit: NASA / GSFC / LAMBDA)

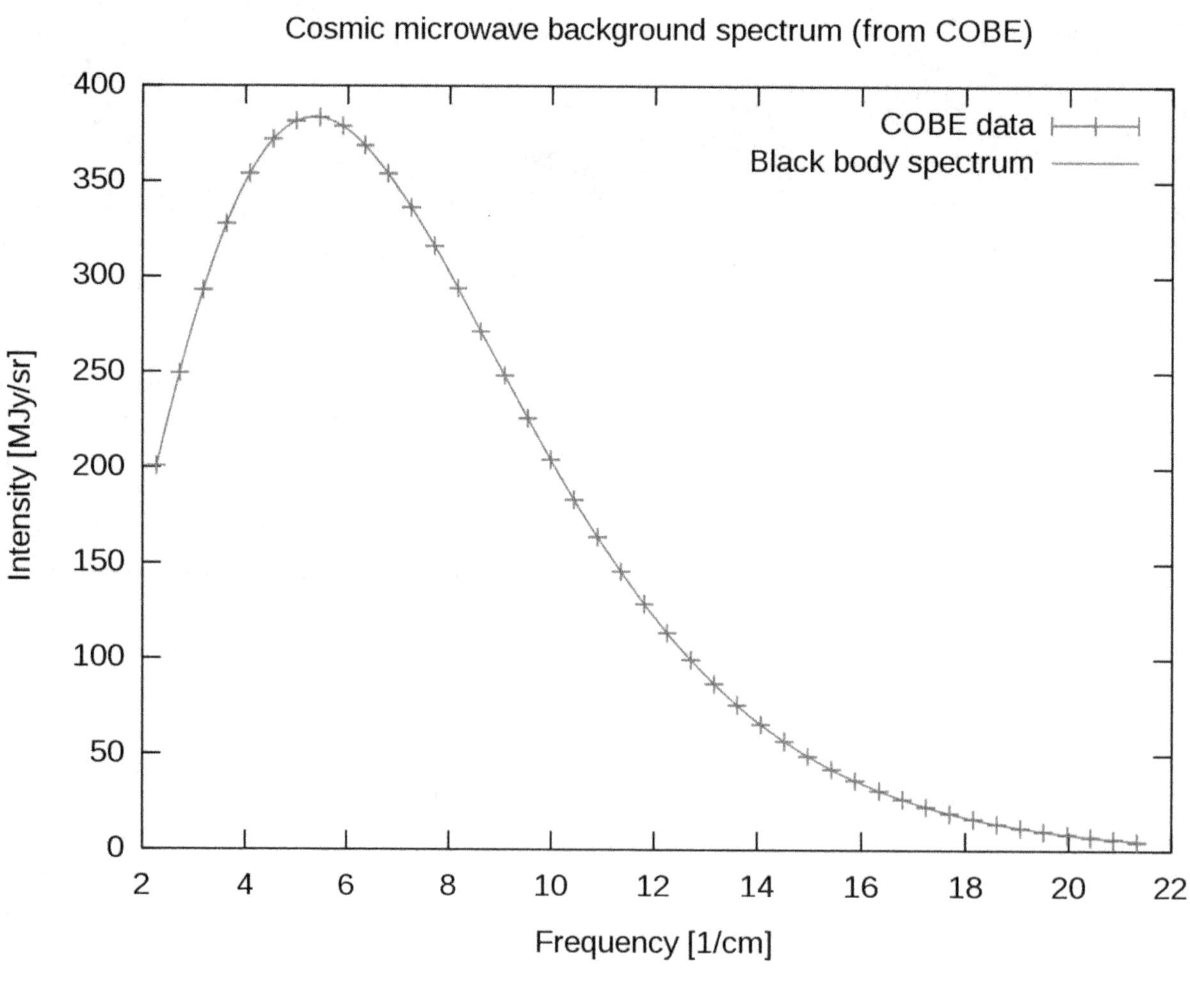

years mapping the sky and created a full sky map of the CMB radiation by subtracting out galactic emissions and dipole at various frequencies. Figure 3 represents the DMR measurements at a frequency of 53 GHz. The top image shows the nearly isotropic brightness of the CMB radiation on a scale of 0 to 4 K. If we wish to see the very subtle variations in the CMB we need to remove the isotropic emission. The middle image reveals the CMB dipole by subtracting the average brightness from the top image, intended to enhance the contrast due to the dipole. The blue color indicates that the CMB radiation is slightly less intense than average and the red color indicates that the radiation is slightly more intense than average. This difference in the intensity in different directions is called the CMB dipole anisotropy, and is due to our motion relative to the radiation. In the direction we are moving, the intensity will be greater and vice versa. The bottom image is the result after subtracting the dipole component from the middle image. The horizontal red band that runs across the bottom image is the foreground emission from our Milky Way galaxy, and the blue and red spots correspond to areas of greater or lesser density respectively in the early universe. DMR has found, for the first time, anisotropies in the CMB radiation, showing how matter and energy were distributed during the very early phase of the universe. It has found the "primordial seeds". The variations in the intensity of the CMB radiation over the sky was found to be very tiny, in the order of one part in 100,000.

The temperature fluctuations found in the CMB maps are the imprint of density variations in the early universe. Before the epoch of recombination, photons and matter were coupled. Due to this coupling, photons will be more abundant in a denser region and vice versa. Since photon density is related to the temperature for a blackbody, the CMB from denser regions will be "hotter" and regions that are less dense will be "colder". Another way for density perturbations to leave signature on the CMB is the gravitational redshift. At denser regions, the gravity is stronger, thus photons have to spend more energy to escape compared to the photons at regions which are less dense. This is translated into an increase in the wavelength and photons leaving a denser region will be redshifted, causing the region to look colder. While these two ways resulted in temperature fluctuations that are opposite of each other, the effects only cancel partially and the net result can still be seen. The last method that can contribute to the temperature variations is the Doppler effect, where a moving source's frequency changes depending on their direction either toward or away from us. All these density fluctuations carry information of the origin of the cosmic structure. Through a process that was still poorly understood at that time, the density ripples seen by the DMR gave rise to structure formation and developed into galaxies, galaxy clusters, and the large-scale structures that populate the universe today.

Apart from the background emission at microwave wavelength, cosmic infrared background (CIB) radiation could potentially contain a total radiant energy density comparable to that of the microwave background. As a result of the cosmic redshift and reprocessing by dust of short-wavelength radiation, CIB radiation is expected to contain the cumulative emissions since the first formation of luminous objects, such as stars and galaxies, in the early epoch. Thus, the DIRBE experiment was designed primarily to do a search for an isotropic CIB radiation and to measure its energy distribution. It is a multi-wavelength infrared detector used to map dust emission. Maps of infrared absolute sky brightness, including zodiacal and galactic components, were obtained by DIRBE in 10 bands ranging in wavelength from 1.25 to 240 microns. DIRBE has also provided maps of linear polarization at 1.25, 2.2, and 3.5 microns. The CIB measurements from DIRBE constrain the cosmological models of star formation and the accumulation of dust and elements heavier than hydrogen over time. DIRBE also had a secondary objective which includes the studies of foreground sources, such as scattered sunlight, thermal emission from interplanetary and interstellar dust, and galactic starlight.

Figure 3. Images of the sky obtained with the Differential Microwave Radiometer (DMR) on the NASA Cosmic Background Explorer (COBE) from the 53 GHz band. The top image shows the nearly isotropic brightness of the CMB radiation, the middle image shows the CMB dipole, and the bottom image shows the evidence for the anisotropies in the CMB radiation. The blue and red spots in the bottom image correspond to areas of greater or lesser density respectively in the early universe (Credit: NASA / GSFC / LAMBDA).

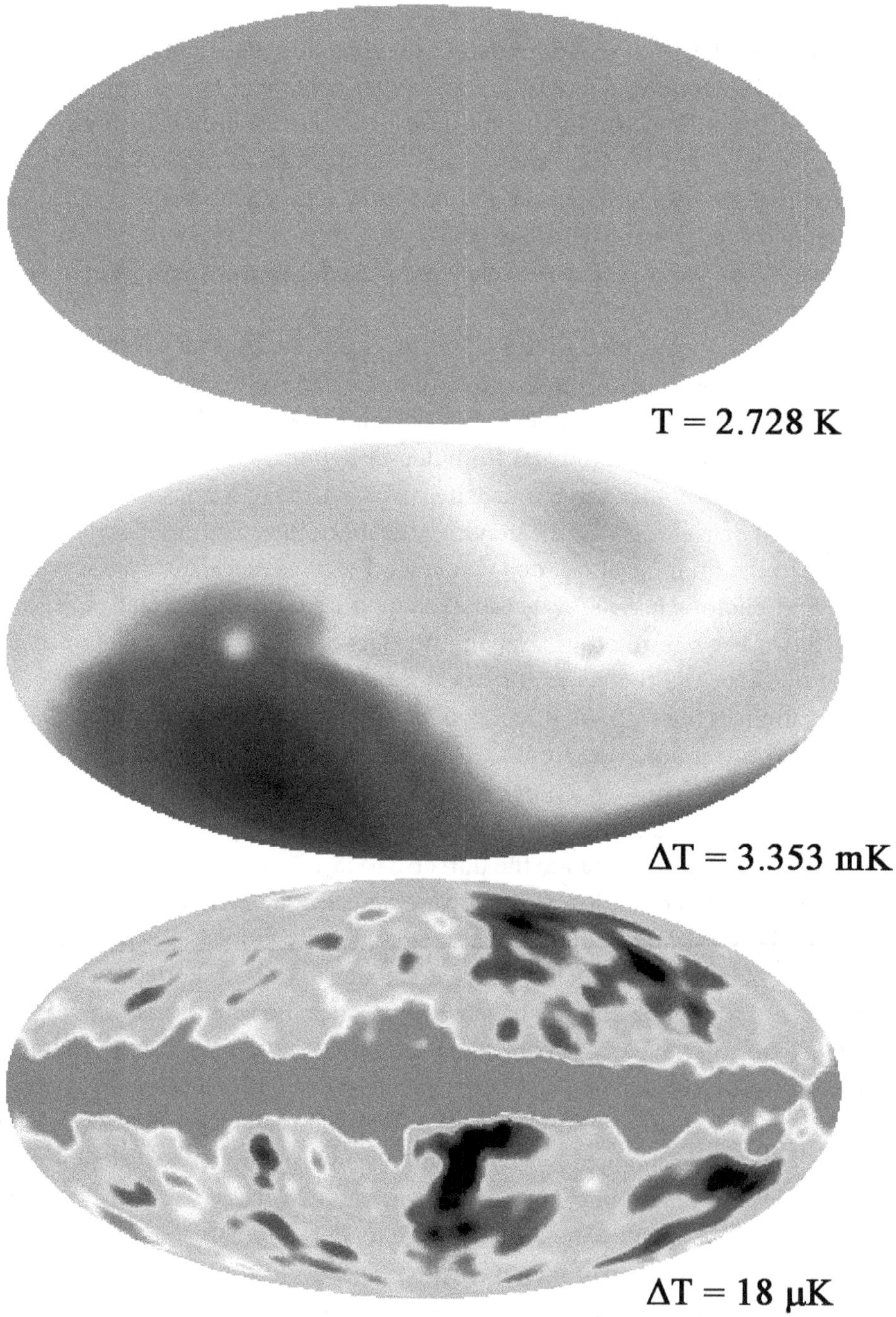

After four years studying the sky, COBE concluded its mission in December 1993. In 2006, Mather and Smoot, the principle investigator for FIRAS and DMR, were jointly awarded the Nobel Prize in Physics for "their discovery of the blackbody form and anisotropy of the cosmic microwave background radiation."

STUDYING THE CMB WITH SPACE MISSIONS

In June 2001, NASA launched another spacecraft, the Microwave Anisotropy Probe (MAP), to succeed COBE in studying the cosmic microwave background radiation. MAP was renamed to Wilkinson Microwave Anisotropy Probe (WMAP) in 2003 to honor David Todd Wilkinson, who was one of the pioneers in the field of cosmology and a member of the mission's science team. WMAP's mission was to make fundamental measurements of cosmology, i.e. revealing the conditions existed in the early universe by measuring the properties of the CMB radiation. It was capable of determining the geometry, content and evolution of the universe. It used five separate observing frequency bands from 22 to 90 GHz to assist in the rejection of foreground signals from our Milky Way galaxy. Previously, COBE has detected tiny variations in temperature, with a difference of only 0.0002 K, in the CMB radiation but with large angular resolution. WMAP had much higher resolution, sensitivity and accuracy compared to COBE and the maps obtained from WMAP observations were 33 times the angular resolution and 45 times the sensitivity of the DMR instrument. These enabled WMAP to measure the anisotropy in the CMB radiation with much finer detail and greater sensitivity (Figure 4).

When we observe distant objects, we are looking into the past since light travels at a finite speed. The CMB radiation contains a lot of information about the properties of the universe since it was emitted at the earliest time and has traveled over such a large distance. By measuring the anisotropy of the CMB radiation, WMAP could measure the basic parameters of the Big Bang model and learn about the origin of cosmic structures. When WMAP released its 9-year results in 2013, it has determined the age of the universe, the time when the first stars began to shine, the density of atoms and all other non-atomic matter, the "lumpiness" of the universe, and how that lumpiness depends on scale size. The age of the universe was determined to be 13.77 billion years old and WMAP has mapped the afterglow of the hot and dense universe when it was only 375,000 years old. After that, when the universe was about 400,000 years old, the first stars started to shine.

WMAP found that the content of the present-day universe consists of only a small fraction (4.6%) of ordinary atoms or also known as baryons. Dark matter, a type of matter that is not made up of atoms, has gravity but does not emit any light, comprises a larger fraction (24%). The largest portion of the universe (71.4%) is dark energy, an anti-gravity force that is driving an acceleration of the universe, causing the increase in the expansion rate of the universe. With the most detailed map of CMB radiation ever produced at that time, WMAP managed to clear the doubt of the existence of dark energy. Its data fit well into a universe dominated by dark energy, in the form of a cosmological constant, and the content points to a Euclidean flat geometry. With no significant anomalies and an adequate goodness of fit, WMAP's data helped in establishing the current standard model of cosmology, i.e. the ΛCDM model. It also supported the inflation theory. Remarkably, WMAP's precision measurement has confirmed specific predictions of the simplest version of inflation, i.e. the distribution of density variations follows a bell curve with the same properties across the sky, and there are equal numbers of hot and cold spots on the map. By measuring the polarization in the CMB it is possible to look at the amplitude of the fluctuations

Figure 4. Cosmic Microwave Background (CMB) radiation maps obtained from Penzias and Wilson's microwave receiver (top), NASA COBE spacecraft (middle) and NASA WMAP spacecraft (bottom). The top right image shows the simulated view of the sky using Penzias and Wilson's microwave receiver. COBE was the first to discover the patterns of the remnant afterglow from the Big Bang and WMAP brought the patterns into much better focus and resolution (Credit: NASA / WMAP Science Team).

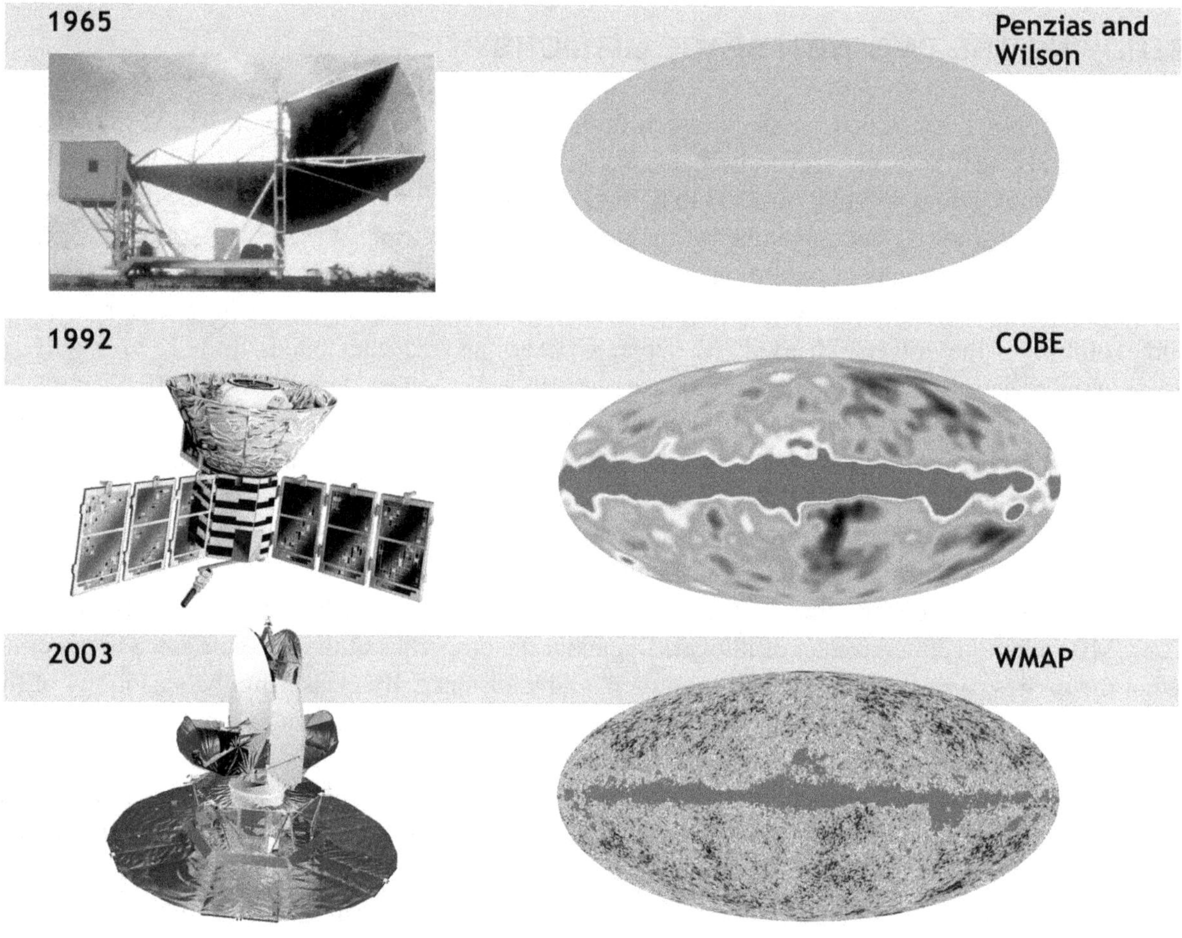

of density in the universe that produced the first galaxies. The reionization phase of the universe was discovered to be earlier than previously thought. WMAP has not only established but provided tighter constrain on the model that describes the history and structure of the universe.

WMAP ended its mission after nine years of operation in 2010. One year before the mission ended, the European Space Agency (ESA) launched a more advanced spacecraft – Planck – in 2009. The Planck mission continued the legacy of COBE and WMAP, studying the CMB radiation with the highest accuracy ever achieved, and helping to provide answers on the past, present and future of our universe. It has the capability to confirm if our current model of the universe was correct to unprecedented levels of accuracy, or if the current model was wrong, Planck would send us back to the drawing board. Planck's sensitive radio receivers operating at extremely low temperatures were capable of distinguishing temperature variations in the CMB radiation in the order of one part in a million. The nine frequency bands observed by Planck range from 27 GHz to 1 THz, and the simultaneous observations of the sky in these wide frequency range allowed us to subtract out the effects of foreground radiations, such as those from our galaxy and others, from the primordial CMB radiation.

The precise measurements by Planck have enabled us to test a huge variety of models that describe the origin and evolution of the universe. The detailed imprint of the early universe as seen by Planck has again confirmed our standard model of cosmology – the CMB temperature fluctuations fitted too well to draw any other conclusions. Apart from measuring the temperature fluctuations of the CMB radiation, Planck also measured its polarization, which indicates if light is vibrating in a preferred direction. The polarized radiation contains information of the last interaction between photons and matter in the early history of the universe, just before the photons started its cosmic journey to us. CMB polarization has two basic patterns – the circular and radial E-modes and the curly B-modes – which are produced by different phenomena on different angular scales. The B-mode polarization is weaker than E-mode polarization, thus it is more difficult to detect. The E-mode polarization arises naturally from Thomson scattering, where photons were scattered off free electron in the surface of last scattering. It was first detected in 2002 by the Degree Angular Scale Interferometer (DASI) located in the South Pole. The B-mode polarization can be generated by two mechanisms – the first is during the cosmic inflationary phase, and the second is by gravitational lensing of the E-modes. Fortunately, these two types of B-mode polarization are distinguishable from each other. The second type of B-mode polarization was successfully detected in 2013 by the South Pole Telescope, but the first type of B-mode polarization still remains elusive.

Planck's team final data was released in 2018. The team has completed a new processing of the data and was confident that both temperature and polarization in the CMB radiation were accurately determined. Figure 5 shows the temperature variations in the mission's final data release and Figure 6 shows the polarization pattern. The results reassured that the standard model of cosmology, where our universe is made up of ordinary matter, dark matter and dark energy, is on the right track. Planck has refined the values of the universe's content found by WMAP a few years back to 4.9% for ordinary matter, 26.8% for dark matter and dark energy accounts for 68.3%. The universe is also found to be a little bit older (13.8 billion years old) than previously determined. The inflationary scenario which describes that our universe underwent an extremely rapid growth in the very early phase of the universe's history, during which the cosmic seeds to build stars and galaxies were sown, was also further confirmed by Planck.

Although Planck is able to confirm almost all our current understanding of the cosmos, it still leaves us some puzzles. One in particular is the Hubble constant, the rate of expansion of the universe, which can be determined through two techniques – either based on the distant, baby universe or based on the local, mature universe. Planck has indicated a value for Hubble Constant of 67.4 km/s for every million parsecs (Mpc) of separation in space, using the cosmological model that fits the CMB. However, another technique to measure the Hubble Constant, i.e. using variable and exploding stars to gauge distances across the universe, has yielded a value of 73.5 km/s/Mpc. These two values do not agree to within their respective uncertainties (Figure 7) and so far, there is still no satisfactory explanation for the differences.

Planck's first data release of the temperature maps in 2013 revealed some anomalies that do not fit very well into the standard model. Although the measurements on small and intermediate angular scales fit very well with the model predictions, there is a deficit in the signal, ten percent weaker than predicted by the model, at large angular scales of about 5 degrees on the sky. This is the most serious anomaly that was found in the CMB temperature maps. Other anomalies include a significant discrepancy of the signal as observed in the two opposite hemispheres of the sky and an abnormally large, low-temperature spot with an unusually steep temperature profile, a so-called "cold spot" in the sky (Figure 8). To provide another almost independent view of the CMB radiation that these anomalies are either statistical flukes

Figure 5. The anisotropies of the CMB radiation, as seen by ESA Planck spacecraft. This remnant radiation from the Big Bang has tiny imperfections and subtle temperature variations that correspond to areas of different densities, representing the primordial seeds for all future cosmic structures (Credit: ESA / Planck Collaboration)

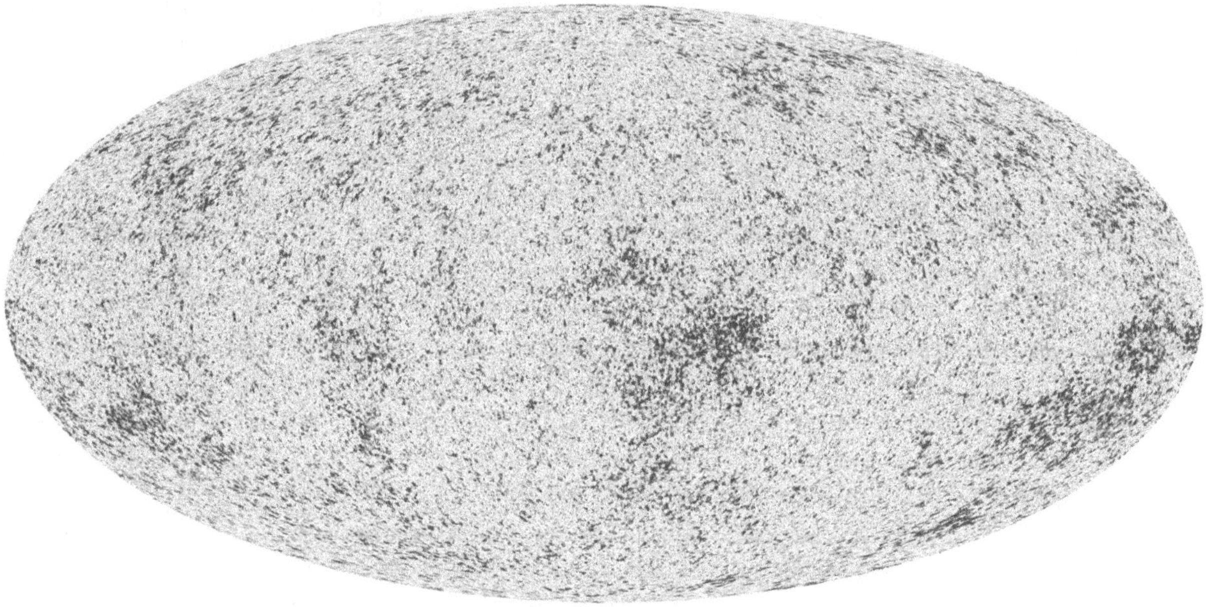

Figure 6. The polarization of the CMB radiation, as seen by ESA Planck spacecraft. The polarized pattern (lines) follows the pattern of the temperature fluctuations (color contours). This polarized light retains the information on the distribution of matter in the surface of last scattering (Credit: ESA / Planck Collaboration)

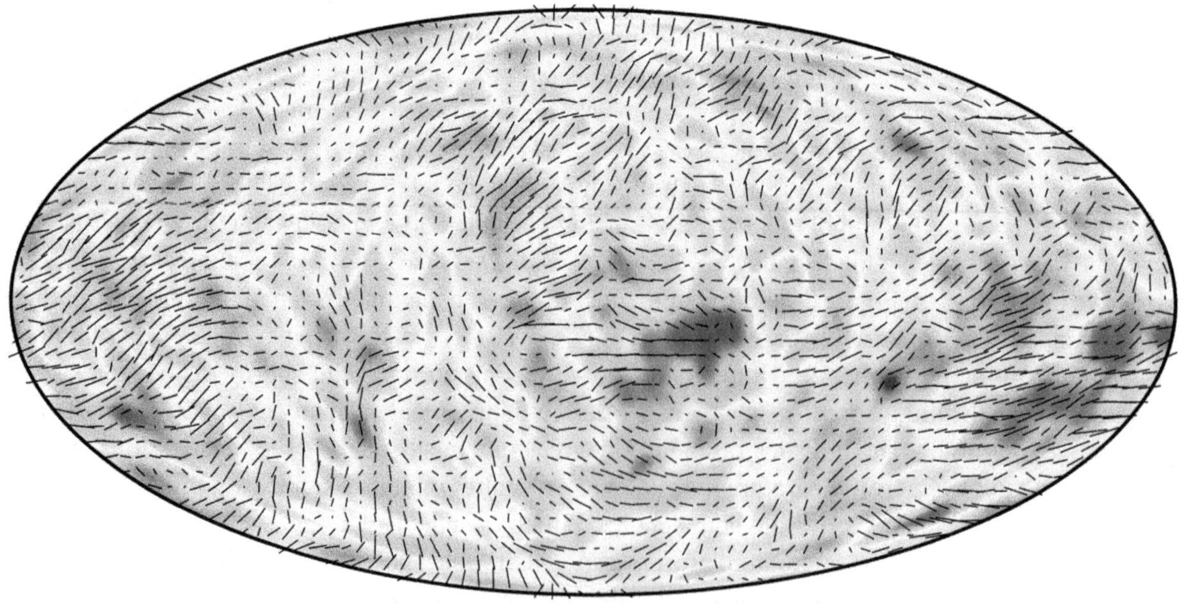

Figure 7. The values of Hubble Constant measured using observations of CMB (green and red) and variable stars or supernovae (blue) over the past two decades. Another method using the observation of gravitational waves is shown in purple (Credit: ESA / Planck Collaboration)

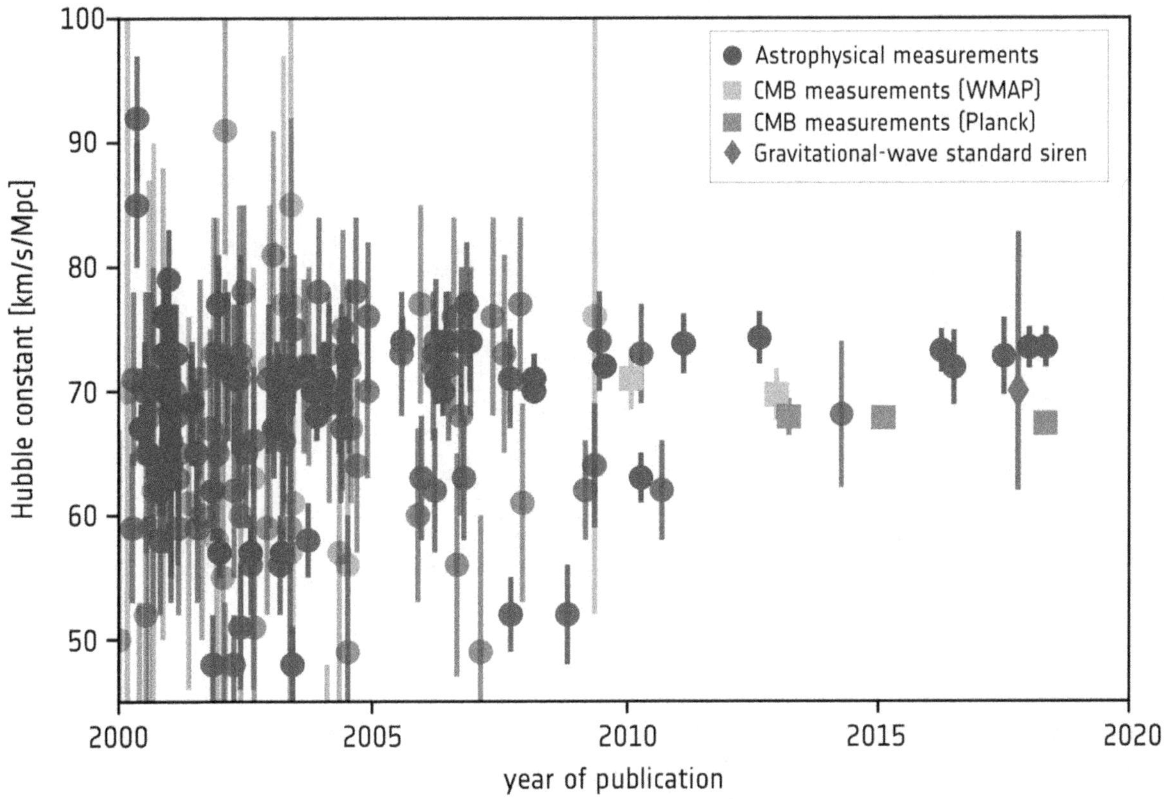

or are due to new physics that we are unaware of, the Planck team turned to the polarization of the CMB radiation. In 2018, the Planck team released the most accurate maps of the CMB E-mode polarisation to date, a great improvement over the first polarization maps published in 2015. The latest polarisation results show no significant sign of anomalies. At most, the analysis only shows some weak hints that the anomalies may be present. Thus, this would seem like the anomalies are statistical flukes, but it does not rule out the potential of new physics in play – a phenomenon that only affects the temperature, but not the polarisation of the CMB radiation.

In 2014, a team of researchers announced the results of their experiment using the BICEP2 instrument and also the preliminary data from the Keck Array located at the South Pole. They detected the CMB polarisation curly B-modes over a small patch of the sky and interpreted it as originated in primordial gravitational waves. When the universe is only a tiny fraction of a second old, it experienced an inflationary phase that would generate tiny perturbations in the fabric of space-time, known as gravitational waves. These perturbations are postulated to leave their signature on the CMB B-mode polarization, whereas the temperature fluctuations only create E-mode polarization. Although the gravitational waves generated by inflation are too weak to be detected with current technology, the polarization imprinted on

Figure 8. Two anomalies in the CMB, as seen by ESA Planck spacecraft. The first anomaly is indicated by the curved line that separates the two hemispheres with asymmetry in the average temperatures – the southern ecliptic hemisphere has a significantly stronger signal on average compared to the north. The second anomaly is indicated by the circle, a cold spot, which is an abnormally large, low-temperature spot (Credit: ESA / Planck Collaboration)

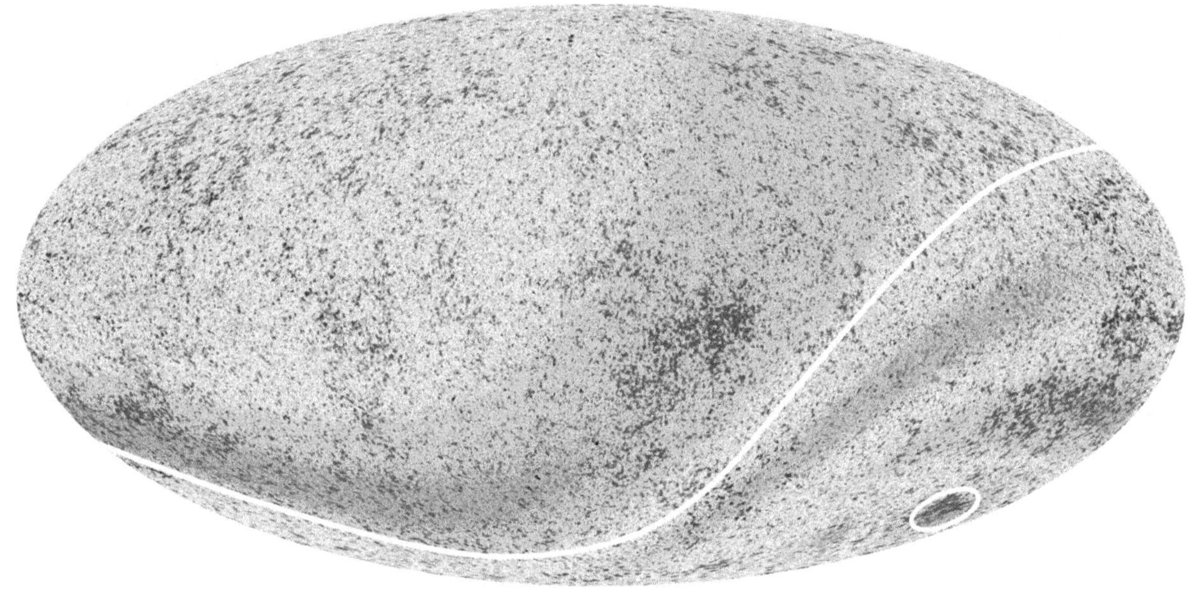

the CMB radiation may be detectable. Thus, by detecting the B-mode polarization in the CMB radiation, the researchers claimed that they have found the first direct evidence of cosmic inflation. Unfortunately, subsequent results from Planck in 2015 found no conclusive evidence of primordial gravitational waves. The results show that cosmic dust accounts for most, or maybe all, of the observed B-mode polarization. While currently we are unable to confirm that the detection is an imprint of inflation, the primordial gravitational waves signal could still be out there, waiting to be found. If and when it is found, this signal can be crucial to make or break the various inflation models that we have put forth.

Planck ended its mission in 2013 after four years scanning the sky at microwave and sub-millimeter frequencies. It has left us the best of both worlds – it helped to set the most precise constraints on some of the parameters of the standard model of cosmology, ranging from the Hubble constant, the densities of the content of the universe and the relative amount of primordial fluctuations that allowed the formation of cosmic structure on different scales, and at the same time, gave us some puzzles to ponder over for many years to come.

REFERENCES

Ade, P. A. R., Aghanim, N., Alves, M. I. R., Armitage-Caplan, C., Arnaud, M., Ashdown, M., ... Zonca, A.Planck Collaboration. (2014). Planck 2013 results. I. Overview of products and scientific results. *Astronomy & Astrophysics*, *571*(A1), 1–49. doi:10.1051/0004-6361/201321529

Alpher, R. A., Bethe, H., & Gamow, G. (1948). The Origin of Chemical Elements. *Physical Review*, *73*(7), 803–804. doi:10.1103/PhysRev.73.803 PMID:18877094

Alves, J., Bertout, C., Combes, F., Ferrara, A., Forveille, T., Guillot, T., ... Walmsley, M.Planck Collaboration. (2014). Planck 2013 results. XVI. Cosmological parameters. *Astronomy & Astrophysics*, *571*(A16), 1–69. doi:10.1051/0004-6361/201425195

Alves, J., Combes, F., Ferrara, A., Forveille, T., & Shore, S.Planck Collaboration. (2016). Planck 2015 results. XIII. Cosmological parameters. *Astronomy & Astrophysics*, *594*(A13), 1–67. doi:10.1051/0004-6361/201629543

Bennett, C. L., Larson, D., Weiland, J. L., Jarosik, N., Hinshaw, G., Odegard, N., ... Halpern, M. (2013). Nine-Year Wilkinson Microwave Anisotropy Probe (WMAP) Observations: Final Maps and Results. *The Astrophysical Journal. Supplement Series*, *208*(20), 1–54.

BICEP2 Collaboration. (2014). Detection of B-Mode Polarization at Degree Angular Scales by BICEP2. *Physical Review Letters*, *112*(241101), 1–26. PMID:24996078

BICEP2/Keck & Planck Collaborations. (2015). Joint Analysis of BICEP2/Keck Array and Planck Data. *Physical Review Letters*, *114*(101301), 1–17. PMID:25815919

Bondi, H., & Gold, T. (1948). The Steady-State Theory of the Expanding Universe. *Monthly Notices of the Royal Astronomical Society*, *108*(3), 252–270. doi:10.1093/mnras/108.3.252

Cosmic Background Explorer Mission PageN. A. S. A. (n.d.). Retrieved from https://lambda.gsfc.nasa.gov/product/cobe/

Dicke, R. H., Peebles, P. J. E., Roll, P. G., & Wilkinson, D. T. (1965). Cosmic Black-Body Radiation. *The Astrophysical Journal*, *142*, 414–419. doi:10.1086/148306

European Space Agency (ESA) Planck Mission Page. (n.d.). Retrieved from https://sci.esa.int/web/planck/home

Fixsen, D. J., Cheng, E. S., Gales, J. M., Mather, J. C., Shafer, R. A., & Wright, E. L. (1996). The Cosmic Microwave Background Spectrum from the Full COBE* FIRAS Data Set. *The Astrophysical Journal*, *473*(2), 576–587. doi:10.1086/178173

Gamow, G. (1946). Expanding Universe and the Origin of Elements. *Physical Review*, *70*(7-8), 572–573. doi:10.1103/PhysRev.70.572.2

Gamow, G. (1948). The Evolution of the Universe. *Nature*, *162*(4122), 680–682. doi:10.1038/162680a0 PMID:18893719

Gamow, G. (1952). Half an hour of creation. *Physics Today*, *3*(8), 16–21. doi:10.1063/1.3066969

Guth, A. H. (1981). Inflationary universe: A possible solution to the horizon and flatness problems. *Physical Review D: Particles and Fields*, *23*(2), 347–356. doi:10.1103/PhysRevD.23.347

Guth, A. H. (1997). *The Inflationary Universe: The Quest for a New Theory of Cosmic Origins*. New York: Perseus Books.

Hanson, D., Hoover, S., Crites, A., Ade, P. A. R., Aird, K. A., Austermann, J. E., ... Zemcov, M. (2013). Detection of B-mode polarization in the Cosmic Microwave Background with data from the South Pole Telescope. *Physical Review Letters*, *111*(141301), 1–6. PMID:24138230

Hinshaw, G., Larson, D., Komatsu, E., Spergel, D. N., Bennett, C. L., Dunkley, J., ... Wright, E. L. (2013). Nine-Year Wilkinson Microwave Anisotropy Probe (WMAP) Observations: Cosmological Parameter Results. *The Astrophysical Journal. Supplement Series*, *208*(19), 1–25.

Hoyle, F. (1948). A New Model for the Expanding Universe. *Monthly Notices of the Royal Astronomical Society*, *108*(5), 372–382. doi:10.1093/mnras/108.5.372

Hubble, E. (1929). A relation between distance and radial velocity among extra-galactic nebulae. *Proceedings of the National Academy of Sciences of the United States of America*, *15*(3), 168–173. doi:10.1073/pnas.15.3.168 PMID:16577160

Legacy Archive for Microwave Background Data Analysis (LAMBDA). (n.d.). Retrieved from https://lambda.gsfc.nasa.gov/

Leitch, E. M., Kovac, J. M., Pryke, C., Carlstrom, J. E., Halverson, N. W., Holzapfel, W. L., ... Sandberg, E. S. (2002). Measurement of polarization with the Degree Angular Scale Interferometer. *Nature*, *420*(6917), 763–771. doi:10.1038/nature01271 PMID:12490940

Lemaître, G. (1931). The Beginning of the World from the Point of View of Quantum Theory. *Nature*, *127*(3210), 706–706. doi:10.1038/127706b0

Mather, J. C., Fixsen, D. J., Shafer, R. A., Mosier, C., & Wilkinson, D. T. (1999). Calibrator Design for the COBE* Far Infrared Absolute Spectrophotometer (FIRAS). *The Astrophysical Journal*, *512*(2), 511–520. doi:10.1086/306805

Peebles, P. J. E. (1982). Large-scale background temperature and mass fluctuations due to scale-invariant primeval perturbations. *The Astrophysical Journal*, *263*, L1–L5. doi:10.1086/183911

Peebles, P. J. E., & Yu, J. T. (1970). Primeval adiabatic perturbation in an expanding universe. *The Astrophysical Journal*, *162*, 815–836. doi:10.1086/150713

Penzias, A. A., & Wilson, R. W. (1965). A Measurement of Excess Antenna Temperature at 4080 Mc/s. *The Astrophysical Journal*, *142*, 419–421. doi:10.1086/148307

Planck Collaboration. (2019). Planck 2018 results. I. Overview and the cosmological legacy of Planck. *Astronomy & Astrophysics*.

Planck Collaboration. (2019). Planck 2018 results. VII. Isotropy and Statistics of the CMB. *Astronomy & Astrophysics*.

Rees, M. (1967). *Physical Processes in Radio Sources and the Intergalactic Medium* (Doctoral dissertation, University of Cambridge, Cambridgeshire, UK). Retrieved from https://discover.libraryhub.jisc.ac.uk/search?q=Physical%20Processes%20in%20Radio%20Sources%20and%20the%20Intergalactic%20Medium%20&rn=1

Sachs, R. K., & Wolfe, A. M. (1967). Perturbations of a Cosmological Model and Angular Variations of the Microwave Background. *The Astrophysical Journal*, *147*, 73–90. doi:10.1086/148982

Silk, J. (1968). Cosmic Black-Body Radiation and Galaxy Formation. *The Astrophysical Journal*, *151*, 459. doi:10.1086/149449

Slipher, V. M. (1917). Radial velocity observations of spiral nebulae. *Observatory*, *40*, 304–306.

Smoot, G. F., Bennett, C. L., Kogut, A., Wright, E. L., Aymon, J., Boggess, N. W., ... Wilkinson, D. T. (1992). Structure in the COBE Differential Microwave Radiometer First-Year Maps. *The Astrophysical Journal. Letters*, *396*, L1–L5. doi:10.1086/186504

Smoot, G. F., Tenorio, L., Banday, A. J., Kogut, A., Wright, E. L., Hinshaw, G., & Bennett, C. L. (1994). Statistics and Topology of the COBE-DMR First-Year Sky Maps. *The Astrophysical Journal*, *437*, 1–11. doi:10.1086/174970

Sunyaev, R. A., & Zel'dovich, Y. B. (1970). Small-scale fluctuations of relic radiation. *Astrophysics and Space Science*, *7*(1), 3–19.

Wilkinson Microwave Anisotropy Probe Mission PageN. A. S. A. (n.d.). Retrieved from https://wmap.gsfc.nasa.gov/

Wright, E. L., Meyer, S. S., Bennett, C. L., Boggess, N. W., Cheng, E. S., Hauser, M. G., ... Wilkinson, D. T. (1992). Interpretation of the Cosmic Microwave Background Radiation Anisotropy Detected by the COBE Differential Microwave Radiometer. *The Astrophysical Journal*, *396*, L13–L18. doi:10.1086/186506

Chapter 16
The Quest for Extraterrestrial Intelligence

Hui Chieh Teoh
Universiti Tunku Abdul Rahman, Malaysia

Katrina Pui Yee Shak
Universiti Tunku Abdul Rahman, Malaysia

ABSTRACT

The constant depictions of contact with extraterrestrial life and their constant basic presence in science fiction shows the deep human desire for connection and transcendence with other life forms. In reality, continuous efforts on the search for aliens are being made by renown not-for-profit research organization such as the Search for Extraterrestrial Intelligence (SETI) since 1984. Over the years, plenty of detected signals were dismissed as noise from transmitters on Earth or orbiting satellites but one – the "Wow!" signal. However, artificial signals from extraterrestrial sources could be the key to detecting extraterrestrial intelligence. Apart from passively searching, some are doing active SETI, or known as METI (Messaging Extraterrestrial Intelligence), where humans create and transmit interstellar messages to aliens instead of waiting for theirs. Substantial effort in many areas – awareness, time, technological advancement, techniques – would be necessary to increase the probability of locating outer space intelligence.

For centuries, we have always been fascinated by the existence of other intelligent beings out in the unknown, or most commonly known as aliens. Judging by the number of science fiction books and movies based on alien contact, alien invasion or alien abduction, our imagination has never run dry on what would happen when we finally meet them. We are curious to know who they are, where do they come from, what would they look like, why would they contact us, if they are friend or foe, or will they conquer and enslave us?

There are even some among us who believe that aliens have actually visited us, abducted us, or that we have found them, captured them, and experimented on them. There is a place in Nevada located in the western region of the United States called Area 51, a highly classified United States Air Force facil-

DOI: 10.4018/978-1-7998-2381-0.ch016

ity, which has long been the center of a conspiracy that the government keeps captured alien spaceships and studies the captured aliens.

But are aliens only in the realm of imagination? Is it possible that we are not alone, and the universe is actually teeming with intelligent beings, waiting to be discovered?

If the universe is actually teeming with intelligent beings, then where is everybody? This is the famous question asked by Enrico Fermi during a lunch session with his fellow physicists in the summer of 1950, now known as the Fermi Paradox. This paradox is a contradiction between the high estimates of probability that favor extraterrestrial civilizations being common in the universe, and the missing evidence that they existed elsewhere other than Earth. Fermi realized that if extraterrestrial civilizations are common in the universe, given the age of our galaxy, aliens should have more than enough time to colonize the Milky Way galaxy. Their presence should be obvious all around us. But we see none of them.

An approach was conceived by Frank Donald Drake in 1961 to estimate the number of communicative extraterrestrial civilizations that may exist in our Milky Way galaxy – the Drake equation. The equation includes seven factors that are considered to be important for the advancement of such civilizations. Firstly, it considers the rate of formation of stars that are suitable for the development of intelligent life, and secondly it considers that among these stars, that percentage of them that have planetary systems. The third factor is the number of planets in a planetary system that have an environment suitable for life, and fourthly, out of these life bearing planets, the fraction of them that life actually appears. The fifth factor represents the fraction on which not only just life, but intelligent life, emerges. The sixth factor is the fraction of civilization with intelligent life that develops a technology that releases detectable signs of their existence into space, and finally, the last factor considers how long such civilizations can release detectable signals into space.

If we really wanted to be certain that we have found convincing signs of alien existence, then we should do it in an organized and systematic way – the scientific way – to ensure that we eliminate all false alarms, and find what we are really looking for.

SETI AND METI

Human's curiosity never stops us from trying anything. We are not just interested in hearing from aliens, but also trying to talk to them. This is the main distinction between the two modes of alien communication that we are currently conducting – the Search for Extraterrestrial Intelligence (SETI) and Messaging Extraterrestrial Intelligence (METI). SETI's main focus is searching for messages from aliens, and to answer the question "are we alone?". METI, on the other hand, is about creating messages to aliens and let everyone knows that "you are not alone!".

A systematic search for life beyond Earth started with an article ("Searching for Interstellar Communications") dated back to 1959, which was published in Nature by two nuclear physicists named Giuseppe Cocconi (cosmic ray specialist) and Philip Morrison. Cocconi and Morrison suggested that a systematic search for life forms from outer space was worth the effort – using electromagnetic waves, specifically the radio region at the proposed frequencies (hydrogen frequency). The duo wrote that "we shall assume that long ago they established a channel of communication that would one day become known to us, and that they look forward patiently to the answering signals from the Sun which would make known to them that a new society has entered the community of intelligence". Before this, not many reliable theories were available to estimate the probabilities of extraterrestrial intelligence. Radio waves have been used in the search since these waves have the longest wavelengths that can penetrate and travel deep into space without interference by thick clouds of gas and dust. With renewed optimism of the space age, it fueled the debut of the search for extraterrestrial intelligence (SETI) program. It was the start of a new era where SETI started utilizing radio telescopes to detect radio waves with special patterns, which could signify the presence of other life forms in outer space.

Throughout the 1960s, SETI has evolved from the realm of speculation to scientific enquiry after the introduction of radio telescopes. Similar to Cocconi and Morrison's notable proposal, Drake had gained a funding, with the help from a renowned astronomer named Otto Struve, for a project on the radio search for natural radio emissions from interstellar hydrogen gas clouds. Drake was at an advantage to prove his formal professional quest of finding aliens. Shortly in 1960, Drake moved on to perform the first modern SETI experiment known as Project Ozma, inspired by his supposed radio contact with Princess Ozma from The Marvelous Land of Oz – a faraway place which is inhabited by strange beings. The Green Bank telescope was used by Drake's team to observe sun-like stars such as Tau Ceti and Epsilon Eridani. The team picked up some signals from the second star, Epsilon Eridani, but a thought came to Drake's mind which was of importance to direct SETI towards real science in action: was the signal confirmed to be artificial and not of human origin, and if not, what could they do to avoid false data? Unfortunately, they concluded that the mysterious signals could be of human origin since he constantly picked up the same signal pulsing through the airwaves when he aimed towards another antenna at a wide region of the sky. However, Project Ozma went on to become the model for future radio-based SETI projects. Although the project failed to locate extraterrestrial (E.T.) intelligence, the observation of sun-like stars has made a big impact/impression on us as seen through the 1960's science fiction which featured E.T. life forms such as (1) Mr. Spock from Star Trek who was from planet Vulcan which orbits around Epsilon Eridani, and (2) Barbarella, a space traveler who crash landed on Tau Ceti's 16th planet.

Later on, Drake went on to form a formidable team of scientists of sorts at a private meeting (Green Bank conference in 1961) to further the investigation on the search of E.T. intelligence, after being encouraged by the US National Academy of Sciences. At this point, the newly established National Aeronautics and Space Administration (NASA) was planning to explore Mars, where many scientists were expecting to find life forms. Drake's newly formed team was keen to find and communicate with aliens. They were known as the "Order of the Dolphin", which was inspired by a fellow member's work with dolphins (John Lilly) – in which communication with an E.T. intelligence could be built upon his work on translating non-human communication from dolphins. At the very same meeting, Drake presented a formula to tackle the meeting's biggest question – the probability of humans being in contact with E.T. intelligence. This formula is now known as the Drake Equation. In the same decade, scientists from Struve's backyard (Soviet Union) have also tried to pick up radio signal from E.T. intelligence by utilizing radio communication, specifically omnidirectional antennas.

In 1967, Jocelyn Bell (now Dame Susan Jocelyn Bell Burnell) detected a pulsing signal with great regularity while studying objects using a radio telescope located at the Mullard Radio Astronomy Observatory, under the supervision of Antony Hewish. The radio telescope was originally intended to be used for studying quasars or big galaxies with bright centers via the interplanetary scintillation technique. After eliminating all possible sources, both terrestrial and cosmic, the pulsing signal was still unable to be accounted for. This led them to the idea of the possibility of picking up E.T. intelligence signals. The source was playfully referred to as Little Green Man (LGM-1). However, after studying more data from the radio array, similar repeating signals were obtained from a different part of the galaxy. As a result, it was concluded that the signals came from a family of objects instead of a signal from a foreign civilization. The discovery was still of significance as it was considered a new type of star, known as a pulsar.

In the following decade, a different attempt was taken in conjunction to the celebration for the upgrade of the Arecibo radio telescope in 1974. Instead of searching for signals from E.T. intelligence, Drake and Carl Sagan prepared a message, carrying basic information about Earth and humanity (Figure 1), to be sent towards the M13 star cluster using frequency modulated radio waves. The message was sent to a position 25,000 light years away but M13 might not be in position when it arrived at its targeted point. A subsequent attempt to beam a song – The Beatles' "Across the Universe" – into space was done in 2008 by NASA. The latest broadcast received negative criticisms stating that the broadcast may be mistaken as an invitation for an intergalactic attack by aggressive E.T. intelligence.

The first long-term SETI project took place at the Ohio State University which housed the Big Ear radio telescope (Kraus-type). The telescope is as big as three football fields and is instrumental to one of SETI's biggest mysteries: The "Wow!" Signal (Figure 2). The unusual signal detected by Jerry Ehman in 1977 was logged as a grid of numbers on a rolled-out paper. It came through the same wavelength of 21 centimeters (this frequency can easily penetrate the atmosphere, where atoms of hydrogen absorb and emit energy), where most SETI scientists consider favorable to perform their search for E.T intelligence. However, to no avail, the countless trials following the discovery of the "Wow!" signal presented no support to the presence of E.T. intelligence. Despite that, Ehmer and fellow scientists still could not account for a terrestrial signal although terrestrial broadcasters try to avoid transmitting on the hydrogen line. At present, Antonio Paris and his team from the Center of Planetary Sciences (CPS) have suggested that the hydrogen clouds from Comets 266/P Christensen and P/2008 Y2 (Gibbs) could be sources of the "Wow!" signal based on the noteworthy and comparable data obtained from the comets. However, Ehmer and several astronomers still believes that it is unlikely that the signals from a hydrogen comet can be that strong although it might be able to extend quite far. Today, the Big Ear radio telescope has been dismantled due to financial pressure as funding came merely from small NASA grants until the 1990s after funding from the Congress was cut in 1972.

In order for a more dedicated research and education in the search for E.T. intelligence around the factors of the Drake equation, the SETI Institute was established in 1984. SETI Institute went on to fund Project Phoenix in 1995. Being the largest telescope in the Southern hemisphere, the Parkes radio telescope, was used throughout 1995. Subsequently, the project was based at the National Radio Astronomy Observatory (NRAO) between 1996 to 1998. The Arecibo telescope took its turn from 1998 to 2004. Unlike previous projects, the telescopes were aimed to concentrate on sun-like stars only (800 stars within 200 light-years) and searching for narrow to broad bandwidth (1 Hz to 3,000 MHz) in the radio region. Unfortunately, a conclusion was made that there were no signals of E.T. origin detected.

In early 2001, the co-founder of Microsoft, Paul Allen decided to pledge millions on an idea to build a one-hectare telescope for SETI and to perform astronomical observations. The array is now known as the Allen Telescope Array (ATA). It is considered as the first implantation of a large-N, small-D radio telescope. The ATA is expected to cover frequencies from 1,000 to 15,000 MHz, which covers at least five times larger than the range possible in Project Phoenix. At present, the telescope is only partially completed. However, in April 2018, the ATA commenced its service to search for signals from intelligent life beyond Earth. The galactic centre over the entire "waterhole" frequency range, from 1,420 to 1,720 MHz was studied to detect dissociation products of water, which is known as the essence to life.

In 1999, SETI@Home was released to the public and now everyone can participate in the search for aliens. A free program can be installed and using home computers, the observational data from the

Figure 1. The Arecibo message as sent from the Arecibo Observatory in 1974. Left: Original binary transmission of 1679 bits with 23 columns and 73 rows. Right: The picture generated by the bits. The original transmission has no color information. The colors were added to highlight the separate sections.

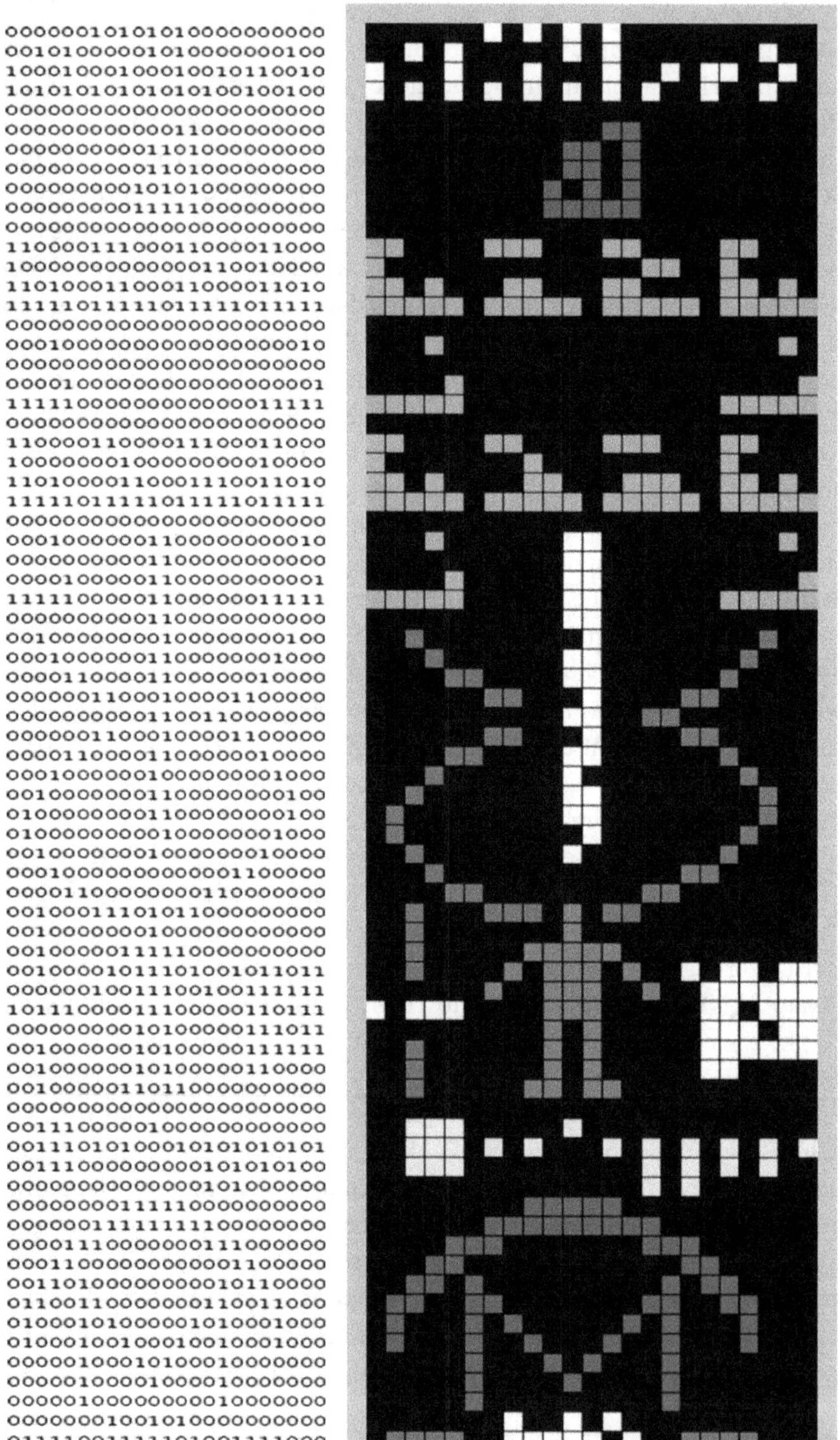

Figure 2. The Wow! Signal (Credit: Big Ear Radio Observatory and North American AstroPhysical Observatory (NAAPO))

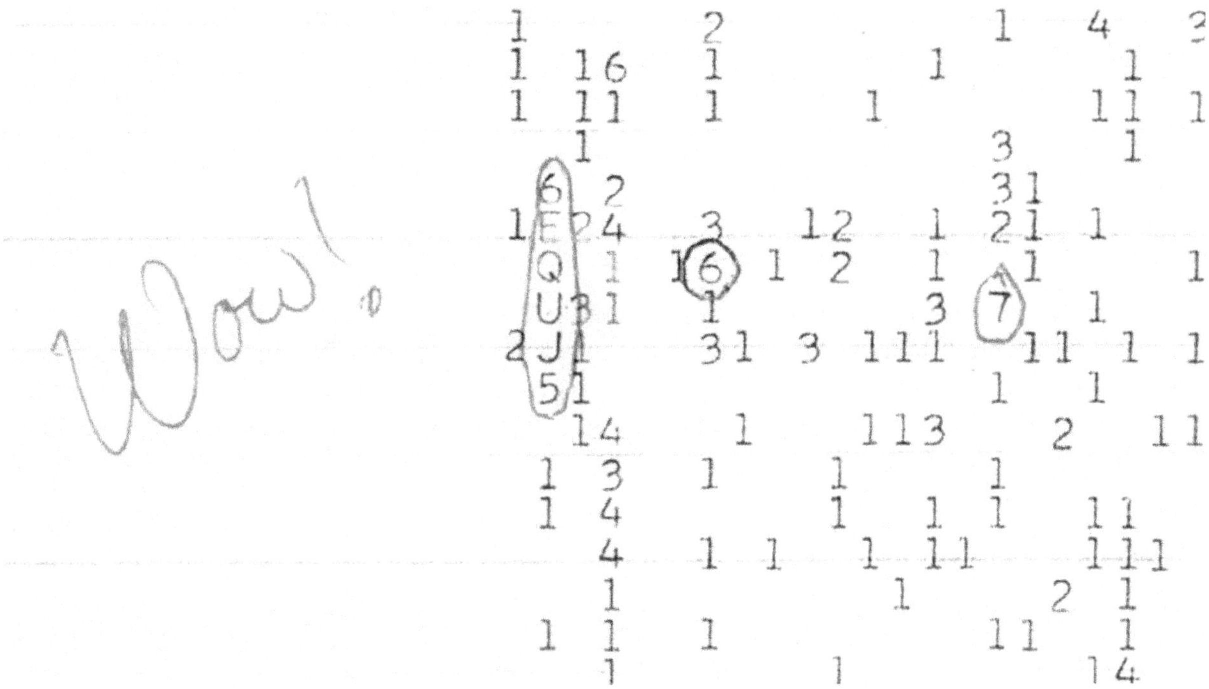

Arecibo radio telescope and the Green Bank Telescope can be downloaded and analyzed to search for abnormal signals that cannot be associated to noise.

In the early 1960s, we have been sending probes to explore our Solar System. These artifacts of ours, even after their operational life, will still be floating in space and would one day escape our Solar System, wandering into interstellar space. Who knows that our artifacts would one day be intercepted by aliens. If so, do we want to include a message in this "bottle" that we throw into the cosmic "ocean"? With this in mind, we attached messages from mankind onto both of the Pioneer spacecrafts and Voyager spacecrafts that we sent to explore the outer Solar System. The plaques on the Pioneers showed nude figures of a male and a female human along with our Solar System makeup and its position relative to the center of our galaxy and 14 pulsars (Figure 3). On both Voyagers, we have attached a golden record that contained selected images and sounds to show the diversity of life and culture on Earth. The record included an instruction on how to playback the record and showed the same pulsar map as in the Pioneer plaques (Figure 4). Besides just SETI, we have in fact been doing active SETI, i.e. creating and transmitting interstellar messages to aliens, for many decades now.

In 2015, METI International was formed to do active SETI. One of its missions is to "foster multidisciplinary research on the design and transmission of interstellar messages", which includes disciplines from natural and social sciences, arts and humanities. METI's first message, known as Sónar Calling GJ273b, was transmitted using a radio telescope facility in Tromsø, Norway to a potentially habitable exoplanet called GJ237b at a distance of 12 light-years away.

At present, the Arecibo observatory has been replaced by the newly built five-hundred-meter Aperture Spherical Telescope (FAST) for the title of the world's largest single-dish telescope. The telescope is considered the most sensitive single-dish telescope, which could detect possible signals from other civilizations. There are plans to search for extrasolar planets (exoplanets) by first trying to detect the radio radiation of exoplanets. This discovery will help immensely to confirm the magnetic field of exoplanets and study the possibility for interstellar migration. Furthermore, signs of artificial modification by anthropogenic activity in an exoplanet's atmosphere could provide information of a possible past or present civilization.

THE ROAD AHEAD

Trying to listen or talk to alien is no easy task – firstly, is the vast distances involved, then there is a great deal of exoplanets out there making it difficult to know where to start. Apart from that, in order for us to communicate, we need a language. If aliens wanted to communicate with us, or if we wanted to communicate with them, what language shall we speak? Since the very beginning of the search for E.T. intelligence, we have been focusing on radio waves as the medium for interstellar communication.

Figure 3. The pioneer plague (Credit: NASA Ames)

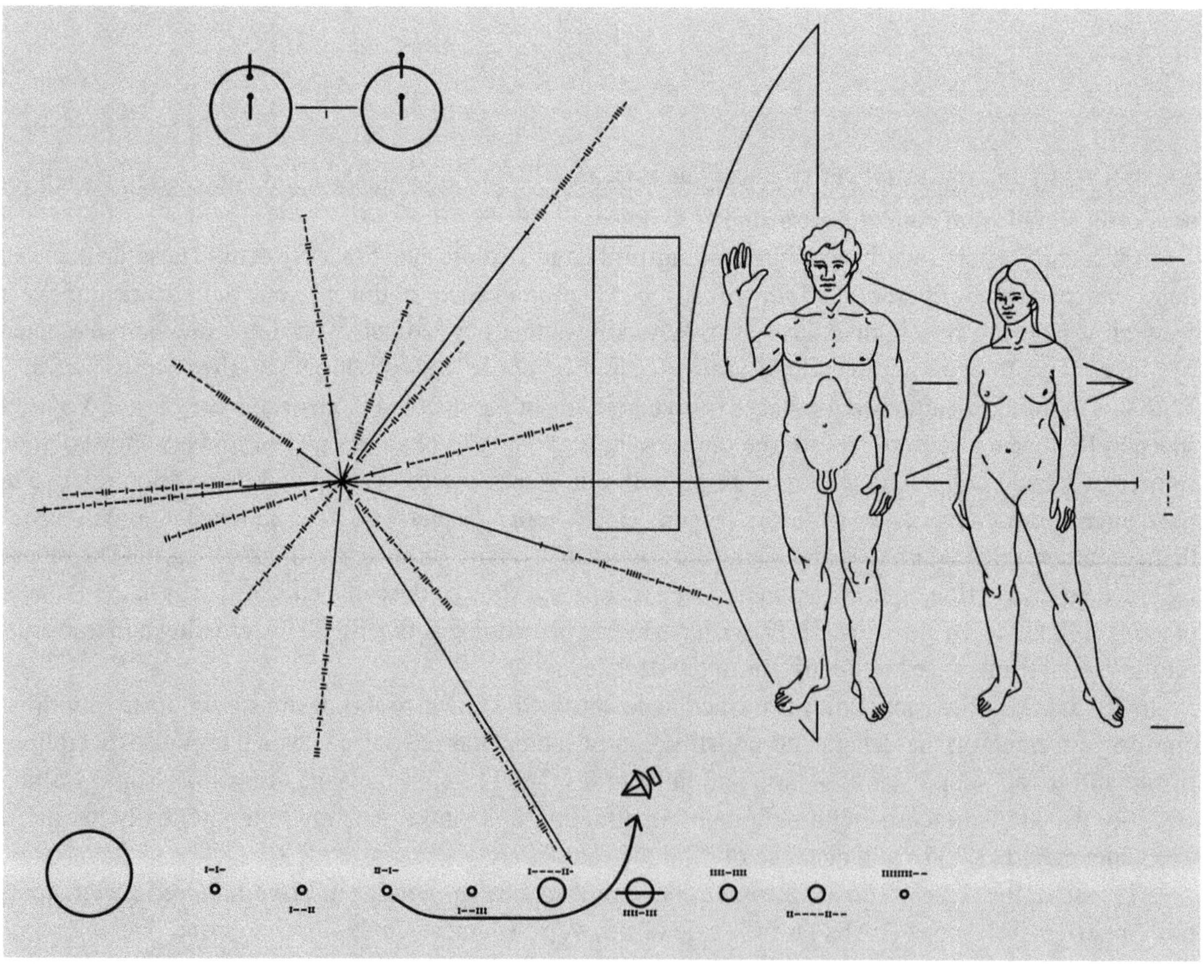

Figure 4. Voyager golden record (Credit: NASA / JPL)

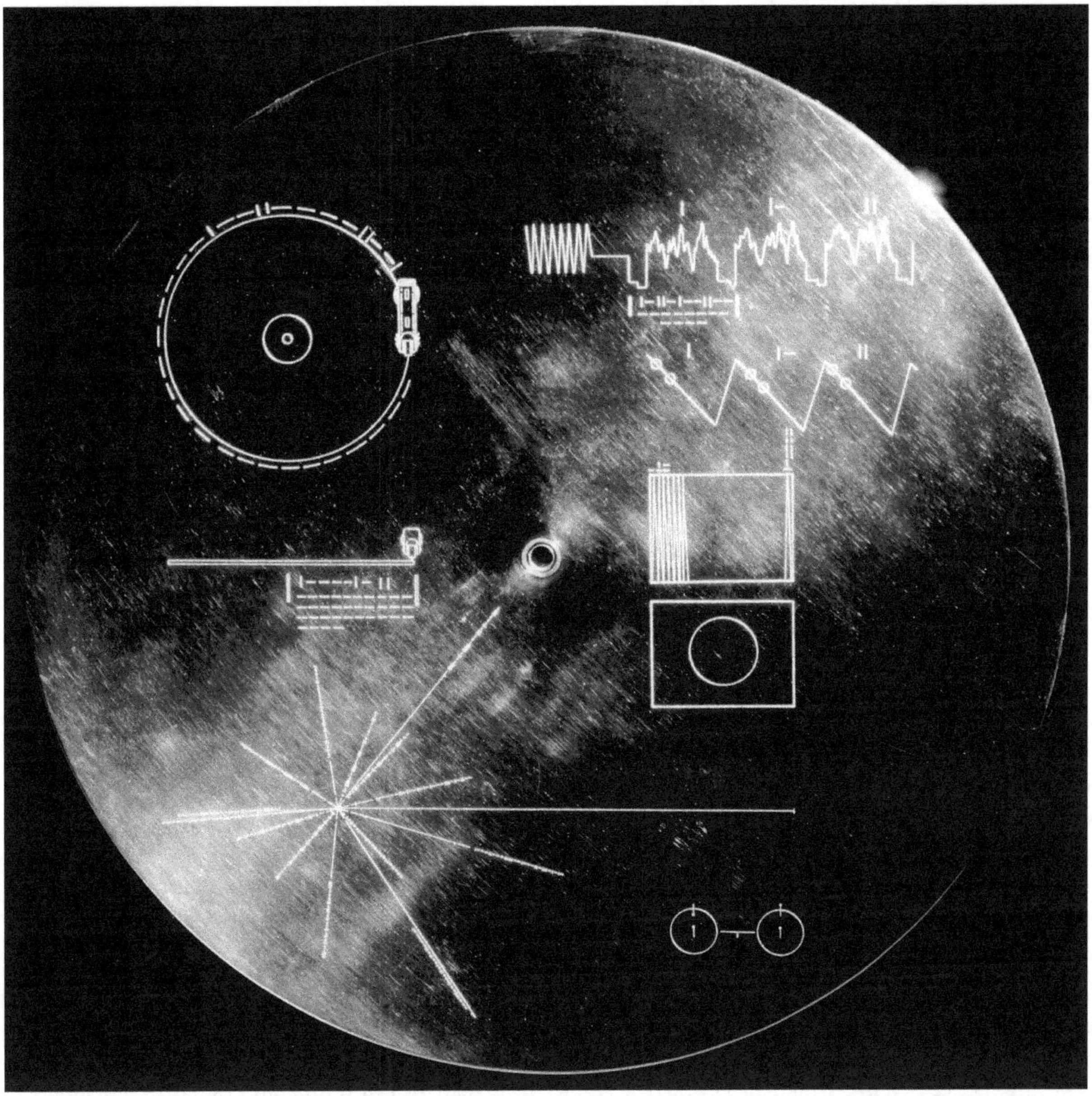

However, after so many decades of search, it still turns up as nothing. This makes some of us to wonder, are we using the correct language?

The range of electromagnetic spectrum is very broad, from the shortest wavelength of gamma rays and x-ray, to ultraviolet and visible and infrared, and to the longest wavelength of microwaves and radio. Therefore, it is logical to think that the options for communication are not only limited to radio waves, for example using optical laser as a mode of communication. Beside electromagnetic waves, the detection of gravitational waves in 2015 might provide us another means of communication. Nevertheless, the gravitational waves are extremely difficult to generate because a very strong gravitation field which involves neutron stars and black holes is required. Some have even proposed neutrino-based communication.

Apart from the difficulties of knowing which channels the aliens might be broadcasting, the timing is also an important factor. We have to tune in at the time when the aliens are talking, else even if we are pointing at the right direction, we might not get any signals.

Finally, we always have this dilemma of alien contact: are they going to be friendly or hostile? Is it going to be dangerous exposing ourselves to aliens?

We still have a long road ahead of us in this scientific exploration. We do not stop at just thinking of how to search for them, but also what we should do if aliens really make contact? These are something for us to ponder in the years to come.

As for the question "do aliens really exist?", we do not know. But the quest for E.T. intelligence is certainly on.

REFERENCES

Andersen, R. (2017, December). What Happens if China Makes First Contact? *The Atlantic*. Retrieved from https://www.theatlantic.com/magazine/archive/2017/12/what-happens-if-china-makes-first-contact/544131/

Backus, P. (1998). The Phoenix search results at Parkes. *Acta Astronautica*, *42*(10-12), 651–654. doi:10.1016/S0094-5765(98)00020-4

Big Ear Radio Observatory. (n.d.). The Big Ear Wow! Signal (30th Anniversary Report). Retrieved from http://www.bigear.org/wowmenu.htm

Bizony, P. (2012). *The Search for Aliens: A Rough Guide to Life on Other Worlds*. London, UK: Rough Guides Ltd.

Chivers, T. (2009, September 29). SETI at 50: 10 key moments in the search for extraterrestrial life. *The telegraph*. Retrieved from https://www.telegraph.co.uk/news/science/science-news/6240998/SETI-at-50-10-key-moments-in-the-search-for-extraterrestrial-life.html

Cofield, C. (2017, November 28). Little Green Men? Pulsars Presented a Mystery 50 Years Ago. *Space.com*. Retrieved from https://www.space.com/38916-pulsar-discovery-little-green-men.html

Darling, D., & Schulze-Makuch, D. (2016). *The Extraterrestrial Encyclopedia* (1st ed.). Sarasota, FL: Design Publishing. doi:10.21899/978-1506901428

Hart, M. H. (1975). Explanation for the Absence of Extraterrestrials on Earth. *Quarterly Journal of the Royal Astronomical Society*, *16*, 128–135.

Horvat, M. (2006). Calculating the probability of detecting radio signals from alien civilizations. *International Journal of Astrobiology*, *5*(2), 143–149. doi:10.1017/S1473550406003004

Huber, P. (2010). Submarine neutrino communication. *Physics Letters. [Part B]*, *692*(4), 268–271. doi:10.1016/j.physletb.2010.08.003

Lamb, D. (1997). Communication with Extraterrestrial Intelligence: SETI and Scientific Methodology. *Issues and Images in the Philosophy of Science*, *192*, 223–251. doi:10.1007/978-94-011-5788-9_13

Learneda, J. G., Pakvasa, S., & Zee, A. (2009). Galactic neutrino communication. *Physics Letters. [Part B]*, *671*(1), 15–19. doi:10.1016/j.physletb.2008.11.057

Nan, R., & Li, D. (2013). The Five-hundred-meter Aperture Spherical radio Telescope (FAST) project. *Materials Science and Engineering*, *4*, 1–4.

NASA, Ames Research Center. (2007, March 3). Pioneer-10 and Pioneer-11. Retrieved from https://www.nasa.gov/centers/ames/missions/archive/pioneer10-11.html

NASA, Jet Propulsion Laboratory (JPL). (n.d.) Voyager: The golden record. Retrieved from https://voyager.jpl.nasa.gov/golden-record/

Paris, A., & Davies, E. (2015). Hydrogen Clouds from Comets 266/P Christensen and P/2008 Y2 (Gibbs) are Candidates for the Source of the 1977 "WOW" Signal. *Journal of the Washington Academy of Sciences*, *101*, 25–31.

Shostak, S. (2019, July 1). Search for space aliens comes up empty, but extraterrestrial life could still be out there. *SETI Institute*. Retrieved from https://www.seti.org/search-space-aliens-comes-empty-extraterrestrial-life-could-still-be-out-there

Stancil, D., Adamson, P., Alania, M., Aliaga, L., Andrews, M., Castillo, C., ... Ziemer, B. (2012). Demonstration of Communication using Neutrinos. *Modern Physics Letters A*, 27.

Walker, J. (n.d.). The Arecibo Message. Retrieved from https://www.bibliotecapleyades.net/circulos_cultivos/chilbolton01.htm

Compilation of References

Abidin, Z. Z., Ibrahim, Z. A., Adnan, S. B. R. S., & Anuar, N. K. (2009). Investigation of radio astronomical windows between 1 MHz – 2060 MHz in University Malaya. *New Astronomy*, *14*(6), 579–583. doi:10.1016/j.newast.2009.02.003

Adana, F. S. D., Gutiérrez, O., Conzález, I., Cátedra, M. F., & Lozano, L. (2011). *Practical applications of asymptotic techniques in electromagnetics* (pp. 73–118). Artech House.

Ade, P. A. R., Aghanim, N., Alves, M. I. R., Armitage-Caplan, C., Arnaud, M., Ashdown, M., ... Zonca, A.Planck Collaboration. (2014). Planck 2013 results. I. Overview of products and scientific results. *Astronomy & Astrophysics*, *571*(A1), 1–49. doi:10.1051/0004-6361/201321529

Adnan, S. B. R. B. S. (2010). The study of radio frequency interference (RFI) for radio astronomy in some remote locations in peninsular Malaysia. Master Dissertation. University of Malaya.

Agrawal, P. K., & Bailey, M. C. (1977). An analysis technique for microstrip antennas. *IEEE Transactions on Antennas and Propagation*, *AP-25*(6), 756–759. doi:10.1109/TAP.1977.1141706

ALMA. (n.d.). Origins. Retrieved from https://www.almaobservatory.org/en/about-alma-at-first-glance/origins/

Alpher, R. A., Bethe, H., & Gamow, G. (1948). The Origin of Chemical Elements. *Physical Review*, *73*(7), 803–804. doi:10.1103/PhysRev.73.803 PMID:18877094

Altschuler, D. R. (2002). The national astronomy and ionosphere's center's (NAIC) Arecibo observatory in Puerto Rico. *Single-Dish Radio Astronomy: Techniques and Applications*, *278*, 1–24.

Alves, J., Bertout, C., Combes, F., Ferrara, A., Forveille, T., Guillot, T., ... Walmsley, M.Planck Collaboration. (2014). Planck 2013 results. XVI. Cosmological parameters. *Astronomy & Astrophysics*, *571*(A16), 1–69. doi:10.1051/0004-6361/201425195

Alves, J., Combes, F., Ferrara, A., Forveille, T., & Shore, S.Planck Collaboration. (2016). Planck 2015 results. XIII. Cosmological parameters. *Astronomy & Astrophysics*, *594*(A13), 1–67. doi:10.1051/0004-6361/201629543

Andersen, R. (2017, December). What Happens if China Makes First Contact? *The Atlantic*. Retrieved from https://www.theatlantic.com/magazine/archive/2017/12/what-happens-if-china-makes-first-contact/544131/

Andrews, J. W., & Hall, P. S. (2002). Phase-locked-loop control of active microstrip patch antennas. *IEEE Transactions on Microwave Theory and Techniques*, *50*(1), 201–206. doi:10.1109/22.981266

ASTRON. (n.d.). *About LOFAR*. Retrieved from http://www.lofar.org/about-lofar/about-lofar

Baade, W., & Zwicky, F. (1934). Remarks on Super-Novae and Cosmic Rays. *Physical Review*, *46*(1), 76–77. doi:10.1103/PhysRev.46.76.2

Baars, J. W. M., & Karcher, H. J. (2017). *Radio telescope reflectors: Historical development of design and construction*. Switzerland: Springer.

Backus, P. (1998). The Phoenix search results at Parkes. *Acta Astronautica, 42*(10-12), 651–654. doi:10.1016/S0094-5765(98)00020-4

Bailey, M. C. (1979) Analysis of the properties of microstrip antennas using strips embedded in a grounded dielectric slab. In *Proceedings IEEE Antennas and Propagation Society International Symposium*, pp. 370-373. June 1979, Seattle. IEEE. 10.1109/APS.1979.1148143

Balanis, C. A. (1982). *Antenna Theory -Analysis and Design*. NY: Harper & Row.

Balanis, C. A. (1989). *Advanced Engineering Electromagnetics*. US: John Wiley & Sons.

Balanis, C. A. (2005). *Antenna Theory, Analysis and Design*. New York: John Wiley & Sons.

Balanis, C. A. (2005). *Antenna Theory: Analysis and Design*. US: Wiley-Interscience.

Balanis, C. A. (2008). *Modern Antenna Handbook*. John Wiley & Sons. doi:10.1002/9780470294154

Bañados, E., Venemans, B. P., Mazzucchelli, C., Farina, E. P., Walter, F., Wang, F., ... Winters, J. M. (2018). An 800-million-solar-mass black hole in a significantly neutral Universe at a redshift of 7.5. *Nature, 553*(7689), 473–476. doi:10.1038/nature25180 PMID:29211709

Bardeen, J., Cooper, L. N., & Schrieffer, J. R. (1957). Theory of superconductivity. *Physical Review, 108*(5), 1175–1204. doi:10.1103/PhysRev.108.1175

Bartholome, P. J., & Vogt, I. M. (1968). COMET - a new meteor-burst system incorporating ARQ and diversity reception. *IEEE Transactions on Communications, 16*(2), 268–278. doi:10.1109/TCOM.1968.1089833

Bell Burnell, S. J. (1979, Jan). Little Green Men, White Dwarfs, or Pulsars? *Cosmic Search, 1*(1), 16. Retrieved from http://www.bigear.org/vol1no1/burnell

Bennett, A. S., & Simth, F. G. (1962). The preparation of the revised 3C catalogue of radio sources. *Monthly Notices of the Royal Astronomical Society, 125*(1), 75–86. doi:10.1093/mnras/125.1.75

Bennett, C. L., Larson, D., Weiland, J. L., Jarosik, N., Hinshaw, G., Odegard, N., ... Halpern, M. (2013). Nine-Year Wilkinson Microwave Anisotropy Probe (WMAP) Observations: Final Maps and Results. *The Astrophysical Journal. Supplement Series, 208*(20), 1–54.

Bhartia, P., & Bahl, I. J. (1982) A frequency agile microstrip antenna. In *Proceedings IEEE Antennas and Propagation Society International Symposium*, pp. 304-307. May 1982, Albuquerque. IEEE.

Bhartia, P., & Bahl, I. J. (1982). Frequency agile microstrip antennas. *Microwave Journal, 25*, 67–70.

Bhattacharya, A., & Raha, B. (2013). Detection of Radio Emissions from Jupiter. *Oriental Journal of Physics, 5*, 1–12.

BICEP2 Collaboration. (2014). Detection of B-Mode Polarization at Degree Angular Scales by BICEP2. *Physical Review Letters, 112*(241101), 1–26. PMID:24996078

BICEP2/Keck & Planck Collaborations. (2015). Joint Analysis of BICEP2/Keck Array and Planck Data. *Physical Review Letters, 114*(101301), 1–17. PMID:25815919

Big Ear Radio Observatory. (n.d.). The Big Ear Wow! Signal (30th Anniversary Report). Retrieved from http://www.bigear.org/wowmenu.htm

Bird, T. S. (2009). Definition and misuse of return loss. *IEEE Antennas & Propagation Magazine*, *51*(2), 166–167. doi:10.1109/MAP.2009.5162049

Birkeland, J., & Itoh, T. (1989). FET-based planar circuits for quasi-optical sources and transceivers. *IEEE Transactions on Microwave Theory and Techniques*, *37*(9), 1452–1459. doi:10.1109/22.32230

Bizony, P. (2012). *The Search for Aliens: A Rough Guide to Life on Other Worlds*. London, UK: Rough Guides Ltd.

Bolton, J. G., Stanley, G. J., & Slee, O. B. (1949). Positions of Three Discrete Sources of Galactic Radio-Frequency Radiation. *Nature*, *164*(4159), 101–102. doi:10.1038/164101b0

Bondi, H., & Gold, T. (1948). The Steady-State Theory of the Expanding Universe. *Monthly Notices of the Royal Astronomical Society*, *108*(3), 252–270. doi:10.1093/mnras/108.3.252

Born, M., & Wolf, E. (1964). *Principles of Optics. 2nd ed.* New York: Pergamon Press.

Borovikov, V. A. (1994). Uniform Stationary Phase Method. IEE Electromagnetic Waves Series 40, UK.

Borovikov, V. A., & Kimber, B. Y. (1994). Geometrical theory of diffraction. IEE Electromagnetic Waves Series 37, UK. doi:10.1049/PBEW037E

Boualleg, A., & Merabtine, N. (2005). Analysis of radiation patterns of rectangular microstrip antennas with uniform substrate. Semiconductor Physics. *Quantum Electronics & Optoelectronics.*, *8*(3), 88–91. doi:10.15407pqeo8.03.088

Bowen, E. G. (1981). History of Australian astronomy. *Publications of the Astronomical Society of Australia*, *4*(2), 267–273. doi:10.1017/S132335800001660X

Bowman, J. J., Senior, T. B. A., & Uslenghi, P. L. E. (1969). *Electromagnetic and Acoustic Scattering by Simple Shapes*. Amsterdam, The Netherlands: North-Holland Publishing Company.

Brauner, T., Kung, R., Vogt, R., & Barchtold, W. (2003) 5-6 GHz low-noise active antenna array for multi-dimensional channel-sounding. In *Proceedings of the 2003 SBMO/IEEE MTT-S International Microwave and Optoelectronics Conference*, pp. 297-301. September 2003, Foz do Iguacu. IEEE. 10.1109/IMOC.2003.1244874

Breed, G. (2009). *The fundamentals of patch antenna design and performance* (pp. 48–51). High Frequency Electronics.

Brown, L. W. (1975). Saturn Radio Emission Near 1 MHz. *The Astrophysical Journal*, *198*, L89–L92. doi:10.1086/181819

Buckley, D. A. H., Meintjes, P. J., Potter, S. B., Marsh, T. R., & Gänsicke, B. T. (2017). Polarimetric evidence of a white dwarf pulsar in the binary system AR Scorpii. *Nature Astronomy*, *1*(2), 1–30. doi:10.103841550-016-0029

Burke, B. F., & Franklin, K. L. (1955). Observations of a variable radio source associated with the planet Jupiter. *Journal of Geophysical Research*, *60*(2), 213–217. doi:10.1029/JZ060i002p00213

Burke, B. F., Graham-Smith, F., & Wilkinson, P. N. (2019). *An Introduction to Radio Astronomy*. Cambridge University Press. doi:10.1017/9781316987506

Candotti, M., Baryshev, A. M., & Trappe, N. (2009). Quasi-optical assessment of the ALMA band 9 frond-end. *Infrared Physics & Technology*, *52*(5), 174–179. doi:10.1016/j.infrared.2009.07.007

Cane, H. V., & Reames, D. V. (1988). Some statistics of solar radio bursts of spectral types II and IV. *The Astrophysical Journal*, *325*, 901–904. doi:10.1086/166061

Carpenter, R. J., & Ochs, G. R. (1959). The NBS meteor burst communication system, *IRE Trans.*, CS-7, 4, pp. 263-271.

Chang, K. (1989). Handbook of Microwave and Components. Vol. I, Microwave Passive & Antenna Components, Wiley Interscience.

Chang, K., Bahl, I., & Nair, V. (2002). *RF and Microwave Circuit and Component Design for Wireless Applications*. New York: John Wiley & Sons.

Chang, K., York, R. A., Hall, P. S., & Itoh, T. (2002). Active integrated antennas. *IEEE Transactions on Microwave Theory and Techniques*, *50*(3), 937–944. doi:10.1109/22.989976

Cheng, D. K. (1989). *Field and Wave Electromagnetics*. US: Addison-Wesley.

Chew, S., & Itoh, T. (1995) A 2x2 beam-switching active antenna array. In *Proceedings IEEE MTT-S International Microwave Symposium Digest*, pp. 925-928. May 1995, Orlando. IEEE.

Chivers, T. (2009, September 29). SETI at 50: 10 key moments in the search for extraterrestrial life. *The telegraph*. Retrieved from https://www.telegraph.co.uk/news/science/science-news/6240998/SETI-at-50-10-key-moments-in-the-search-for-extraterrestrial-life.html

Chou, H. T., & Pathak, P. H. (2004). Fast Gaussian beam-based synthesis of shaped reflector antennas for contoured beam applications. *IEEE Proceedings Microwaves, Antennas, and Propagation,* 151, 13 – 20. 10.1049/ip-map:20040250

Chung, Y., Hang, C. Y., Cai, S., Qian, Y., Wen, C.-P., Wang, K. L., & Itoh, T. (2003). AlGaN/GaN HFET power amplifier integrated with microstrip antenna for RF front-end applications. *IEEE Transactions on Microwave Theory and Techniques*, *51*(2), 653–659. doi:10.1109/TMTT.2002.807685

Cocke, W. J., Disney, M., & Taylor, D. J. (1969). Discovery of Optical Signals from Pulsar NP 0532. *Nature*, *221*(5180), 525–527. doi:10.1038/221525a0

Cofield, C. (2017, November 28). Little Green Men? Pulsars Presented a Mystery 50 Years Ago. *Space.com*. Retrieved from https://www.space.com/38916-pulsar-discovery-little-green-men.html

Cohen, M. H. (1994). The Owens Valley radio observatory: Early years. *Engineering and Science*, *57*, 8–23.

Cook, J. J., Cross, L. G., Bair, M. E., & Arnold, C. B. (1960). Radio Detection of the Planet Saturn. *Nature*, *188*(4748), 393–394. doi:10.1038/188393a0

Cooper, L. N. (1956). Bound electrons pairs in a degenerate Fermi gas. *IEEE Transactions on Microwave Theory and Techniques*, *104*, 1189–1190.

Copeland, J. R., & Robertson, W. J. (1961). Antennaverters and antennafiers. *Electronics (Basel)*, 68–71.

Copeland, J. R., Robertson, W. J., & Verstraete, R. G. (1964). Antennafier arrays. *IEEE Transactions on Antennas and Propagation*, *AP-2*(2), 227–233. doi:10.1109/TAP.1964.1138196

Cosmic Background Explorer Mission PageN. A. S. A. (n.d.). Retrieved from https://lambda.gsfc.nasa.gov/product/cobe/

Covington, A. E. (1947). Micro-Wave Solar Noise Observations During the Partial Eclipse of November 23, 1946. *Nature*, *159*(4038), 405–406. doi:10.1038/159405a0

Cryan, M. J., Hall, P. S., Tsang, S. H., & Sha, J. (1997). Integrated active antenna with full duplex operation. *IEEE Transactions on Microwave Theory and Techniques*, *45*(10), 1742–1748. doi:10.1109/22.641721

Czech, D., Mishra, A. K., & Inggs, M. (2016). *Identifying radio frequency interference with hidden Markov model. Radio Frequency Interference* (pp. 17–20). Socorro, NM: RFI.

Dąbrowski, B. P., Krankowski, A., Błaszkiewicz, L., & Rothkaehl, H. (2016). Prospects for Solar and Space Weather Research with Polish Part of the LOFAR Telescope. *Acta Geophysica*, *64*(3), 825–840. doi:10.1515/acgeo-2016-0028

Darling, D., & Schulze-Makuch, D. (2016). *The Extraterrestrial Encyclopedia* (1st ed.). Sarasota, FL: Design Publishing. doi:10.21899/978-1506901428

Deal, W. R., Radisic, V., Qian, Y., & Itoh, T. (1999). Integrated-antenna push-pull power amplifiers. *IEEE Transactions on Microwave Theory and Techniques*, *47*(8), 1901–1909. doi:10.1109/22.780389

Derneryd, A. G. (1976). Linearly polarized microstrip antennas. *IEEE Transactions on Antennas and Propagation*, *AP-24*(6), 846–851. doi:10.1109/TAP.1976.1141445

Derneryd, A. G. (1978). A theoretical investigation of the rectangular microstrip antenna element. *IEEE Transactions on Antennas and Propagation*, *AP-26*(4), 532–535. doi:10.1109/TAP.1978.1141890

Derneryd, A. G. (1979a). Analysis of the microstrip disk antenna element. *IEEE Transactions on Antennas and Propagation*, *AP-27*(5), 660–664. doi:10.1109/TAP.1979.1142159

Derneryd, A. G., & Lind, A. (1979b). Extended analysis of rectangular microstrip resonator antennas. *IEEE Transactions on Antennas and Propagation*, *AP-27*(6), 846–849. doi:10.1109/TAP.1979.1142206

Deschamps, G. A. (1953). Microstrip microwave antennas. In *Proceedings of the Third Symposium on the USAF Antenna Research and Development Program, Oct* (pp. 18-22).

Dicke, R. H., Peebles, P. J. E., Roll, P. G., & Wilkinson, D. T. (1965). Cosmic Black-Body Radiation. *The Astrophysical Journal*, *142*, 414–419. doi:10.1086/148306

Drake, F. D. (1961). Radio emission from the planets. *Physics Today*, *14*(4), 30–34. doi:10.1063/1.3057498

Drake, F. D., & Ewen, H. I. (1958). A Broad-Band Microwave Source Comparison Radiometer for Advanced Research in Radio Astronomy. *Proceedings of the IRE*, *46*(1), 53–60. doi:10.1109/JRPROC.1958.286710

Dyudina, U. A., Ingersoll, A. P., Ewald, S. P., Porco, C. C., Fischer, G., & Yair, Y. (2013). Saturn's visible lightning, its radio emissions, and the structure of the 2009–2011 lightning storms. *Icarus*, *226*(1), 1020–1037. doi:10.1016/j.icarus.2013.07.013

Edge, D. O., Shakeshaft, J. R., McAdam, W. B., Baldwin, J. E., & Archer, S. (1959). A survey of radio sources at frequency of 159 Mc/s. *Memoirs of the Royal Astronomical Society*, *68*, 37–60.

Encrenaz, P., Gomez-Gonzalez, J., Lequeux, J., & Orchiston, W. (2011). Highlighting the history of French radio astronomy. 7: The genesis of the institute of radioastronomy at millimeter wavelengths (IRAM). *Journal of Astronomical History and Heritage*, *14*, 83–92.

Eshleman, V. R., & Manning, L. A. (1954). Radio communication by scattering from meteoric ionization. *Proceedings of the IRE*, *42*(3), 530–536. doi:10.1109/JRPROC.1954.274813

European Southern Observatory (ESO). (n.d.). Centaurus A. Retrieved from https://www.eso.org/public/images/eso0903a/

European Space Agency (ESA) Planck Mission Page. (n.d.). Retrieved from https://sci.esa.int/web/planck/home

European Space Agency (ESA). (n.d.). Best image of bright quasar 3C 273. Retrieved from https://www.spacetelescope.org/images/potw1346a/

Fanaroff, B. L., & Riley, J. M. (1974). The morphology of extragalactic radio sources of high and low luminosity. *Monthly Notices of the Royal Astronomical Society*, *167*(1), 31–36. doi:10.1093/mnras/167.1.31P

Fan, X., Wang, F., Yang, J., Keeton, C. R., Yue, M., Zabludoff, A., ... Wu, X.-B. (2019). The Discovery of a Gravitationally Lensed Quasar at z = 6.51. *The Astrophysical Journal*, *870*(2), L11. doi:10.3847/2041-8213/aaeffe

Fath, E. A. (1909). The Spectra of Some Spiral Nebulae and Globular Star Clusters. *Lick Observatory Bulletins*, *149*, 71–77. doi:10.5479/ADS/bib/1909LicOB.5.71F

Feain, I., Cornwell, T., Ekers, R., Morganti, R., & Junkes, N. (n.d.). The Centaurus A Galaxy. CSIRO. Retrieved from https://www.scienceimage.csiro.au/image/10850/the-centaurus-a-galaxy

Fisher, J. R. (2002). Back-ends. In Single-Dish Radio Astronomy: Techniques and Applications, 278, 113 – 122.

Fixsen, D. J., Cheng, E. S., Gales, J. M., Mather, J. C., Shafer, R. A., & Wright, E. L. (1996). The Cosmic Microwave Background Spectrum from the Full COBE* FIRAS Data Set. *The Astrophysical Journal*, *473*(2), 576–587. doi:10.1086/178173

Flynt, R. A., Fan, L., Navarro, J. A., & Chang, K. (1996). Low cost and compact active integrated antenna transceiver for system application. *IEEE Transactions on Microwave Theory and Techniques*, *44*(10), 1642–1649. doi:10.1109/22.538955

Forsythe, P. A., Vogan, E. L., Hansen, D. R., & Hines, C. O. (1957). The principles of JANET: A meteor burst communication system. *Proceedings of the IRE*, *45*(12), 1642–1657. doi:10.1109/JRPROC.1957.278296

Frater, R. H., Goss, W. M., & Wendt, H. W. (2017). *Four Pillars of Radio Astronomy: Mills, Christiansen, Wild, Bracewell*. Switzerland: Springer. doi:10.1007/978-3-319-65599-4

Fukuda, A., Mukumoto, K., Yoshihiro, Y., Nagasawa, M., Yamagishi, H., Sato, N., ... & Jin, L. J. (2003). Experiments on meteor burst communications in the Antarctic, *Advances in polar upper atmosphere research*, 17, 120-136.

Furse, C., & Bhatia, R. S. (2006). A Simple Radio Telescope Operating at Ku Band for Educational Purposes. *IEEE Antennas & Propagation Magazine*, *48*(5), 144–152. doi:10.1109/MAP.2006.277116

Galerkin, B. G. (1915). On electrical circuits for the approximate solution of the Laplace equation. *Vestnik Inzh.*, *19*, 897–908.

Gamow, G. (1946). Expanding Universe and the Origin of Elements. *Physical Review*, *70*(7-8), 572–573. doi:10.1103/PhysRev.70.572.2

Gamow, G. (1948). The Evolution of the Universe. *Nature*, *162*(4122), 680–682. doi:10.1038/162680a0 PMID:18893719

Gamow, G. (1952). Half an hour of creation. *Physics Today*, *3*(8), 16–21. doi:10.1063/1.3066969

Garg, R., Bhartia, P., Bahl, I., & Ittipiboon, A. (2001). *Microstrip Antenna Design Handbook*. Boston, MA: Artech House.

Gary, D. E., & Hurford, G. J. (1994). Coronal temperature, density, and magnetic field maps of a solar active region using the Owens Valley Solar Array. *The Astrophysical Journal*, *420*, 903–912. doi:10.1086/173614

Goldsmith, P. F. (1998). *Quasioptical systems; Gaussian beam, quasioptical propagation and applications*. New York: IEEE Press.

Gold, T. (1968). Rotating Neutron Stars as the Origin of the Pulsating Radio Sources. *Nature*, *218*(5143), 731–732. doi:10.1038/218731a0

Gonzalez, A., Uzawa, Y., Fujii, Y., & Kaneko, K. (2011). ALMA band 10 tertiary optics. *Infrared Physics & Technology*, *54*(6), 488–496. doi:10.1016/j.infrared.2011.07.007

Goubau, G. (1969). Optical or quasi-optical transmission schemes. In F. A. Benson (Ed.), *Millimetre and Submillimetre Waves*. London, UK: Illife Press.

Green, K. A. (1963). Modified Cassegrain antenna for arbitrary aperture illumination. *IEEE Transactions on Antennas and Propagation, AP-11*(5), 589–590. doi:10.1109/TAP.1963.1138084

Greenstein, J. L., & Matthews, T. A. (1963). Red-Shift of the Unusual Radio Source: 3C 48. *Nature, 197*(4872), 1041–1042. doi:10.1038/1971041a0

Gurnett, D. A. (2005). Radio and Plasma Wave Observations at Saturn from Cassini's Approach and First Orbit. *Science, 307*(5713), 1255–1259. doi:10.1126cience.1105356 PMID:15604362

Gurnett, D. A., Kurth, W., Kirchner, D., Hospodarsky, G., Averkamp, T., Zarka, P., ... Pedersen, A. (2004). The Cassini Radio and Plasma Wave Investigation. *Space Science Reviews, 114*(1-4), 395–463. doi:10.100711214-004-1434-0

Guru, B. S., & Hiziroglu, H. R. (2004). *Electromagnetic Field Theory Fundamentals*. US: Cambridge University Press. doi:10.1017/CBO9781139165297

Guth, A. H. (1981). Inflationary universe: A possible solution to the horizon and flatness problems. *Physical Review D: Particles and Fields, 23*(2), 347–356. doi:10.1103/PhysRevD.23.347

Guth, A. H. (1997). *The Inflationary Universe: The Quest for a New Theory of Cosmic Origins*. New York: Perseus Books.

Gutton, H., & Baissinot, G. (1955) Flat aerial for ultra-high frequencies. French Patent No. 703 113.

Halbach, K. (1964). Matrix representation of Gaussian optics. *American Journal of Physics, 32*(2), 90–108. doi:10.1119/1.1970159

Hall, P. S., Gardner, P., & Ma, G. (2002). Active integrated antennas. *IEICE Transactions on Communications, E85-B*(9), 1661–1667.

Halpern, P., & Tomasello, N. (2016). Size of the observable universe. *Advances in Astrophysics, 1*(3), 135–137. doi:10.22606/adap.2016.13001

Hammerstad, E. O. (1975) Equations for microstrip circuit design. In *Proceedings of the 5th European Microwave Conference*, pp. 268-272. September 1975, Hamburg. IEEE.

Hannan, P. W. (1961). Microwave antennas derived from the Cassegrain telescope. *IRE Transactions on Antennas and Propagation, 9*(2), 140–153. doi:10.1109/TAP.1961.1144976

Hanson, D., Hoover, S., Crites, A., Ade, P. A. R., Aird, K. A., Austermann, J. E., ... Zemcov, M. (2013). Detection of B-mode polarization in the Cosmic Microwave Background with data from the South Pole Telescope. *Physical Review Letters, 111*(141301), 1–6. PMID:24138230

Harrington, R. F. (1961). *Time-Harmonic Electromagnetic Fields* (p. 183). New York: McGraw-Hill.

Hart, M. H. (1975). Explanation for the Absence of Extraterrestrials on Earth. *Quarterly Journal of the Royal Astronomical Society, 16*, 128–135.

Hazard, C., Mackey, M. B., & Shimmins, A. J. (1963). Investigations of the Radio Source 3C 273 By The Method of Lunar Occultations. *Nature, 197*(4872), 1037–1039. doi:10.1038/1971037a0

Hessels, J. W. T., Ransom, S. M., Stairs, I. H., Freire, P. C., Kaspi, V. M., & Camilo, F. (2006). A Radio Pulsar Spinning at 716 Hz. *Science, 311*(5769), 1901–1904. doi:10.1126cience.1123430 PMID:16410486

Hewish, A., Bell, S. J., Pilkington, J. D. H., Scott, P. F., & Collins, R. A. (1968). Observation of a Rapidly Pulsating Radio Source. *Nature, 217*(5130), 709–713. doi:10.1038/217709a0

Hey, J. S. (1946). Solar Radiations in the 4–6 Metre Radio Wave-Length Band. *Nature*, *157*(3976), 47–48. doi:10.1038/157047b0

Hildebrandt, J., Krüger, A., Chertok, I. M., Fomichev, V. V., & Gorgutsa, R. V. (1998). Solar Microwave Bursts from Electron Populations with a 'Broken' Energy Spectrum. *Solar Physics*, *181*(2), 337–349. doi:10.1023/A:1005007321040

Hines, C. O., & Forsyth, P. A. (1957). The forward scattering of radio waves from overdense meteor trails. *Canadian Journal of Physics*, *35*(9), 1033–1041. doi:10.1139/p57-115

Hinshaw, G., Larson, D., Komatsu, E., Spergel, D. N., Bennett, C. L., Dunkley, J., ... Wright, E. L. (2013). Nine-Year Wilkinson Microwave Anisotropy Probe (WMAP) Observations: Cosmological Parameter Results. *The Astrophysical Journal. Supplement Series*, *208*(19), 1–25.

Hiramatsu, M. (2017). ALMA and its observational achievements: Unveiling the mysteries of the dark universe. *Fujitsu Scientific and Technical Journal*, *53*, 9–14.

Horvat, M. (2006). Calculating the probability of detecting radio signals from alien civilizations. *International Journal of Astrobiology*, *5*(2), 143–149. doi:10.1017/S1473550406003004

Howard, W. E. III, Barrett, A. H., & Haddock, F. T. (1961). The Measurement of Microwave Radiation from the Planet Mercury. *The Astronomical Journal*, *66*, 287. doi:10.1086/108417

Howard, W. E. III, Barrett, A. H., & Haddock, F. T. (1962). Measurement of Microwave Radiation from the Planet Mercury. *The Astrophysical Journal*, *136*, 995–1004. doi:10.1086/147451

Howell, J. Q. (1975). Microstrip antennas. *IEEE Transactions on Antennas and Propagation*, *AP-23*(1), 90–93. doi:10.1109/TAP.1975.1141009

Hoyle, F. (1948). A New Model for the Expanding Universe. *Monthly Notices of the Royal Astronomical Society*, *108*(5), 372–382. doi:10.1093/mnras/108.5.372

Hubble, E. (1929). A relation between distance and radial velocity among extra-galactic nebulae. *Proceedings of the National Academy of Sciences of the United States of America*, *15*(3), 168–173. doi:10.1073/pnas.15.3.168 PMID:16577160

Hubble, E. P. (1926). Extragalactic nebulae. *The Astrophysical Journal*, *64*, 321–369. doi:10.1086/143018

Huber, P. (2010). Submarine neutrino communication. *Physics Letters. [Part B]*, *692*(4), 268–271. doi:10.1016/j.physletb.2010.08.003

Hulse, R. A., & Taylor, J. H. (1975). Discovery of a pulsar in a binary system. *The Astrophysical Journal*, *195*, L51–L53. doi:10.1086/181708

Huygens, C. (1690). Leyden: Traite de la Lumiere.

Itoh, T., Haddad, G., & Harvey, J. (2001). *RF Technologies for Low-Power Wireless Communications*. New York: Wiley. doi:10.1002/0471221643

James, G. L. (1980). *Geometrical Theory of Diffraction for Electromagnetic Waves* (2nd ed.). London, UK: Peter Peregrinus.

Jansky, K. G. (1932). Directional Studies of Atmospherics at High Frequencies. *Proceedings of the Institute of Radio Engineers*, *20*(12), 1920–1932. doi:10.1109/JRPROC.1932.227477

Jansky, K. G. (1933). Electrical Disturbances Apparently of Extraterrestrial Origin. *Proceedings of the Institute of Radio Engineers*, *21*(10), 1387–1398.

Jansky, K. G. (1933). Radio Waves from Outside the Solar System. *Nature*, *132*(3323), 66–66. doi:10.1038/132066a0

Jansky, K. G. (1935). A Note on the Source of Interstellar Interference. *Proceedings of the Institute of Radio Engineers*, *23*(10), 1158–1163.

Johnson, D. E. (1987) Ten years experience with the SNOTEL meteor burst data acquisition system, In *Proceedings of MBC Symposium*, pp. SII.5-SII.20

Kaur, J., & Khanna, R. (2013). Co-axial fed rectangular microstrip patch antenna for 5.2 GHz WLAN application. *Universal Journal of Electrical and Electronic Engineering*, *1*(3), 94–98. doi:10.13189/ujeee.2013.010306

Kellermann, K. I. (1970). Thermal Radio Emission From the Major Planets. *Radio Science*, *5*(2), 487–493. doi:10.1029/RS005i002p00487

Kellermann, K. I., & Moran, J. M. (2001). The development of high-resolution imaging in radio astronomy. *Annual Review of Astronomy and Astrophysics*, *39*(1), 457–509. doi:10.1146/annurev.astro.39.1.457

Kempen, T. V., Corder, S., Lucas, R., & Mauersberger, R. (2012). How ALMA is calibrated: I. Antenna-based pointing, focus and amplitude calibration. *ALMA Newsletter*, *9*, 8–16.

Kerton, R. (1989). Parkes radio telescope. CSIRO. Retrieved from https://www.scienceimage.csiro.au/image/2720/parkes-radio-telescope/

Kerton, R. (2009). The 70 metre antenna (DSS 43) at the Canberra Deep Space Communication Complex. CSIRO. Retrieved from https://www.scienceimage.csiro.au/image/10808/the-70-metre-antenna-dss-43-at-the-canberra-deep-space-communication-complex/

Kim, H., Yoon, I. J., & Yoon, Y. I. (2005). A novel fully integrated transmitter front-end with high power added efficiency. *IEEE Transactions on Microwave Theory and Techniques*, *53*(4), 3206–3214.

Kirkman-Bey, M., Cintron-Tirado, P., Sebsebie, R., Dogan, N., & Xie, Z. (2014). Design and simulation of Ku-band low-noise block down-converter in 0.18 micrometer CMOS technology. In *Proceedings of the IEEE SoutheastCon*, 13-16. Lexington, KY. 10.1109/SECON.2014.6950648

Kitchen, J. (1987). Octave bandwidth varactor-tuned oscillators. *Microwave Journal*, *30*(5), 347–353.

Kobayashi, H. (2012). Geometrical Theory of Diffraction and Related Topics. Press of Niigata University. (in Japanese)

Kobayashi, H. (2018). A Simple Expression for Curved Rectangular Patch Antenna Pattern. International Symposium on Antennas and Propagation (ISAP2018), A8_1007.

Kobayashi, H., Kawamura, R., Yamada, Y. (2014). Fabrication of Horn Antenna using 3D Printer and its Evaluation. *IEICE Tech. Rep., 114*(193), AP2014-72, pp. 1-6. (in Japanese)

Kobayashi, H. (2011). *Electromagnetic Wave in Space-Domain.* Tokyo, Japan: Sakura-tech. (in Japanese)

Kobayashi, H., & Hongo, K. (1996). Scattering of Electromagnetic Plane Wave by Conducting Convex Plate. In *Proceedings Symposium of Antenna Technology and Applied Electromagnetics (ANTEM96)*, pp. 663-667.

Kobayashi, H., & Tokumitsu, Y. (1998). The Element Antenna, Phase Shifter and Amplifier for Wide-band Systems. In *Proceedings of 1998 China-Japan Joint Meeting on Microwaves (CJMW'98)*, pp. 283-286.

Kobayashi, H., Tokumitsu, Y., & Hongo, K. (1999). RCS Calculation Software System using PTD-TC. In *Proceedings International Conference on Computational Electromagnetics and Its Applications (ICCEA'99)*, pp. 569-572.

Kontar, E. P., Yu, S., Kuznetsov, A. A., Emslie, A. G., Alcock, B., Jeffrey, N. L. S., ... Subramanian, P. (2017). Imaging spectroscopy of solar radio burst fine structures. *Nature Communications*, *8*(1), 1–9. doi:10.103841467-017-01307-8 PMID:29142220

Kraus, J. D. (1986). *Radio Astronomy*. Ohio: Cygnus-Quasar Books.

Krüger, A. (1979). *Introduction to Solar Radio Astronomy and Radio Physics*. The Netherlands: Springer. doi:10.1007/978-94-009-9402-7

Kumar, G., & Gupta, K. C. (1985). Non-radiating edges and four edges gap coupled multiple resonator broadband microstrip antennas. *IEEE Transactions on Antennas and Propagation*, *AP-33*(2), 173–178. doi:10.1109/TAP.1985.1143563

Kumar, G., & Ray, K. P. (2003). *Broadband Microstrip Antennas* (p. 8). Norwood: Artech House.

Kundu, M. R. (1965). *Solar radio astronomy*. New York: Interscience Publishers, John Wiley & Sons.

Kundu, M. R., & Vlahos, L. (1982). Solar microwave bursts - A review. *Space Science Reviews*, *32*(4), 405–462. doi:10.1007/BF00177449

Kurth, W. S., & Zarka, P. (2001). Saturn Radio Waves. In *"Planetary Radio Emissions V, Proceedings of the 5th International Workshop"* pp. 247–259. Vienna, Austria: Austrian Academy of Sciences Press.

Lamb, J. W. (1999). Optimized optical layout for MMA 12-m antennas. *MMA Memo* 246.

Lamb, D. (1997). Communication with Extraterrestrial Intelligence: SETI and Scientific Methodology. *Issues and Images in the Philosophy of Science*, *192*, 223–251. doi:10.1007/978-94-011-5788-9_13

Lamb, J. W. (1986). Quasioptical coupling of Gaussian beam systems to large Cassegrain antennas. *International Journal of Infrared and Millimeter Waves*, *7*(10), 1511–1536. doi:10.1007/BF01010755

Learneda, J. G., Pakvasa, S., & Zee, A. (2009). Galactic neutrino communication. *Physics Letters. [Part B]*, *671*(1), 15–19. doi:10.1016/j.physletb.2008.11.057

Lee, J. (2007). Radio Emissions from Solar Active Regions. *Space Science Reviews*, *133*(1-4), 73–102. doi:10.100711214-007-9206-2

Lee, K. F., & Chen, W. (1997). *Advances in Microstrip and Printed Antennas* (pp. 53–62). New York: John Wiley & Sons.

Legacy Archive for Microwave Background Data Analysis (LAMBDA). (n.d.). Retrieved from https://lambda.gsfc.nasa.gov/

Lehpamer, H. (2008). *RFID Design Principles*. Boston, MA: Artech House.

Leitch, E. M., Kovac, J. M., Pryke, C., Carlstrom, J. E., Halverson, N. W., Holzapfel, W. L., ... Sandberg, E. S. (2002). Measurement of polarization with the Degree Angular Scale Interferometer. *Nature*, *420*(6917), 763–771. doi:10.1038/nature01271 PMID:12490940

Lemaître, G. (1931). The Beginning of the World from the Point of View of Quantum Theory. *Nature*, *127*(3210), 706–706. doi:10.1038/127706b0

Leverington, D. (2017). *Observatories and Telescopes of Modern Times: Ground-based Optical and Radio Astronomy Facilities since 1945*. UK: Cambridge University Press. doi:10.1017/9781139051507

Lindenmeier, H. (1976, October). Design of electrically small broadband receiving antennas under consideration of nonlinear distortions in amplifier elements. In *Proceedings IEEE Antennas and Propagation Society International Symposium*, pp. 242-245. Amherst, MA. IEEE. 10.1109/APS.1976.1147544

Long, S. A., & Walton, M. D. (1979). A dual-frequency stacked circular-disc antenna. *IEEE Transactions on Antennas and Propagation*, *AP-27*(2), 270–273. doi:10.1109/TAP.1979.1142078

Lo, Y. T., & Lee, S. W. (1988). *Antenna Handbook*. New York: Van Nostrand Reinhold. doi:10.1007/978-1-4615-6459-1

Lo, Y. T., Solomon, D., & Richards, W. F. (1979). Theory and experiment on microstrip antennas. *IEEE Transactions on Antennas and Propagation*, *AP-27*(2), 137–145. doi:10.1109/TAP.1979.1142057

Mach, E. (1926). *The Principle of Physical Optics*. Dover Pub.

Margenau, H., & Murphy, G. (1956). *The mathematics of physics and chemistry* (pp. 120–126). New York: Van Nostrand.

Marques, M. S., Zarka, P., Echer, E., Ryabov, V. B., Alves, M. V., Denis, L., & Coffre, A. (2017). Statistical analysis of 26 yr of observations of decametric radio emissions from Jupiter. *Astronomy and Astrophysics*, *604*, 1–18. doi:10.1051/0004-6361/201630025

Martinez, R. D., & Compton, R. C. (1994). A quasi-optical oscillator/modulator for wireless transmission. pp. 839-842. IEEE. 10.1109/MWSYM.1994.335226

Masterson, J. (n.d.). Australia Telescope Compact Array. CSIRO. Retrieved from https://www.scienceimage.csiro.au/image/3881/five-antennas-at-narrabri/

Mather, J. C., Fixsen, D. J., Shafer, R. A., Mosier, C., & Wilkinson, D. T. (1999). Calibrator Design for the COBE* Far Infrared Absolute Spectrophotometer (FIRAS). *The Astrophysical Journal*, *512*(2), 511–520. doi:10.1086/306805

Mather, J. C., & Hinshaw, G. F. (2008). Cosmic background explorer. *Scholarpedia*, *3*(3), 4732. doi:10.4249cholarpedia.4732

Matthews, T. A., & Sandage, A. R. (1963). Optical Identification of 3C 48, 3C 196, and 3C 286 with Stellar Objects. *The Astrophysical Journal*, *138*, 30–56. doi:10.1086/147615

Matveenko, L. I. (1968). Position of a Source of Small Angular Size in the Crab Nebula. *Soviet Astronomy*, *12*, 552–553.

Mayer, C. H., McCullough, T. P., & Sloanaker, R. M. (1958). Observation of Mar and Jupiter at a Wave Length of 3.15 cm. *The Astrophysical Journal*, *127*, 11–16. doi:10.1086/146434

Mayer, C. H., McCullough, T. P., & Sloanaker, R. M. (1958). Observations of Venus at 3.15-CM Wave Length. *The Astrophysical Journal*, *127*, 1–10. doi:10.1086/146433

Meinke, H., & Landstorfer, F. (1968). Noise and bandwidth limitations with transistorized antennas. In *Proceedings IEEE Antennas and Propagation Society International Symposium*, pp. 245-246. September 1968, Boston, MA. IEEE. 10.1109/APS.1968.1150608

Melrose, D. B. (1975). Plasma emission due to isotropic fast electrons, and types I, II, and V solar radio bursts. *Solar Physics*, *43*(1), 211–236. doi:10.1007/BF00155154

Melrose, D. B. (1985). Plasma emission mechanisms. In *Solar Radiophysics: Studies of Emission from the Sun at Metre Wavelengths* (pp. 177–210). Cambridge, New York: Cambridge University Press.

Mitrovic, B., & Rozema, L. A. (2008). On the correct formula for the lifetime broadened superconducting density of states. *Journal of Physics Condensed Matter*, *20*(1). doi:10.1088/0953-8984/20/01/015215

Mizugutch, Y., Akagawa, M., & Yokoi, H. (1976). Offset dual reflector antenna, *Int. Symp. of Antennas and Propagation Society*, 14, 2 – 5.

Moran, J. M. (1998). Thirty years of VLBI: Early days, successes, and future. *International Astronomical Union Colloquium*, 164, 1–10.

Mukumoto, K., Fukuda, A., Nagasawa, M., Yoshihiro, Y., Nakano, K., Ohichi, S., ... & Yao, M. W. (2005). VHF data transmission experiments using MBC equipment conducted during the period from JARE-43 to JARE-45. National Institute of Polar Research. *Advances in polar upper atmosphere research*, 19, 89-105.

Munson, R. E. (1974). Conformal microstrip antennas and microstrip phased arrays. *IEEE Transactions on Antennas and Propagation*, *AP-22*(1), 74–78. doi:10.1109/TAP.1974.1140723

Nan, R., & Li, D. (2013). The Five-hundred-meter Aperture Spherical radio Telescope (FAST) project. *Materials Science and Engineering*, *4*, 1–4.

NASA, Ames Research Center. (2007, March 3). Pioneer-10 and Pioneer-11. Retrieved from https://www.nasa.gov/centers/ames/missions/archive/pioneer10-11.html

NASA, Jet Propulsion Laboratory (JPL). (n.d.) Voyager: The golden record. Retrieved from https://voyager.jpl.nasa.gov/golden-record/

NASA, Jet Propulsion Laboratory, University of Iowa. (2005). PIA07966: Eerie Sounds of Saturn's Radio Emissions. Retrieved from https://photojournal.jpl.nasa.gov/catalog/PIA07966

NASA, Jet Propulsion Laboratory. (2002). PIA03478: Inner Radiation Belts of Jupiter. Retrieved from https://photojournal.jpl.nasa.gov/catalog/PIA03478

NASA, Space Telescope Science Institute. (STScI), HubbleSite (n.d.). Dust Disk Fuels Black Hole in Giant Elliptical Galaxy NGC 4261. Retrieved from https://hubblesite.org/image/82

NASA, Space Telescope Science Institute. (STScI), HubbleSite. (n.d.). Combined x-ray and optical images of the crab nebula. Retrieved from https://hubblesite.org/contents/media/images/2002/24/1248-Image.html

Nather, R. E., Warner, B., & Macfarlane, M. (1969). Optical Pulsations in the Crab Nebula Pulsar. *Nature*, *221*(5180), 527–529. doi:10.1038/221527a0

National Radio Astronomy Observatory. (n.d.). Green Bank Solar Radio Burst Spectrometer. Retrieved from http://www.astro.umd.edu/~white/gb/

National Radio Astronomy Observatory. (n.d.). Retrieved from https://public.nrao.edu/

Navarro, J., Shu, Y.-H., & Chang, K. (1992). Broadband electronically tunable planar active radiating elements and spatial power combiners using notch antennas. *IEEE Transactions on Microwave Theory and Techniques*, *40*(3), 323–328. doi:10.1109/22.120105

Nelson, G. J., & Melrose, D. B. (1985). Type II bursts. In *Solar Radiophysics: Studies of Emission from the Sun at Metre Wavelengths* (pp. 333–359). Cambridge: Cambridge University Press.

Netzer, H. (2013). *The Physics and Evolution of Active Galactic Nuclei* (1st ed.). UK: Cambridge University Press. doi:10.1017/CBO9781139109291

Newkirk, G. Jr. (1961). The solar corona in active regions and the thermal origin of the slowly varying component of solar radio radiation. *The Astrophysical Journal*, *133*, 983–1012. doi:10.1086/147104

Newman, E. H., & Tulyathan, P. (1981). Analysis of microstrip antennas using Moment Methods. *IEEE Transactions on Antennas and Propagation*, *AP-29*(1), 47–53. doi:10.1109/TAP.1981.1142532

Nobel Media, A. B. (2018). All Nobel prizes in physics. Retrieved from https://www.nobelprize.org/prizes/lists/all-nobel-prizes-in-physics/

Noguchi, T., Naruse, M., & Sekimoto, Y. (2012). RF conductivity and surface impedance of a superconductor taking into account the complex superconducting gap energy. *Physics Procedia*, *36*, 318–323. doi:10.1016/j.phpro.2012.06.166

Noguchi, T., Naruse, M., & Sekimoto, Y. (2013). Contribution of quasiparticles in the subgap states to the surface impedance of superconductors. *IEEE Transactions on Applied Superconductivity*, *23*(3). doi:10.1109/TASC.2013.2240761

Noguchi, T., Suzuki, T., Endo, A., & Tamura, T. (2009). Contribution of the imaginary part of the superconducting gap energy on the SIS tunneling current. *Physica. C, Superconductivity*, *469*(15-20), 1585–1588. doi:10.1016/j.physc.2009.05.218

Noguchi, T., Suzuki, T., & Tamura, T. (2011). Subgap tunneling current at low temperature in Nb/Al-AlN/Nb SIS junctions. *IEEE Transactions on Applied Superconductivity*, *21*(3), 756–759. doi:10.1109/TASC.2010.2089033

Oke, J. B. (1963). Absolute Energy Distribution in the Optical Spectrum of 3C 273. *Nature*, *197*(4872), 1040–1041. doi:10.1038/1971040b0

Oort, J. H., Kerr, F. T., & Westerhout, G. (1958). The galactic system as a spiral nebula. *Monthly Notices of the Royal Astronomical Society*, *118*(4), 379–389. doi:10.1093/mnras/118.4.379

Ormiston, T. D., Gardner, P., & Hall, P. S. (1998). Compact low noise receiving antenna. *Electronics Letters*, *34*(14), 1367–1368. doi:10.1049/el:19980524

Osipov, A. V., & Tretyakov, A. A. (2017). *Modern Electromagnetic Scattering Theory with Applications*. John Wiley & Sons. doi:10.1002/9781119004639

Pacini, F. (1967). Energy Emission from a Neutron Star. *Nature*, *216*(5115), 567–568. doi:10.1038/216567a0

Padman, R., Murphy, J., & Hills, R. (1987). Gaussian mode analysis of Cassegrain antenna efficiency. *IEEE Transactions on Antennas and Propagation*, *35*(10), 1093–1103. doi:10.1109/TAP.1987.1143983

Palmer, J., & Davenhall, A. C. (2001). The CCD Photometric Calibration Cookbook. Retrieved from http://star-www.rl.ac.uk/star/docs/sc6.htx/sc6se5.html

Paris, A., & Davies, E. (2015). Hydrogen Clouds from Comets 266/P Christensen and P/2008 Y2 (Gibbs) are Candidates for the Source of the 1977 "WOW" Signal. *Journal of the Washington Academy of Sciences*, *101*, 25–31.

Pathak, P. H. (1988). *Techniques for High-Frequency Problems. Antenna Handbook* (Y. T. Lo, & S. W. Lee, Eds.). Van Nostrand Reinhold.

Peebles, P. J. E. (1982). Large-scale background temperature and mass fluctuations due to scale-invariant primeval perturbations. *The Astrophysical Journal*, *263*, L1–L5. doi:10.1086/183911

Peebles, P. J. E., & Yu, J. T. (1970). Primeval adiabatic perturbation in an expanding universe. *The Astrophysical Journal*, *162*, 815–836. doi:10.1086/150713

Penzias, A. A., & Wilson, R. W. (1965). A Measurement of Excess Antenna Temperature at 4080 Mc/s. *The Astrophysical Journal*, *142*, 419–421. doi:10.1086/148307

Pickard, G. W. (1931). A note on the relation of meteor showers and radio reception. *Proceedings of the IRE*, *19*(7), 1166–1170. doi:10.1109/JRPROC.1931.222442

Planck Collaboration. (2019). Planck 2018 results. I. Overview and the cosmological legacy of Planck. *Astronomy & Astrophysics*.

Planck Collaboration. (2019). Planck 2018 results. VII. Isotropy and Statistics of the CMB. *Astronomy & Astrophysics*.

Planelles, S., & Biffi, V. (2017). A mysterious universe: Revealing the bright and dark sides of the cosmos. *Metode Science Studies Journal*, *7*, 153–161.

Pobanz, C., & Itoh, T. (1994) A microwave non-contact identification transponder using subharmonic interrogation. In *Proceedings IEEE MTT-S International Microwave Symposium Digest, pp.* 753-756. May 1994, San Diego, CA. IEEE. 10.1109/MWSYM.1994.335246

Pozar, D. M. (1990). *Microwave Engineering*. New York: Addison-Wesley.

Prestage, R. M., Constantikes, K. T., Hunter, T. R., King, L. J., Lacasse, R. J., Lockman, F. J., & Norrod, R. D. (2009). The Green Bank telescope. *Proceedings of the IEEE*, *97*(8), 1382–1390. doi:10.1109/JPROC.2009.2015467

Purchine, M. P., & Aberle, J. T. (1994). A tunable L-band circular microstrip patch antenna. *Microwave Journal*, 80–88.

Radisic, V., Chew, S. T., Qian, Y., & Itoh, T. (1997a). High-efficiency power amplifier integrated with antenna. *IEEE Microwave and Guided Wave Letters*, *7*(2), 39–41. doi:10.1109/75.553052

Radisic, V., Qian, Y., & Itoh, T. (1997b) Class F power amplifier integrated with circular sector microstrip antenna. In *Proceedings IEEE MTT-S International Microwave Symposium Digest, pp.* 687-690. June 1997, Denver. IEEE. 10.1109/MWSYM.1997.602884

Ramsdale, P., & Maclean, T. (1971). Active loop-dipole aerials. *Proceedings of the Institution of Electrical Engineers*, *118*(12), 1698–1710. doi:10.1049/piee.1971.0313

Reber, G. (1944). Cosmic static. *The Astrophysical Journal*, *100*, 279–287. doi:10.1086/144668

Rees, M. (1967). *Physical Processes in Radio Sources and the Intergalactic Medium* (Doctoral dissertation, University of Cambridge, Cambridgeshire, UK). Retrieved from https://discover.libraryhub.jisc.ac.uk/search?q=Physical%20Processes%20in%20Radio%20Sources%20and%20the%20Intergalactic%20Medium%20&rn=1

Reifenstein, E. C. III, Staelin, D. H., & Brundage, W. D. (1969). Crab Nebula Pulsar NPO527. *Physical Review Letters*, *22*(7), 311. doi:10.1103/PhysRevLett.22.311

Richards, W. F. (1988). Microstrip Antennas, Chapter 7 in Antennas Handbook – Theory, Applications, and Design. New York: Van Nostrand Reinhold Co.

Richards, W. F., & Long, S. A. (1986a) Adaptive pattern control of a reactively loaded, dual-mode microstrip antenna. In *Proceedings of the International Telemetering Conference*, pp. 291-296. October 1986, Las Vegas.

Richards, W. F., & Long, S. A. (1986b) Impedance control of microstrip antennas utilizing reactive loading. In *Proceedings of the International Telemetering Conference*, pp. 285-290. October 1986, Las Vegas.

Richards, W. F., & Lo, Y. T. (1983). Theoretical and experimental investigation of a microstrip radiator with multiple lumped linear loads. *Electromagnetics*, *3*(3), 371–385. doi:10.1080/02726348308915194

Richards, W. F., Lo, Y. T., & Harrison, D. D. (1981). An improved theory of microstrip antennas with application. *IEEE Transactions on Antennas and Propagation*, *AP-29*(1), 38–46. doi:10.1109/TAP.1981.1142524

Roberts, J. A. (1963). Radio emission from the planets. *Planetary and Space Science*, *11*(3), 221–259. doi:10.1016/0032-0633(63)90026-6

Robertson, P. (2011). An Australian icon: Planning and construction of the Parkes telescope. Science with Parks @ 50 Years Young, Australia.

Rowe, J. (n.d.). An artist's impression of magnetar XTE J1810-197 showing the radio emissions and the magnetic field. CSIRO. Retrieved from https://www.scienceimage.csiro.au/image/3210/an-artist-s-impression-of-magnetar-xte-j1810-197-showing-the-radio-emissions-and-the-magnetic-field-/

Ruck, G. T., Barrick, D. E., Stuwart, W. D., & Kirchbaum, C. K. (1970). *Radar Cross Section Handbook* (Vol. I). Plenum Press. doi:10.1007/978-1-4899-5324-7

Rudge, A. W., Milne, K., Olver, A. D., & Knight, P. (1986). *The handbook of antenna design*. IEE UK.

Ryle, M., Smith, F. G., & Elsmore, B. (1950). A preliminary survey of the radio stars in the northern hemisphere. Monthly Notices of the Royal Astronomical Society, 110, 508–523. doi:10.1093/mnras/110.6.508

Ryle, M., & Vonberg, D. D. (1948). An investigation of radio-frequency radiation from the sun. *Proceedings of the Royal Society of London. Series A, Mathematical and Physical Sciences*, *193*(1032), 98–120. doi:10.1098/rspa.1948.0036 PMID:18865816

Sachs, R. K., & Wolfe, A. M. (1967). Perturbations of a Cosmological Model and Angular Variations of the Microwave Background. *The Astrophysical Journal*, *147*, 73–90. doi:10.1086/148982

Sadiku, M. N. O. (2007). *Elements of Electromagnetics*. US: Oxford University Press.

Schmidt, M. (1963). 3C 273: A Star-Like Object with Large Red-Shift. *Nature*, *197*(4872), 1040. doi:10.1038/1971040a0

Schmidt, M. (1965). Large Redshifts of Five Quasi-Stellar Sources. *The Astrophysical Journal*, *141*, 1295–1300. doi:10.1086/148217

Schmidt, M., & Matthews, T. A. (1964). Redshift of the Quasi-Stellar Radio Sources 3C 47 and 3C 147. *The Astrophysical Journal*, *139*, 781–785. doi:10.1086/147815

Schneider, P. (2015). *Extragalactic Astronomy and Cosmology: An Introduction* (2nd ed.). Springer-Verlag Berlin and Heidelberg GmbH & Co. KG.

Senior, T. B. A., & Volakis, J. L. (1995). Approximate Boundary Conditions in Electromagnetics. IEE Electromagnetic Wave Series 41, UK. doi:10.1049/PBEW041E

Shakeshaft, J. R., Ryle, M., Baldwin, J. E., Elsmore, B., & Thomson, J. H. (1955). A survey of radio sources between declinations -38° and +83°. *Memoirs of the Royal Astronomical Society*, *67*, 106–154.

Shen, L. C., Long, S. A., Allerding, M. R., & Walton, M. D. (1977). Resonant frequency of a circular disc, printed-circuit antenna. *IEEE Transactions on Antennas and Propagation*, *AP-25*(4), 595–596. doi:10.1109/TAP.1977.1141643

Shibasaki, K., Alissandrakis, C. E., & Pohjolainen, S. (2011). Radio Emission of the Quiet Sun and Active Regions (Invited Review). *Solar Physics*, *273*(2), 309–337. doi:10.100711207-011-9788-4

Shields, G. A. (1999). A Brief History of Active Galactic Nuclei. *Publications of the Astronomical Society of the Pacific*, *111*(760), 661–678. doi:10.1086/316378

Shostak, S. (2019, July 1). Search for space aliens comes up empty, but extraterrestrial life could still be out there. *SETI Institute*. Retrieved from https://www.seti.org/search-space-aliens-comes-empty-extraterrestrial-life-could-still-be-out-there

Silk, J. (1968). Cosmic Black-Body Radiation and Galaxy Formation. *The Astrophysical Journal*, *151*, 459. doi:10.1086/149449

Silver, S. (1949). *Microwave Antenna Theory and Design. Radiation Laboratories Series*. McGraw Hill.

SKA. (n.d.). *The SKA project*. Retrieved from https://www.skatelescope.org/

Slipher, V. M. (1917). Radial velocity observations of spiral nebulae. *Observatory*, *40*, 304–306.

Slipher, V. M. (1917). The spectrum and velocity of the nebula N.G.C. 1068 (M 77). *Lowell Observatory Bulletin*, *3*, 59–62.

Smoot, G. F., Bennett, C. L., Kogut, A., Wright, E. L., Aymon, J., Boggess, N. W., ... Wilkinson, D. T. (1992). Structure in the COBE Differential Microwave Radiometer First-Year Maps. *The Astrophysical Journal. Letters*, *396*, L1–L5. doi:10.1086/186504

Smoot, G. F., Tenorio, L., Banday, A. J., Kogut, A., Wright, E. L., Hinshaw, G., & Bennett, C. L. (1994). Statistics and Topology of the COBE-DMR First-Year Sky Maps. *The Astrophysical Journal*, *437*, 1–11. doi:10.1086/174970

Southworth, G. C. (1945). Microwave radiation from the sun. *Journal of the Franklin Institute*, *239*(4), 285–297. doi:10.1016/0016-0032(45)90163-3

Staelin, D. H., & Reifenstein, E. C. III. (1968). Pulsating radio sources near the Crab Nebula. *Science*, *162*(3861), 1481–1483. doi:10.1126cience.162.3861.1481 PMID:17739779

Stancil, D., Adamson, P., Alania, M., Aliaga, L., Andrews, M., Castillo, C., ... Ziemer, B. (2012). Demonstration of Communication using Neutrinos. *Modern Physics Letters A*, 27.

Stephan, K. D., Camilleri, N., & Itoh, T. (1983). A quasi-optical polarization-duplexed balanced mixer for millimeter-wave applications. *IEEE Transactions on Microwave Theory and Techniques*, *31*(2), 164–170. doi:10.1109/TMTT.1983.1131452

Steven, J. F., & Joseph, H. T. (2017). *The MSK144 Protocol for Meteor-Scatter Communication.* Retrieved from http://www.arrl.org/files/file/QEX_Next_Issue/SeptOct2017/FrankeTaylor.pdf

Stratton, J. A. (1941). *Electromagnetic Theory*. McGraw-Hill Book Co.

Strom, R. G. (2013). How was atomic HI (λ = 21 cm line) in space discovered? *International Journal of Modern Physics: Conference Series*, *23*, 474–477.

Stutzman, W. L., & Thiele, G. A. (1998). *Antenna Theory and Design*. New York: John Wiley & Sons.

Sullivan, W. T., III. (1991). Some highlights of interferometry in early radio astronomy. *International Astronomical Colloquium*, 131, 132–149.

Sunyaev, R. A., & Zel'dovich, Y. B. (1970). Small-scale fluctuations of relic radiation. *Astrophysics and Space Science*, *7*(1), 3–19.

Tayler, J. (2001). *WSJT: New Software for VHF Meteor-Scatter Communication.* Retrieved from http://physics.princeton.edu/pulsar/K1JT/WSJT_QST_Dec2001.pdf

Tham, C. Y., & Withington, S. (2003). *Receiver optics design electromagnetic analysis, second report (Bands 3, 6, 7, and 9). Technical Report*. Cambridge University.

Tham, C. Y., Yassin, G., & Carter, M. (2007). Analysis techniques for the optics in millimetre/submillimetre wave radio telescopes receivers. *JurnalFizik Malaysia*, *28*, 49–53.

Thomas, H., Fudge, D., & Morris, G. (1985). Gunn source integrated with a microstrip patch. *Microwaves & RF*, *24*(2), 87–89.

Tranter, Z. (1995). *Principles of Communications*. New Jersey: John Wiley & Sons.

Ufimtsev, P. Y. (1971). *Method of Edge Waves in the Physical Theory of Diffraction, Translated from Russian*. Prepared by the U.S. Air Force Foreign Tech. Div.

Ufimtsev, P. Y. (2007). *Fundamentals of the Physical Theory of Diffraction*. IEEE Press. doi:10.1002/0470109017

Urry, C. M., & Padovani, P. (1995). Unified Schemes for Radio-Loud Active Galactic Nuclei. *Publications of the Astronomical Society of the Pacific*, *107*, 803–845. doi:10.1086/133630

Uzunoglu, N. K., Alexopoulos, N. G., & Fikioris, J. G. (1979). Radiation properties of microstrip dipoles. *IEEE Transactions on Antennas and Propagation*, *AP-27*(2), 853–858. doi:10.1109/TAP.1979.1142173

Vincent, W. R., Wolfram, R. T., Sifford, B. M., Jaye, W. E., & Peterson, A. M. (1957). A Meteor-Burst System for Extended Range VHF Communications. *Proceedings of the IRE*, *45*(12), 1693–1700. doi:10.1109/JRPROC.1957.278324

Walker, J. (n.d.). The Arecibo Message. Retrieved from https://www.bibliotecapleyades.net/circulos_cultivos/chilbolton01.htm

Walker, C. K., Kooi, J. W., Chan, M., LeDuc, H. G., Schaffer, P. L., Carlstrom, J. E., & Phillips, T. G. (1992). A low-noise 492 GHz SIS waveguide receiver. *International Journal of Infrared and Millimeter Waves*, *13*(6), 785–798. doi:10.1007/BF01011595

Wang, Z., Gurnett, D. A., Fischer, G., Ye, S.-Y., Kurth, W. S., Mitchell, D. G., ... Russell, C. T. (2010). Cassini observations of narrowband radio emissions in Saturn's magnetosphere. *Journal of Geophysical Research: Space Physics*, *115*, 1–12.

Warmuth, A., & Mann, G. (2005). Lecture Notes in Physics: Vol. 656. *The application of radio diagnostics to the study of the solar drivers of space weather* (pp. 51–70). doi:10.1007/978-3-540-31534-6_3

Warwick, J. W. (1964). Radio Emission from Jupiter. *Annual Review of Astronomy and Astrophysics*, *2*(1), 1–22. doi:10.1146/annurev.aa.02.090164.000245

Warwick, J. W., Pearce, J. B., Evans, D. R., Carr, T. D., Schauble, J. J., Alexander, J. K., ... Barrow, C. H. (1981). Planetary radio astronomy observations from Voyager 1 near Saturn. *Science*, *212*(4491), 239–243. doi:10.1126cience.212.4491.239 PMID:17783837

Weisberg, J. M., Nice, D. J., & Taylor, J. H. (2010). Timing Measurements of the Relativistic Binary Pulsar PSR B1913+16. *The Astrophysical Journal*, *722*(2), 1030–1034. doi:10.1088/0004-637X/722/2/1030

Weiss, L. A. A., & Stewart, R. T. (1965). Solar radio bursts of spectral type V. *Australian Journal of Physics*, *18*(2), 143–166. doi:10.1071/PH650143

Weiss, M. D., & Popovic, Z. (1999) A 10 GHz high-efficiency active antenna. In *Proceedings IEEE MTT-S International Microwave Symposium Digest*, pp. 663-666. June 1999, Anaheim, CA. IEEE.

Weitzen, J. A. (1987). A data base approach to analysis of meteor burst data. *Radio Science*, *22*(1), 133–140. doi:10.1029/RS022i001p00133

Wendt, H., Orchiston, W., & Slee, B. (2008). W. N. Christiansen and the development of the solar grating array. *Journal of Astronomical History and Heritage*, *11*, 173–184.

Wengler, M. J. (1992). Submillimeter-wave detection with superconducting tunnel diodes. *Proceedings of the IEEE*, *80*(11), 1810–1825. doi:10.1109/5.175257

Wheeler, H. A. (1975). Small antennas. *IEEE Transactions on Antennas and Propagation*, *AP-12*(4), 462–469. doi:10.1109/TAP.1975.1141115

White, S. M. (2007). Solar Radio Bursts and Space Weather. *Asian Journal of Physics*, *16*, 189–207.

Wielebinski, R., Junkes, N., & Grahl, B. H. (2011). The Effelsberg 100-m radio telescope: Construction and forty years of radio astronomy. *Journal of Astronomical History and Heritage*, *14*, 3–21.

Wild, J. P., Smerd, S. F., & Weiss, A. A. (1963). Solar Bursts. *Annual Review of Astronomy and Astrophysics*, *1*(1), 291–366. doi:10.1146/annurev.aa.01.090163.001451

Wilkinson Microwave Anisotropy Probe Mission PageN. A. S. A. (n.d.). Retrieved from https://wmap.gsfc.nasa.gov/

Wilson, T. L. (2013). Techniques of Radio Astronomy. In T. D. Oswalt, & H. E. Bonds (Eds.), *Planets, stars, and stellar systems* (pp. 283–323). The Netherlands: Springer. doi:10.1007/978-94-007-5618-2_6

Withington, S., Murphy, J., Egan, A., & Hills, R. (1992). On the design of broadband quasioptical systems for submillimter-wave radio-astronomy receivers. *International Journal of Infrared and Millimeter Waves*, *13*(10), 1515–1537. doi:10.1007/BF01009233

Woerden, H. V., & Strom, R. G. (2007). Dwingeloo – the golden radio telescope. *Astronomische Nachrichten: Astronomical Notes*, *328*(5), 376–387. doi:10.1002/asna.200710767

Wolszczan, A., & Frail, D. A. (1992). A planetary system around the millisecond pulsar PSR1257 + 12. *Nature*, *355*(6356), 145–147. doi:10.1038/355145a0

Woltjer, L. (1964). X-rays and Type I Supernovae. *The Astrophysical Journal*, *140*, 1309–1313. doi:10.1086/148028

Wong, K. L. (1999). *Design of Nonplanar Microstrip Antennas and Transmission Lines*. New York: John Wiley & Sons. doi:10.1002/0471200662

Wong, K. L. (2002). *Compact and Broadband Microstrip Antennas* (pp. 232–273). New York: John Wiley & Sons. doi:10.1002/0471221112.ch7

Wood, D. (2015). Radio Telescope Definition, Parts & Facts. [online] Available at https://study.com/academy/lesson/radio-telescope-definition-parts-facts.html [Accessed 1 February 2018]

Wright, E. L., Meyer, S. S., Bennett, C. L., Boggess, N. W., Cheng, E. S., Hauser, M. G., ... Wilkinson, D. T. (1992). Interpretation of the Cosmic Microwave Background Radiation Anisotropy Detected by the COBE Differential Microwave Radiometer. *The Astrophysical Journal*, *396*, L13–L18. doi:10.1086/186506

Wu, X., Wang, F., Fan, X., Yi, W., Zuo, W., Bian, F., ... Beletsky, Y. (2015). An ultraluminous quasar with a twelve-billion-solar-mass black hole at redshift 6.30. *Nature*, *518*(7540), 512–515. doi:10.1038/nature14241 PMID:25719667

Yang, S. L. S., Kishk, A. A., & Lee, K. F. (2008). Frequency reconfigurable U-slot microstrip patch antenna. *IEEE Antennas and Wireless Propagation Letters*, *7*, 127–129. doi:10.1109/LAWP.2008.921330

Yeap, K. H., & Hirasawa, K. (2019). Introductory Chapter: Electromagnetism. In K. H. Yeap, & K. Hirasawa (Ed.), Electromagnetic fields and waves (pp. 1 – 12). UK: InTechOpen.

Yeap, K. H., & Hirasawa, K. (2019). Introductory Chapter: Electromagnetism. In K. H. Yeap, & K. Hirasawa (Ed.), Electromagnetic fields and waves (pp. 1–12). UK: InTechOpen.

Yeap, K. H., Law, Y. H., Rizman, Z. I., Cheong, Y. K., Ong, C. E., & Chong, K. H. (2013). Performance analysis of paraboloidal reflector antennas in radio telescopes. *International Journal of Electronics, Computer, and Communications Technologies*, *4*(1), 21–25.

Yeap, K. H., Loh, M. C., Tham, C. Y., Yiam, C. Y., Yeong, K. C., & Lai, K. C. (2016). Analysis of reflector antennas in radio telescopes. *Advanced Electromagnetics*, *5*(3), 32–38. doi:10.7716/aem.v5i3.402

Yeap, K. H., Ong, S. S., Nisar, H., Lai, K. C., & Ng, C. A. (2016). Attenuation in superconducting circular waveguides. *Advanced Electromagnetics*, *5*(2), 34–38. doi:10.7716/aem.v5i2.388

Yeap, K. H., Teh, J. S. M., Nisar, H., Yeong, K. C., & Hirasawa, K. (2015). Attenuation in superconducting rectangular waveguides. *Frequenz Journal of RF-Engineering and Telecommunications*, *69*, 111–117.

Yeap, K. H., & Tham, C. Y. (2018). Optimization of an offset receiver optics for radio telescopes. *Journal of Infrared, Millimeter, and Terahertz Waves*, *39*(1), 64–76. doi:10.100710762-017-0449-z

Yeap, K. H., Tham, C. Y., Yassin, G., & Yeong, K. C. (2011). Attenuation in rectangular waveguides with finite conductivity walls. *Radioengineering Journal*, *20*, 472–478.

Yeap, K. H., Wong, E. V. S., Nisar, H., Lai, K. C., & Ng, C. A. (2017). Attenuation in circular and rectangular waveguides. *Electromagnetics*, *37*(3), 171–184. doi:10.1080/02726343.2017.1301198

Yeap, K. H., Yeong, K. C., Tham, C. Y., & Nisar, H. (2016). Analysis of Energy Loss in Superconducting Waveguides. In V. Ponnusamy, N. Zaman, T. J. Low, & A. H. M. Amin (Eds.), *Biologically-Inspired Energy Harvesting through Wireless Sensor Technologies* (pp. 71–83). US: IGI Global. doi:10.4018/978-1-4666-9792-8.ch004

Yeap, K. H., Yiam, C. Y., Lai, K. C., Loh, M. C., Lim, S. K., & Rizman, Z. I. (2016b). Analysis of offset antennas in radio telescopes. *International Journal on Advanced Science, Engineering. Información Tecnológica*, *6*(6), 997–1004.

York, R. A., & Popovic, Z. B. (1997). *Active and Quasi-Optical Arrays for Solid-State Power Combining*. New York: Wiley.

Yusef-Zadeh, F. (2003). The Origin of the Galactic Center Nonthermal Radio Filaments: Young Stellar Clusters. *The Astrophysical Journal*, *598*(1), 325–333. doi:10.1086/378715

Zarka, P. (1998). Auroral radio emissions at the outer planets: Observations and theories. *Journal of Geophysical Research*, *103*(E9), 20159–20194. doi:10.1029/98JE01323

Zarka, P., & Kurth, W. S. (2005). Radio wave emissions from the outer planets before Cassini. *Space Science Reviews*, *116*(1-2), 371–397. doi:10.100711214-005-1962-2

Zarka, P., Lamy, L., Cecconi, B., Prangé, R., & Rucker, H. O. (2007). Modulation of Saturn's radio clock by solar wind speed. *Nature*, *450*(7167), 265–267. doi:10.1038/nature06237 PMID:17994092

Zheleznyakov, V. V. (1970). International Series of Monographs in Natural Philosophy: Vol. 25. *Radio Emission of the Sun and Planets*. Oxford, UK: Pergamon Press.

Zurcher, J. F., & Gardiol, F. E. (1995). *Broadband Patch Antennas*. Boston, MA: Artech House.

About the Contributors

Kim Ho Yeap received his BEng (Hons) from Universiti Teknologi Petronas in 2004, MSc from Universiti Kebangsaan Malaysia in 2005, and PhD from Universiti Tunku Abdul Rahman in 2011. n 2008 and 2015, he underwent research attachments at University of Oxford (experimental cosmology laboratory) and Nippon Institute of Technology, respectively. Yeap is a senior member of the IEEE and a member of the IET. He is also a Chartered Engineer registered with the UK Engineering Council and a Professional Engineer registered with the Board of Engineers Malaysia. He is currently an associate professor in Universiti Tunku Abdul Rahman. Throughout his career, Yeap has served in various administrative capacities, including the Head of Programme of Master of Engineering Science and the Head of Department of Electronic Engineering. Yeap's research areas of interests are in radio astronomy, electromagnetics, and microelectronics. Since 2006, he has published 64 journal papers, 35 conference proceedings, 4 books, and 8 book chapters. Yeap is the external examiner of Wawasan Open University. He is the Editor in Chief of the i-manager's Journal on Digital Signal Processing. Yeap has been given various awards which include the university teaching excellence award, 4 Kudos awards from Intel Microelectronics and 18 research grants.

Veerendra Dakulagi received the B.E and M.Tech. degrees from Visvesvaraya Technological University, Belagavi, India, in 2007 and 2011 respectively. He received the Ph.D. degree in Adaptive antennas from the same university in 2018. From 2010-2017 he was an Asst. Professor in Dept. of E&CE, Guru Nanak Dev Engineering College, Bidar, India. Since March 2017 he has been an Associate Professor and R&D Dean of same institute. His research interests are in the area of signal processing and communications, and include statistical and array signal processing, adaptive beamforming, spatial diversity in wireless communications, multiuser and MIMO communications. He has published over 40 technical papers (including IEEE, Elsevier, Springer and Taylor and Francis) in these areas. Dr. Veerendra Dakulagi is a member of Institute of Electronics and Telecommunication Engineers (IETE) and currently serves as an editorial board member of Journal of Computational Methods in Sciences and Engineering, Journal on Communication Engineering and Systems, Journal on Electronics Engineering and International Journal of Scientific Research and Development. Currently he is also working as a Post Doctoral Fellow (PDF) of faculty of Electronics & Communication Engineering, Lincoln University College, Petaling Jaya, Malaysia.

Magdalene Wan Ching Goh obtained her Bachelor of Engineering in Electrical Engineering and Electronics and PhD at the University of Liverpool. She is a senior lecturer at Wawasan Open University and a member of the Penang Science Cluster, radio astronomy team. While in the academic, she has served as a judge for the Innovate Malaysia competitions. She has worked for several years in the electronics/semiconductor industry prior to joining the academia. Magdalene's research interests are in the areas of semiconductor physics and electromagnetics. She also has strong interests in astronomy and a member of the Astronomical Society of Penang. She is working with many volunteers who share the same interests and passion in astronomy education, to create curricula for radio astronomy, for the purpose of introducing school pupils and college students to the concepts of radio astronomy and radio telescopes.

Kazuhiro Hirasawa received his Ph.D. degree in Electrical Engineering from Syracuse University, Syracuse, NY, in 1971. From 1967 to 1975, he was with the department of Electrical and Computer Engineering, Syracuse University. From 1975 to 1977, he was a consultant on research and development of various antennas. Since 1978, he has been with the University of Tsukuba, Ibaraki, Japan. Currently, he is an Emeritus Professor in University of Tsukuba and an International Collaborative Partner of Universiti Tunku Abdul Rahman, Malaysia. He is also an IEEE Life Fellow and IEICE Fellow.

Hirokazu Kobayashi was born in Hokkaido, Japan. He received the B.E.E. and M.E.E. de-grees from Shizuoka University, Hamamatsu, Japan, in 1978 and 1980, respectively, and the Doctorate Eng. from University of Tsukuba, Tsukuba, Japan, in 2000. He joined Fujitsu Ltd., Kawasaki, Japan, in 1980. Since 1981, he has been with the Fujitsu System Integration Laboratories as a researcher on developments for micro/millimeter wave wide band antennas, active phased array radar systems, and theoretical investigation of electromagnetic scattering and antenna. From 1999 to 2010, he served as a General Manager at Fujitsu Ltd. In 2010, he joined the Faculty of Engineering as a Professor with Niigata University, Niigata, Japan. From 2013, he has been currently a professor in the Department of Electronics and Information Systems Engineering, Osaka Institute of Technology, Osaka, Japan. His current research interests include high-frequency scattering field theoretical analysis, computing of radar cross-section (RCS) for large objects, electromagnetic scattering analysis using Physical Optics, Physical Theory of Diffraction and Geometrical Theory of Diffraction (PO/PTD/GTD), RCS near-ðeld to far-ðeld transformation technology for electrical large ob-ject such as airplane based on microwave radar imaging theory, radar SAR and ISAR imaging, radiation analysis such as conformal array with curved patch antenna, and recognizing analysis by micro-Doppler information. Dr. Kobayashi has published around two hundred papers including international and domestic conference paper, and three books. Dr. Kobayashi is a Senior Member of the Institute of Electrical and of Electronic Engineers (IEEE) and a Chair of Technical Committee on Space, Aeronautical and Navigational Elec-tronics (SANE) of the Institute of Electronics, Information and Communication Engineers (IEICE) during 2015-16. He is an Editor in Chief of the section: Microwave and Wireless Communications, Electronics in MDPI journal from 2019.

Wei Gie Lim received his Bachelor of Engineering (Honours) Electronic Engineering from Universiti Tunku Abdul Rahman in 2018. He is currently working as an engineer in Penang, Malaysia.

Mey Chern Loh received both her Bachelor of Engineering and Master of Engineering Science qualifications from the Universiti Tunku Abdul Rahman, Malaysia. She is currently working on her PhD research in the field of radio astronomy.

Humaira Nisar received her BE (Honors) in Electrical Engineering from University of Engineering and Technology, Lahore, Pakistan. She received her MS degree in Nuclear Engineering from Quaide-Azam University, Islamabad, Pakistan. She received her MS degree in Mechatronics and Ph.D. in Information and Mechatronics from Gwangju Institute of Science and Technology, Republic of Korea. Currently, she is an Associate Professor at the Department of Electronic Engineering, Universiti Tunku Abdul Rahman, Kampar, Malaysia. She is also Senior Member of IEEE.

Katrina Shak received her BEng (Hons) and PhD from Monash University Malaysia in 2012 and 2017, respectively. Currently, Dr. Shak is serving the Chemical Engineering Department in Universiti Tunku Abdul Rahman as a Lecturer and researcher.

Tee Tang is an adjunct associate professor in Queensland University of Technology. His areas of expertise are in antennas, electromagnetic compatibility (EMC) and RF circuits. He is a member of the Penang Science Cluster, radio astronomy team.

Hui Chieh Teoh is currently the advisor of the astronomy club in Universiti Tunku Abdul Rahman. She is a Professional Engineer registered with the Board of Engineers Malaysia.

Ryoji Wakabayashi was born in Japan on April 1, 1962. He received the B.E. and M.E. degree from the Department of Electrical and Electronics Engineering, Sophia University in 1987 and 1989 respectively, and became a Research Associate of Tokyo Metropolitan College of Aeronautical Engineering, where he was promoted to Lecturer, Associate professor, Professor in 1992, 1995 and 2004 respectively. In 2006, he became a Professor of Tokyo Metropolitan College of Industrial Technology. He has been engaged in research on near field of wire antennas, EMC, satellite communication and meteor burst communication. In 1993, he received a URSI YS Award.

Kim Huat Yeap received his B.Eng (First Class Honors) in Electrical and Electronics Engineering from Universiti Tenaga Nasional (Uniten), Malaysia in 2004. He received his M.Sc. and Ph.D. from Universiti Sains Malaysia (U.S.M.), Malaysia in 2012 and 2018, respectively. He was a senior electronics design engineer in Motorola Technology from 2004 to 2011. His research interest focuses in the areas of antenna, oscillator, mixer, as well as microwave circuitries.

Index

F

G

H

I

K

L

M

N

O

P

Q

R

S

T

V

W

Y

www.ingramcontent.com/pod-product-compliance
Lightning Source LLC
LaVergne TN
LVHW060420300926
849162LV00006B/1477